CISM COURSES AND LECTURES

The series presents lecture notes, monographs, edited works and proceedings in the field of Mechanics, Engineering, Computer Science and Applied Mathematics.
Purpose of the series is to make known in the international scientific and technical community results obtained in some of the activities organized by CISM, the International Centre for Mechanical Sciences.

INTERNATIONAL CENTRE FOR MECHANICAL SCIENCES

COURSES AND LECTURES - No. 346

MODELLING AND ANALYSIS OF REINFORCED CONCRETE STRUCTURES FOR DYNAMIC LOADING

EDITED BY

CHRISTIAN MEYER
COLUMBIA UNIVERSITY

Springer-Verlag Wien GmbH

Le spese di stampa di questo volume sono in parte coperte da
contributi del Consiglio Nazionale delle Ricerche.

This volume contains 183 illustrations

Originally published by Springer-Verlag Wien New York in 1998
SPIN 10681557

In order to make this volume available as economically and as rapidly as possible the authors' typescripts have been reproduced in their original forms. This method unfortunately has its typographical limitations but it is hoped that they in no way distract the reader.

ISBN 978-3-211-82919-6 ISBN 978-3-7091-2524-3 (eBook)
DOI 10.1007/978-3-7091-2524-3

This volume is published in memory of prof. P. Gergely

Peter Gergely (b. February 12, 1936 - d. August 25, 1995)
Professor of Structural Engineering, School of Civil and Environmental Engineering
Cornell University, Ithaca, N.Y., USA

PREFACE

Nonlinear dynamic analysis of reinforced concrete structures is generating considerable interest, which manifests itself in the numerous research and development projects, conferences, and literature published in recent years. The call for reliable analytical tools comes from many directions: the nuclear and defence industries, the research community, designers of complex structures, and especially from the earthquake engineering community. The task is complex, and many problems remain unsolved. Dynamic analysis can be helpful in estimating the nonlinear response of concrete structures in some cases, but in other cases the predictions are rather unreliable. For that reason it is not advisable to leave the task of nonlinear analysis to engineers who are not familiar with its many fine points and potential pitfalls.

The main purpose of this short course was, therefore, to provide an overview of the modelling and analysis of reinforced concrete structures for dynamic loads and to illustrate the subject through examples. The course presents a mix of fundamental (scientific) knowledge that is still limited mainly to the research arena, as well as information ready for application in engineering practice.

Dynamic problems in structural engineering may be subdivided into four tasks:

1. definition of the input (dynamic load);

2. modelling (idealization) of the structure-foundation system;

3. dynamic analysis;

4. interpretation of the results for design.

The first task is typically subject to great uncertainties, especially in earthquake engineering. The fourth task is not trivial at all and can be done only by experienced engineers who understand the dynamic behavior of concrete structures. This short course deals primarily with the second task. The dynamic analysis techniques of a mathematical model are fairly well established. Therefore they have been relegated to the Appendix of these lecture notes.

During the week of June 28 to July 2, 1993, the International Centre for Mechanical Sciences held a Short Course on "Modelling and Analysis of Reinforced Concrete Structures for Dynamic Loading" at its home, the beautiful Palazzo del Torso in Udine, Italy. In early 1991, Professor Sandor Kaliszky, then the Centre's Rector, had asked Professor Peter Gergely of Cornell University to coordinate and organize the course. With my assistance he assembled a slate of well-qualified lecturers, namely Professors Hans Reinhardt of the University of Stuttgart, Filip Filippou of the University of California at Berkeley, Johan Blaauwendraad of Delft University, and T.P. Tassios of the University of Athens.

Although all lecturers had agreed after the successful conclusion of the course to participate in the publication of the lecture notes, their geographic dispersion and various professional commitments allowed only slow progress in rewriting the lecture notes that had been handed out to the course participants. The process was dealt a serious setback by the untimely death of Peter Gergeley on August 25, 1995. During his long illness he had requested me to assume the editorship, but progress was much slower than had been desirable. Finally, the lecture notes are complete, and I hope they are a fitting memorial to Peter, whom we all hold in fondest memory as a superb researcher and teacher as well as a wonderful human being.

I wish to express my deepest gratitude to the Springer-Verlag and its Scientific Editor, Professor Carlo Tasso, for their patience in seeing this project come to fruition. It is hoped that the effort was not in vain and that the lessons taught in these pages will be noted wherever the threat of destructive earthquakes to reinforced concrete structures reminds engineers to ceaselessly improve their design methods.

Christian Meyer

CONTENTS

MATERIAL MODELLING

H.W. Reinhardt
University of Stuttgart, Stuttgart, Germany

C. Meyer
Columbia University, New York, NY, USA

1.1 Main Characteristics of Concrete

1.1.1 Composition of Concrete

Fresh concrete consists of cement, water, aggregates, admixtures and maybe more additives. In the fresh state, concrete behaves like a fluid when it is vibrated, or like a cohesive soil when it is at rest. However, this state is only relevant during construction and not the topic of this chapter.

As cement hydrates and concrete gains strength, the fresh state changes to the hardened state. Concrete is now an artificial rock with distinct mechanical, physical, and chemical properties. Here, we are dealing with mechanical properties and, even more specifically, with short-term properties like compressive and tensile strength, Young's modulus, ultimate strain, fracture energy, and resistance to crack propagation. Creep and relaxation are not considered.

To understand the bulk properties we have to step down one level of observation from the macro- to the meso-level. It is common practice to distinguish between macro-, meso-, and micro-level. When an engineer utilizes a material he characterizes it in terms such as elastic, plastic, strong, ductile etc., i.e. it is modelled as a homogeneous and (mostly) isotropic continuum, and elasticity or some other theory is used to relate stresses and strains. If the bulk behavior is to be understood the composition of the material has to be considered (Fig. 1.1).

macro - isotropic homogeneous continuum; structures,
structural parts
meso - inhomogeneous composite material with flaws and cracks;
explanation of bulk behavior
micro - molecules, atoms; scientific foundation of
meso and micro behavior

Fig. 1.1 Levels of observation

It is on the meso-level where aggregate particles, hydrated cement paste, pores, cracks, and interfaces play their part. This is the level which is addressed mainly by the materials engineer when he tries to optimize products. The micro-level deals with atoms, molecules and their interaction - a field dealt with by physicists and chemists.

The meso-level is the most appropriate one for the engineer to describe and explain mechanical phenomena. Here, normal weight concrete contains about 70% by volume natural aggregates like sand and coarse grains, 11% by vol. cement, 17% water, and 2% air. After hydration we can distinguish between about 30% by vol. matrix and 70% by vol. aggregates (Fig. 1.2). Concrete is a typical two-phase material, the two phases being bonded together by interparticle forces. The interface between aggregate and matrix is an important part of the whole and may be considered a third phase.

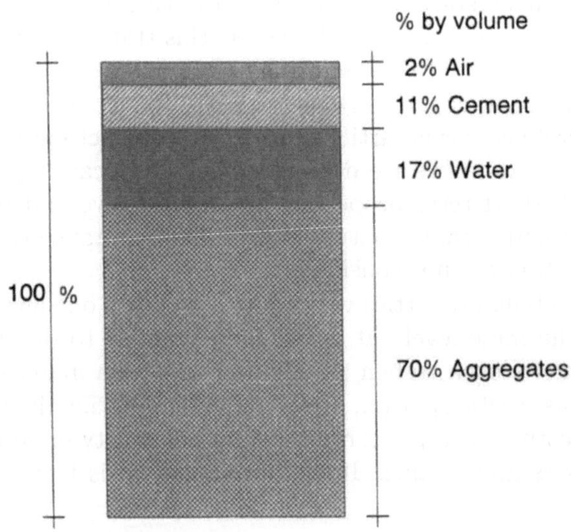

Fig. 1.2 Volume partition of fresh concrete (average example)

The solid aggregates are strong and elastic. Depending on the origin of the material (rounded quartz, crushed limestone, basalt etc.) Young's modulus ranges between 50 and 100 GPa, compressive strength between 80 and 300 MPa, tensile strength between 10 and 30 MPa. In contrast to the aggregate, the matrix is porous and less strong. Whereas the properties of the aggregate are mostly given, the properties of the matrix depend strongly on mix design, curing and age. Thus, they are influenced by the materials engineer.

1.1.2 Pores in Concrete

Hydration of cement means a chemical reaction with water. New products are formed which contain chemically bound water (calcium silicate hydrates, CSH, calcium aluminate hydrates, CAH, calcium hydroxide, CH, etc.) and which have a very large specific surface of about 200 m^2/g. A few layers of water molecules are absorbed on this immense surface. This is the so-called physically bound water. 100 g Portland cement bind about 25 g water chemically and 15 g physically. The ratio between mass of water and mass of cement is called water/cement ratio, which is the technical key parameter for most properties of concrete.

When water is chemically bound it decreases its volume (chemical shrinkage). If the new material is stiff enough the free shrinkage is restrained and so-called capillary pores develop. If the water/cement ratio is larger than about 0.40 there is more water available than the cement can react with and more capillary pores will develop. Fig. 1.3a shows the partition of solid material and pores in hydrated cement paste. A typical water/cement ratio of 0.50 leads to 23% gel pores, which, at usual ambient conditions, are filled with physically bound water and 17% capillary pores. If hydration is stopped half-way ($\alpha = \frac{1}{2}\alpha_{max}$) the capillary pores take 39% (Fig. 1.3b).

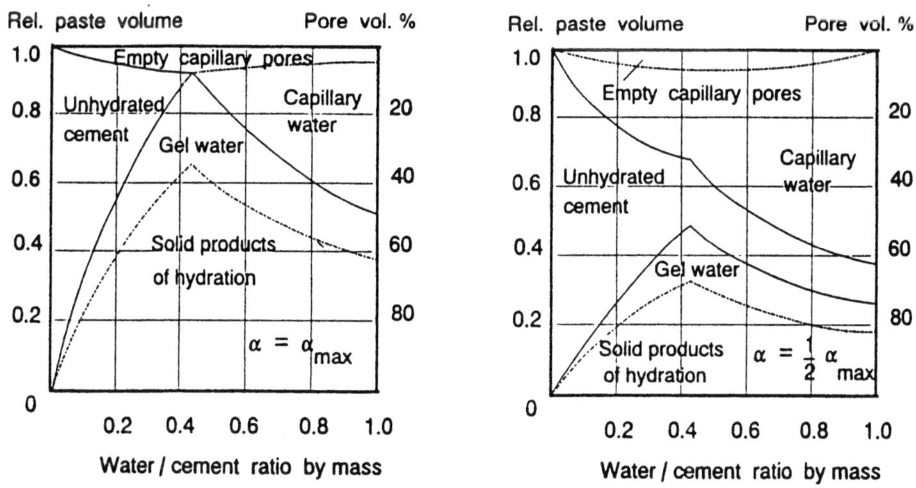

a) complete hydration b) incomplete hydration

Fig. 1.3 Volume partition of hydrated cement paste

It is evident that the matrix in concrete is a rather porous material. Usual concretes have a paste porosity of 40 to 60%. However, since paste occupies only about 30% of concrete, the concrete porosity is about 12 to 18%. Entrapped air increases the porosity by 1 to 2% by volume. The pore sizes cover a wide range. Gel pores reach a few nanometer, capillary pores have radii between 5 and 5000 nm, air voids, cracks and flaws are larger. According to a RILEM definition, a micro-crack has a width of 10 μm (=10,000 nm). Fig. 1.4 shows a pore-size distribution in cement mortar with the typical effect of water/cement ratio.

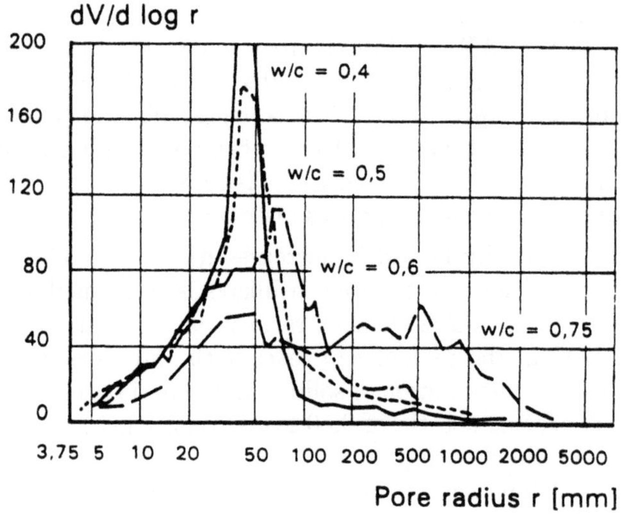

Fig. 1.4 Pore size distribution of cement mortar with various water-cement ratios

1.1.3 Interfaces

The interfacial region between aggregate and matrix is different from the bulk cement paste in terms of morphology and density. Due to bleeding of cement (water segregation) there is space around the aggregate particles which facilitates precipitation of calcium hydroxide crystals. These are less strong than CSH. The porosity of this zone is higher than that of the bulk paste. Different morphology and higher porosity lead to lower strength than in the bulk cement paste (Fig. 1.5).

Addition of silica fume and other fine grain additives can reduce the porosity of the interfacial layer and increase the strength.

Chemical shrinkage (or autogenous shrinkage), drying shrinkage and heat of hydration are three causes for cracks in the interface. There are cracks emanating from the aggregate and there are cracks running along the interface. Both are due to the lack of fit between the shrunk paste and the aggregate. Stereologic pictures show that many cracks exist in concrete before first mechanical loading.

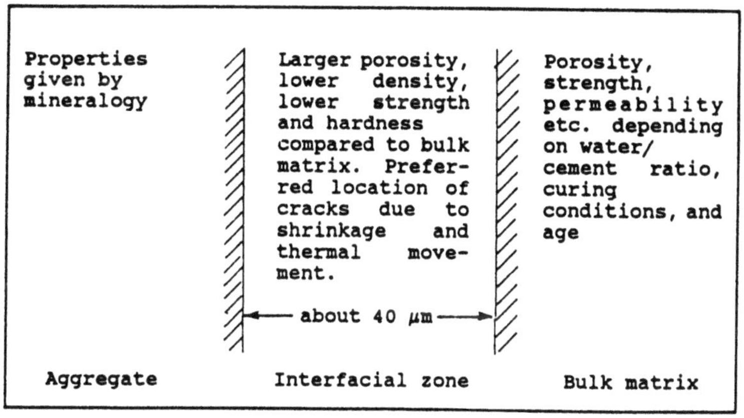

Fig. 1.5 Schematic of interfacial zone

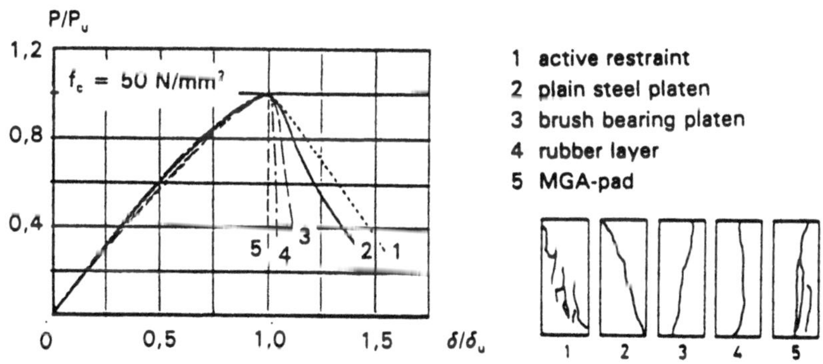

Fig. 1.6 Compressive force vs displacement as function of specimen boundary condition

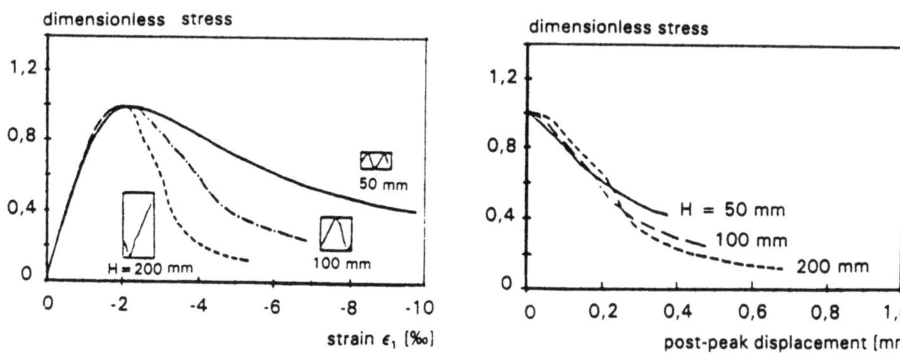

a) in terms of strain b) in terms of displacement

Fig. 1.7 Compressive stress vs displacement diagram

1.1.4 Stress-Strain Response

If a specimen is loaded in compression, three states can be distinguished: a quasi-elastic state at the beginning, a state with stable growth of existing and new cracks, and finally a state with unstable crack growth and subsequent fracture. New cracks are mainly due to the different stiffness of matrix and aggregate. This difference causes tensile stresses between grains which lead to debonding. In a strain-controlled test the softening of concrete can be measured. The final debris of concrete have the shape of large grains with mortar (or matrix) caps on them.

Tests have clearly demonstrated that the boundary conditions and the length (or height) of the specimen influence the test results. Active restraint leads to a more ductile behavior, while no restraint leads to a more sudden failure. Furthermore, a longer specimen shows less strain than a shorter one (Figs. 1.6, 1.7). However, the post-peak displacements are almost identical. Looking at the specimens, a localization of the fracture zone can be identified (shear band). Localized failure is very likely under biaxial and triaxial compressive loading.

A prismatic concrete bar loaded in tension fails by a crack which runs mainly at a right angle to the loading direction. Observed on a meso-scale, debonding of aggregate and matrix starts first, and various cracks coalesce into a discrete crack. Measurements in displacement-controlled tests show an increase of stress and strain, the formation of a cracking zone, and finally the decay of stress by increasing displacement (Fig. 1.8). The so-called softening is due to crack bridging and aggregate matrix friction. The stress-strain response is almost linear up to about 60% of the strength, then non-linearity occurs. The stress decay after peak stress is steepest for high-strength concrete when aggregates fracture, and it is rather gradual for fiber-reinforced concrete with the fibers causing a large crack bridging effect.

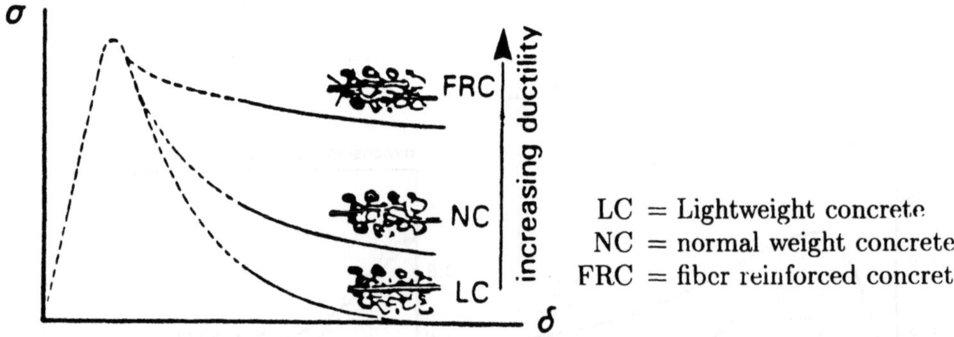

Fig. 1.8 Stress-displacement diagram

1.1.5 Fracture Mechanics Terms

The behavior of a cracked body under load can be approached by fracture mechanics. Following Griffith theory, the elastic energy release is compared with th surface energy gain during crack extension. As long as the latter is larger tha the former one, there is no crack propagation; otherwise, unstable (or catastrophic) failure occurs. The critical stress is given by

$$\sigma_c = \sqrt{\frac{2\gamma E}{\pi c}} \cdot f\ (geom) \tag{1.1}$$

with γ = surface energy, E = Young's modulus, c = half crack length, f = a function taking account of the geometry. For an infinite plate with σ = remote stress, $f = 1$.

Westergaard, Sneddon, and Irwin analysed the stress field near the crack tip and defined a parameter which is a measure for the intensity of the stresses,

$$K_I = \sigma\sqrt{\pi c} \cdot g\ (geom) \tag{1.2}$$

with K_I = stress intensity factor for mode I (crack opening), and g = geometrical function. Failure occurs when K_I approaches the critical stress intensity factor K_{Ic}, which is a material property.

For an ideal brittle material the two approaches converge to

$$K_{Ic} = \sqrt{2\gamma E} = \sqrt{G_c E} = \sqrt{RE} \tag{1.3}$$

with G_c = critical energy release rate, R = crack resistance. For other materials, G_c and R may include also other energy contributions due to plastic, viscous, and frictional actions.

Tensile stresses or strains are required for crack extension. However, remote compressive stresses can cause tensile stresses around pores (cavities, flaws) and can propagate a crack. The same is true for heterogeneous materials with two phases of different elastic stiffness. This is certainly true for concrete on a meso-scale.

On a macro-scale, concrete does not follow the linear elastic fracture mechanics concept. It is a softening material instead of a purely brittle material. A crack causes a process zone ahead of the crack tip with cohesive stresses, or a crack band develops with dissipation of energy. To relate stresses and displacements, the following quantities have to be known (Fig. 1.9):

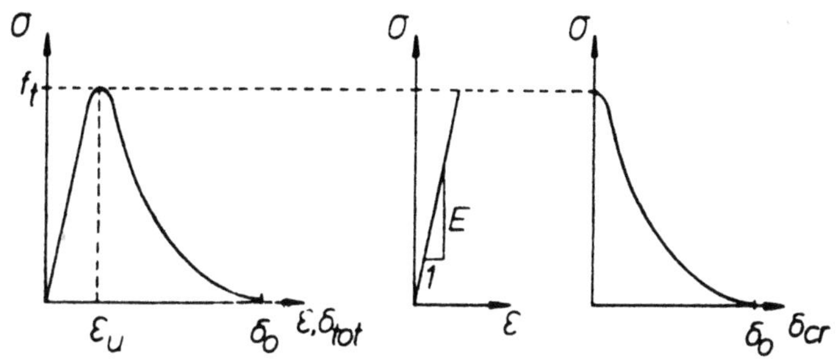

a) total displacement b) elastic strain c) crack opening

Fig. 1.9 Tensile stress vs displacement diagram

- Young's modulus, E
- tensile strength, f_t
- specific fracture energy, G_F
- stress-free crack opening, δ_0
- shape of the softening curve, $f(w)$.

All quantities depend on concrete composition, age, loading rate, temperature, humidity, etc.

Hillerborg has introduced the characteristic length,

$$l_{ch} = EG_F/f_t^2 \tag{1.4}$$

The smaller this length is, the more brittle the material. If a typical dimension of a structure is 10 to 20 times l_{ch}, linear elastic fracture mechanics (LEFM) may apply, otherwise elastoplastic (EPFM) or nonlinear fracture mechanics (NLFM) are appropriate. Concretes yield characteristic lengths l_{ch} between 0.1 and 0.5 m.

1.1.6 Aggregate Interlock

A crack in a concrete section follows an irregular path. Aggregate particles are pulled out of a low-strength matrix or fractured in a high-strength matrix. The higher the strength the smoother the surface of a crack plane. When two crack faces slide on each other, frictional resistance develops which depends on the roughness of the crack and the normal force on the crack plane. Simultaneously, the crack width changes because the asperities of the crack plane tend to override each other. There are at least four quantities involved which interact, i.e. shear stress, normal stress, shear displacement (crack sliding), normal displacement (crack opening). Their relation can be written in an incremental form

$$\begin{pmatrix} \Delta\sigma_{nn} \\ \Delta\sigma_{nt} \end{pmatrix} = \begin{pmatrix} A_{11} & A_{12} \\ A_{21} & A_{22} \end{pmatrix} \begin{pmatrix} \Delta\delta_n \\ \Delta\delta_t \end{pmatrix}$$

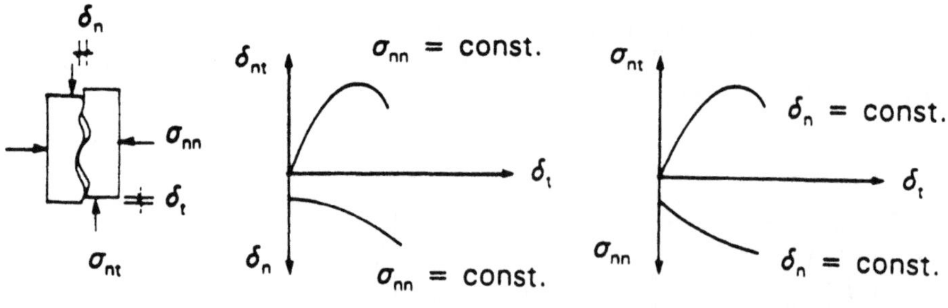

a) with constant normal stress b) with constant normal displacement

Fig. 1.10 Interaction between shear stress, normal stress, shear displacement, and normal displacement

A schematic representation of this behavior is shown in Fig. 1.10. The upper part of this figure shows that shear stress increases with increasing shear displacement and that normal displacement increases also. There is a maximum shear stress which depends on concrete composition and strength. The lower part of the figure indicates that normal stress increases with increasing shear displacement. This is the case for a stiff confinement (δ_n = const) which suppresses additional crack opening. In actual structures, the degree of confinement is governed by boundary conditions and the amount of steel reinforcement in the cross-section. The coefficients A_{ij} in the matrix given above are not constants but functions of stress and displacement. There is a large number of publications available which cover or summarize this phenomenon [1.1, 1.2].

The total shear behavior is a more complex interaction of aggregate interlock, dowel action of the reinforcing bar, bond between steel and concrete, and splitting of concrete cover. This chapter is not aimed at details, and the reader is referred to the literature (e.g. [1.1, 1.2]).

1.2 Modeling of Concrete Material Behavior for Dynamic Analysis

1.2.1 Introduction

Nonlinear finite element analysis is now a well established tool for analyzing a variety of problems in engineering mechanics as well as other physical sciences. In nonlinear analysis, two separate problems arise, which often appear interrelated: 1) the specification of material properties, including failure criteria, at the element integration points; and 2) numerical problems associated with the solution of the nonlinear equations of state. Once constitutive relationships have been defined, it is relatively straightforward to compute element stiffnesses, to assemble the structure stiffness, mass and load vectors, and then to determine the state of stress and strain throughout the time domain, unless numerical problems cause difficulties.

The main difference between concrete and "simple" materials like many metals is that there do not exist simple models or theories that can realistically reproduce all the important aspects of concrete behavior. It is the intent of this overview to summarize what we know about concrete material behavior and to survey some of the theories and models that have been proposed to numerically simulate this behavior for purposes of nonlinear finite element analysis of reinforced concrete structures. We shall comment also on some questions of implementation and numerical considerations. Most engineers are not likely to be developers but rather users of software systems. Therefore their "need to know" is limited to a basic understanding of the theory that such programs are based on, to enable them to make prudent use of these truly powerful systems.

1.2.2 Concrete Behavior Under Static Load

The uniaxial response of plain concrete to monotonically applied load can be subdivided into three or four different phases, Fig. 1.11. It is completely controlled by the evolution of various crack systems. Microcracks and crack-like voids typically exist even prior to the application of any external load and are found typically at the interfaces between coarse aggregate particles and the mortar matrix. Some of these cracks are caused by early volume changes in the concrete due to hydration, drying or carbonation. Other preexisting cracks are caused by settlement of coarse aggregate particles during the placing process and the accompanying "bleeding". These cracks are the cause of an initial anisotropy of the material, with the result that it responds differently to loading parallel to the direction of casting than to perpendicular loading.

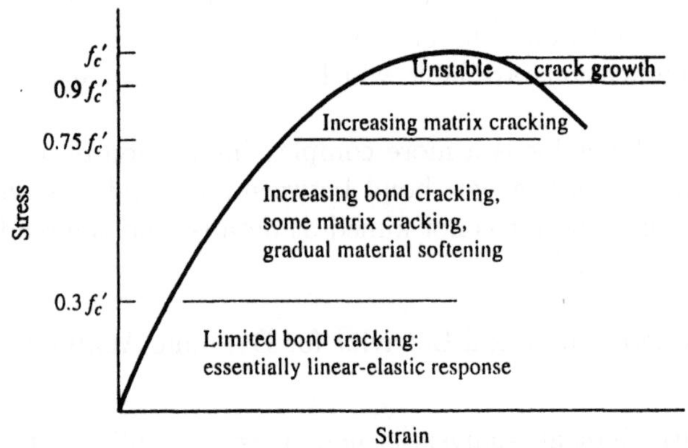

Fig. 1.11 Phases of uniaxial stress-strain response

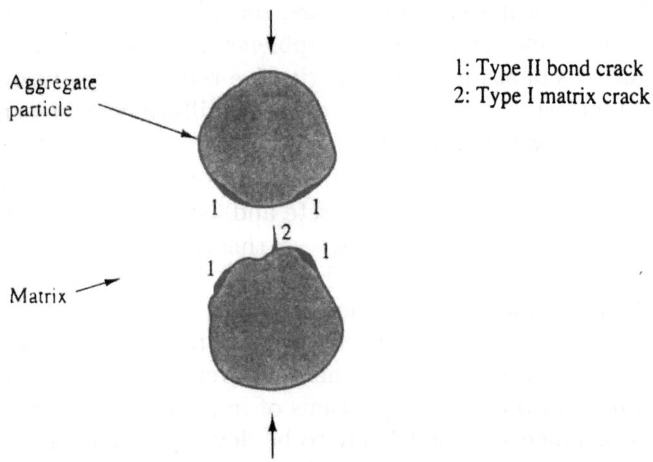

Fig. 1.12 Tensile splitting under compressive stress

Up to about 30% of compressive strength f_c', the extent of bond cracking is very limited. As a result, the stress-strain response is near-linear and basically elastic. From 0.3 f_c' to about 0.75 f_c', formation of new internal free surfaces occurs mostly along the interfaces between large aggregate particles and the mortar matrix. This growth of bond cracks is accompanied by increased softening of the stress-strain response. From 0.75 f_c' to about 0.9 f_c', mortar cracks begin to increase noticeably and to form continuous crack patterns by connecting separate bond cracks along the larger aggregate particles. The stress-strain curve is very flat now and has two components - a diminishing linear elastic one and an increasing inelastic one. Under loading beyond the 90% level crack growth eventually becomes unstable and causes rapidly increasing inelastic deformations. This final phase is accompanied by significant volume dilatation as a result of wholesale cracking and the opening of relatively large voids within the material. Under purely uniaxial compression, failure occurs by tensile splitting. This mechanism can be explained by the resistance of aggregate particles to axial strain, Fig. 1.12.

The uniaxial stress-strain curve is strongly affected by the concrete strength, Fig. 1.13. The higher the standard strength, f_c', the steeper the ascending curve (Young's modulus) and the more linear this branch will be. Also the descending branch is much steeper for high-strength concrete, reflecting a more brittle failure mode and reduced ductility. The increased linearity of the loading branch is due to the reduced bond cracking along the matrix-aggregate interfaces. Failure is sudden because of the reduced redundancy of the material, as compared to normal-strength concrete, in which the gradual formation of crack systems prior to failure provides for the higher ductility.

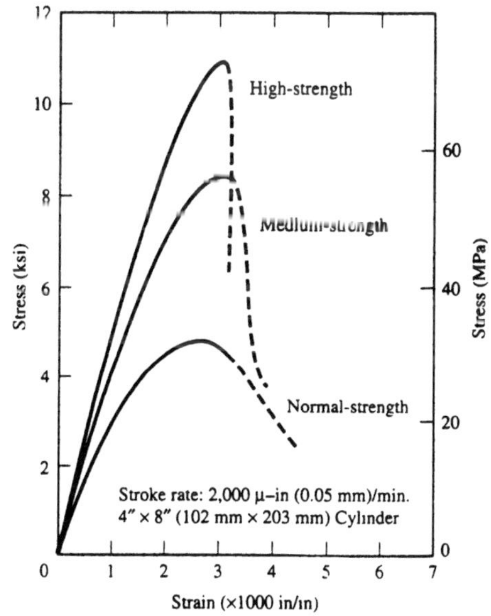

Fig. 1.13 Effect of strength on stress-strain curve [1.88]

Under positive strains, cracks of type 1 dominate. When loading conditions do not cause any extensions, only cracks of type 2 or 3 develop. Type 1 cracks opening up in response to tension may cause unstable material behavior. Under compressive load, extension is caused by Poisson's ratio effect, and splitting cracks occur parallel to the load but open up more slowly. In that case also cracks of type 2 and 3 are present, with friction in the crack surfaces. This friction and the slower growth of the type 1 cracks make concrete more ductile in compression as compared to tension and contribute to the other dissimilarities between tension and compression behavior.

The response of plain concrete to proportional biaxial loading is illustrated in Fig. 1.14, taken from the classical paper by Kupfer, Hilsdorf and Rüsch [1.3]. A second compressive stress component increases both strength and stiffness (Young's modulus), because of the confinement effect, through which, for example, a horizontal compressive stress component retards the opening of vertical concrete cracks (splitting) due to vertical load (see Fig. 1.12). It is important to note that a well-developed crack system with a clear directional bias caused by a tensile stress or a large compressive stress, increases the material's anisotropy well beyond the initial one mentioned earlier. Once the material is cracked, it does not get uncracked in the affected plane, regardless of subsequent loading. This fact complicates the construction of numerical models to predict the biaxial and triaxial behavior of concrete under non-proportional loading.

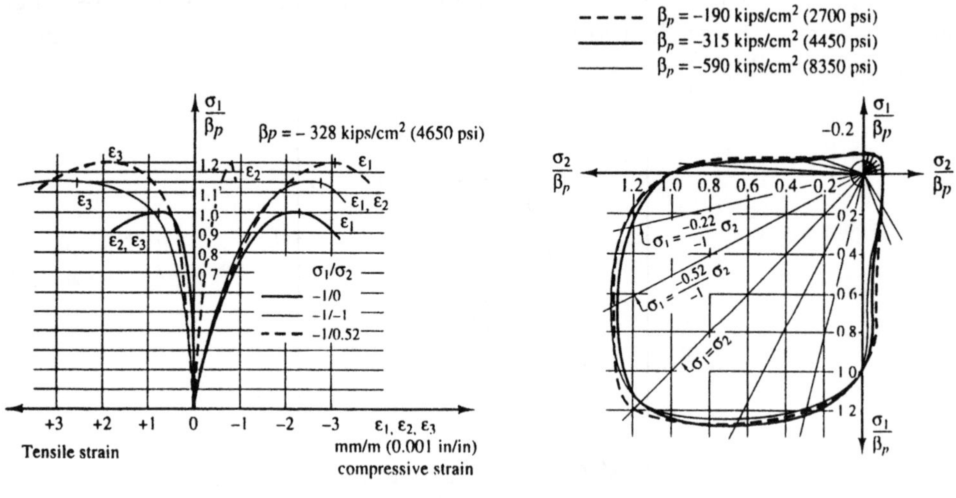

Fig. 1.14 Biaxial strength and stress-strain behavior [1.4]

In three dimensions, both the stress-strain and failure behavior become rather complicated, Fig. 1.15. Most significant is the large increase in strength, especially for stress states that do not deviate too much from the hydrostatic case. Under hydrostatic pressure, one is tempted to claim that concrete cannot fail. The corresponding stress-strain curve does indeed maintain a positive slope presumably indefinitely. In reality, however, the microporous structure of the mortar matrix collapses under high enough pressure, and the concrete consolidates. This manifests itself in a loss of residual uniaxial compressive strength. After applying a hydrostatic pressure of about $6f_c'$, a 25% drop in uniaxial strength has been reported [1.4]. While the failure surface can be constructed experimentally using monotonic proportional stress histories, nonproportional load paths introduce numerous complications for any constitutive model.

It suffices to stress that any confining pressure, caused either actively by load or passively by confinement reinforcement, results in greatly increased strength and stiffness and particularly ductility, as the cracks are prevented from opening and linking up to form the mechanisms associated with nonductile failure.

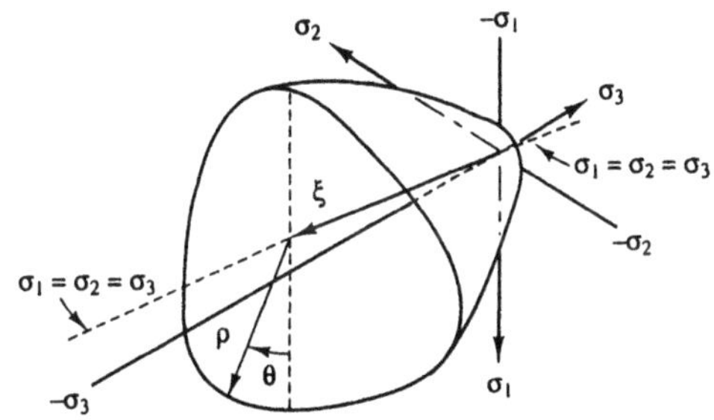

Fig. 1.15 Triaxial failure surface

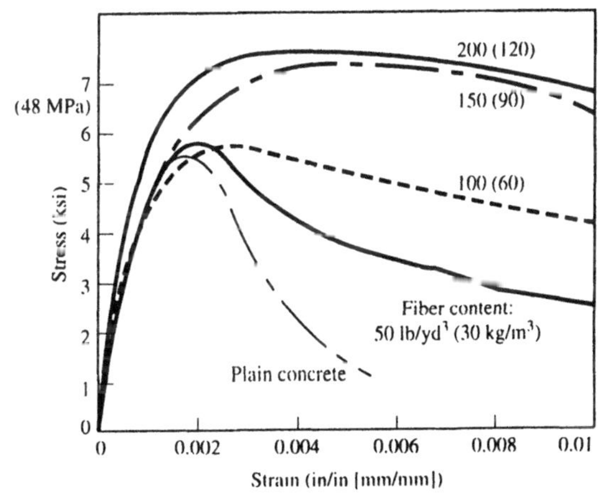

Fig. 1.16 Effect of fiber reinforcement [1.5]

Figure 1.16 illustrates a point made earlier, from a different viewpoint. It shows the effect of fiber reinforcement on the stress-strain behavior of concrete. A considerable body of literature exists on this subject [1.5]. The graph is of interest because it illustrates that different percentages of fiber reinforcement lead to different failure modes. The strength of normal-strength concrete is often controlled by the matrix strength, which is subject to major influence of the fiber reinforcement. The higher strength concretes or those that are controlled by the interface bond, are less affected by the fiber reinforcement. As can be seen, fiber reinforcement has a much smaller effect on the static compressive strength than on the ductility and energy absorption capacity. By judiciously devising experiments, one can force the one or the other failure mode and fairly accurately predict the outcome.

1.2.3 Concrete Behavior Under Dynamic Load

When dealing with dynamic loading, we first have to define what constitutes a dynamic load. Basically, a loading cannot qualify as a purely static load, if one of the following three criteria applies: 1) the accompanying strain rate is high enough to markedly affect the material properties; 2) the load history is rich in frequencies that can excite dynamic response; or 3) there are repeated load applications or even load reversals, which can lead to damage accumulation and material deterioration. The third case by itself is not as intrinsically "dynamic" as the other two, as it can be simulated with quasi-static loading.

The so-called strain-rate effect expresses itself primarily in a strength and stiffness increase with faster load application, Fig. 1.17 [1.6]. The explanation is that the crack systems that lead to failure at "static" loading rates are paths of least resistance. This means they may include lengthy detours around large aggregate particles in order to connect prior cracks. At higher strain rates the material has less time to be "choosy" and may have to let cracks propagate along more resistant paths, including cracks passing through (stronger) aggregate particles. This phenomenon is addressed in more detail in later sections of this chapter.

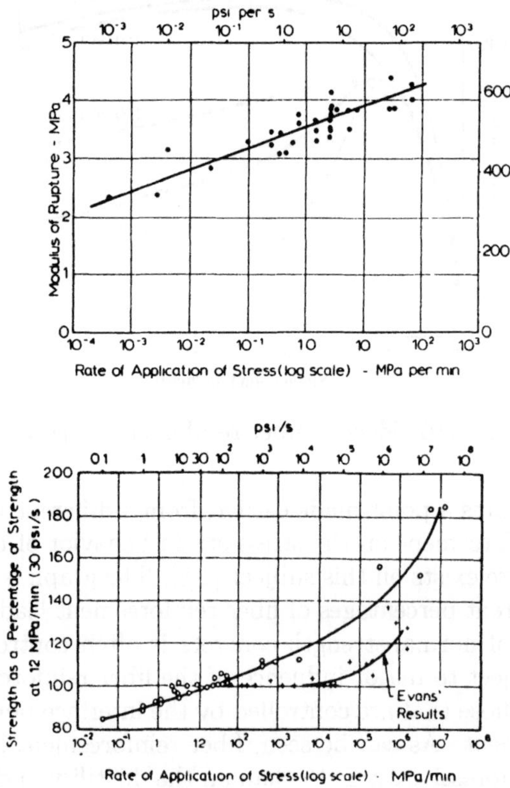

Fig. 1.17 Effect of loading rate on strength [1.6]

The resonant load category is of interest here only if it produces stresses beyond the 0.5 f_c' level. Destructive earthquakes fall into this as well as into the third category that involves load repetitions and reversals. Considering the potential failure modes, there are some striking similarities and differences between the material response to monotonic and repeated load application. For example, the monotonic stress-strain curve has been reported to serve as an envelope to the cyclic stress-strain curve, Fig. 1.18 [1.7]. This claim is somehow an oversimplification, because most of the tests on which it is based involved only small numbers of load cycles and therefore avoided the fatigue phenomenon. Yet, the strain at failure has been shown to be fairly independent of the number of load applications. However, the total length of cracks visible on the surface of specimens subjected to cyclic load with a stress of 0.75 f_c' was 35% greater than that observed on identical specimens loaded up to 0.90 f_c' statically [1.8]. Another indication of fatigue damage is the gradual decrease of the elastic modulus, which may decrease by as much as 40%. This significant stiffness decrease indicates that the system of fatigue cracks is more extensive than the corresponding one for monotonic compression. Thus, fatigue introduces an additional dimension to our problem. It is a process of progressive, permanent internal structural damage that results in complete failure, if the number of cycles is sufficiently large.

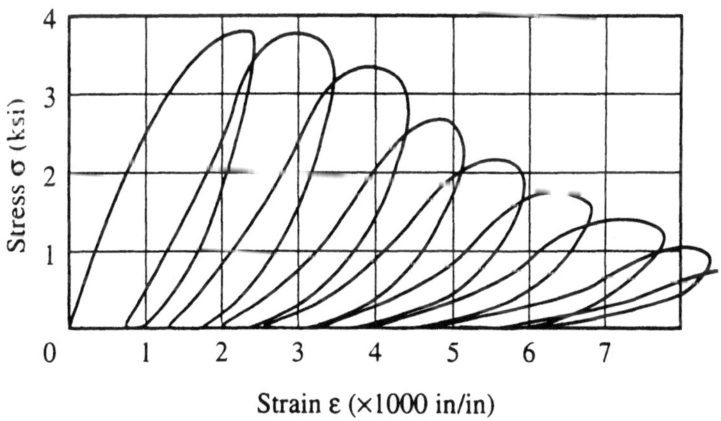

Fig. 1.18 Cyclic stress-strain curve [1.7]

One widely used indicator of damage is the total strain accumulation. Figure 1.19 shows some representative results which display three distinct phases in the fatigue life of a concrete specimen [1.9]. The first one, representing possibly 10% of the fatigue life, appears to be characterized by a relatively rapid increase of damage. This is deceptive, however. The material rather undergoes a period of consolidation, during which preexisting voids are reduced in size, not unlike compaction of granular soils. In the second phase, which covers almost 70% of the material's life, damage increases rather steadily. During the third phase, damage accumulation unmistakably accelerates all the way up to failure.

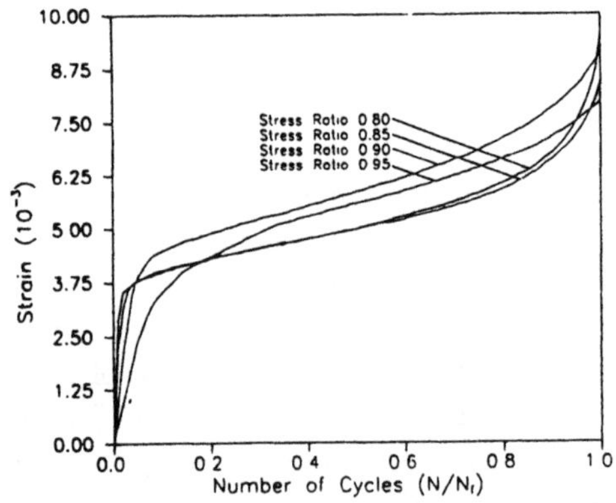

Fig. 1.19 Strain accumulation under cyclic load [1.9]

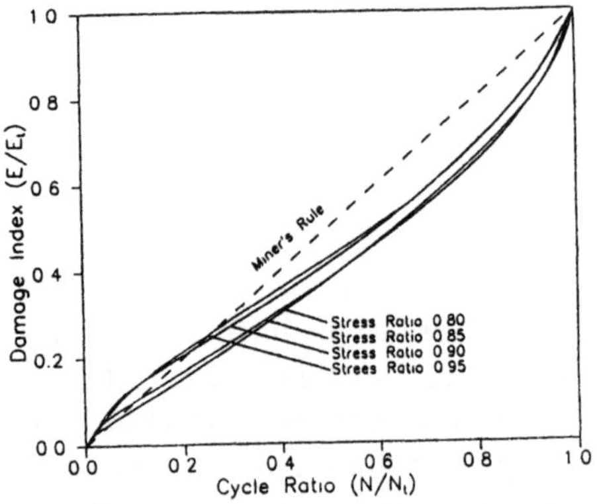

Fig. 1.20 Damage accumulation as measured by energy ratio

If damage is measured as the ratio of dissipated energy to the total energy dissipation capacity per unit volume, then the evolution of damage appears as shown in Fig. 1.20. Here the three phases of life are less distinct and more consistent. The first phase all but disappears, indicating that the amount of work needed in the consolidation phase is not much different than the work needed to create internal fracture. These results also show that the assumption of linear damage accumulation, which underlies the so-called Miner's Rule, is not quite applicable. The degree of deviation from linearity varies with concrete strength, stress level, and other factors.

At this point it is worthwhile to point out the potentially serious consequences of overrelying on "ductility" as a design parameter in the presence of cyclic loads, because the ductility capacity of any concrete material is clearly a function of the number of load applications [1.10]. In other words, if we are dealing with repeated load applications, we cannot rely on material properties derived from monotonic load tests. The material response needs to be treated as a fatigue phenomenon, in either the high- or low-cycle range. It is therefore preferable to use the total energy dissipation capacity (of a unit volume material or a reinforced concrete member) rather than ductility as the important design capacity to be compared with the load demand.

Fatigue data are most commonly presented in S-N diagrams. Whereas for most metals there are distinct stress levels, below which the fatigue life is basically unlimited, concrete does not have such a fatigue limit. Representative S-N curves, if plotted on semi-log paper, consist of two fairly straight line segments with clearly distinct slopes. The first one is associated with low-cycle fatigue, and the second one with high-cycle fatigue [1.11], Fig. 1.21. These two different slopes can be considered to reflect two different types of microcracks. In the low-cycle fatigue range, failure can be attributed to cracking in the mortar matrix, whereas in the high-cycle fatigue range, it is the bond cracks on the mortar-aggregate interface, which ultimately lead to failure. The development of bond cracks and isolated mortar cracks by themselves cannot lead to failure, because of concrete's highly redundant internal structure. This internal redundancy permits a large number of load applications until a gradual crack network develops, which reduces the number of paths, through which the material can transmit load. The remaining load paths have to assume a greater share of the load, thereby hastening the crack propagation until the material fails.

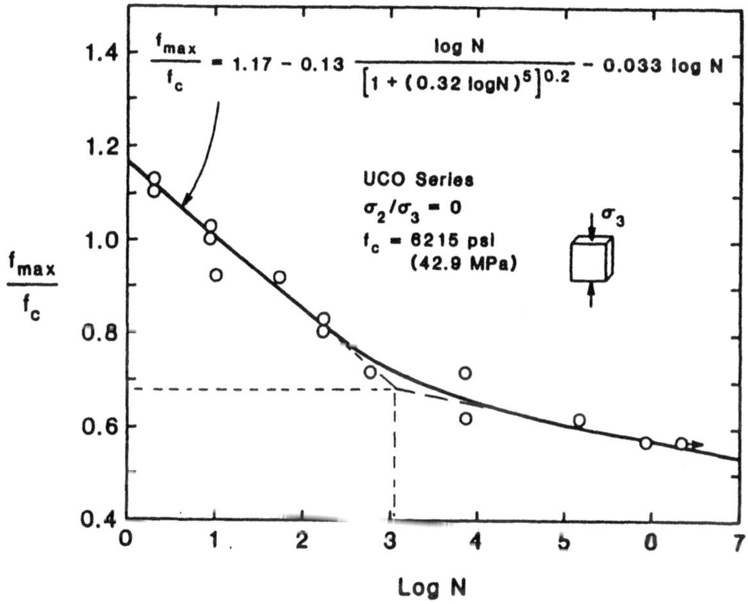

Fig. 1.21 S-N Curve for concrete [1.11]

In the low-cycle fatigue range the stress levels are much higher. Therefore cracking progresses mainly through the mortar, thereby leading to rapid deterioration of the concrete up to failure.

Hsu [1.11] suggests that the point at which the two different straight lines of the S-N diagram intersect, corresponds to the discontinuity stress, which also coincides with the sustained strength limit, i.e. the static stress level that the material is capable of resisting indefinitely, or about 0.75 f_c'. It is also this critical stress level, at which the bond cracks in a monotonic test start propagating into the matrix. This interesting hypothesis is confirmed by tests with fiber reinforced concrete [1.11]. These indicate that fiber reinforcement increases the low-cycle fatigue strength much more than the high-cycle fatigue strength, because the fibers can strengthen the mortar matrix, which controls the low-cycle fatigue strength, but have much less effect on the mortar-aggregate interface bond, which is responsible for the crack initiation and propagation in the high-cycle fatigue range.

Finally, it should be pointed out that cyclic or repeated load application may introduce non-proportional loading. In three-dimensional stress states this complicates the situation tremendously. Some limited results for non-proportional loading in three dimensions are available [1.12]. None of the classical theories to be reviewed below are really capable of reproducing the complicated material response to such loading. But new models based on continuum damage mechanics are now under development, which promise to improve the situation.

1.2.4 Classical Theories

A wide variety of models based on classical elasticity and plasticity theory have been proposed in the past to describe the constitutive behavior of concrete [1.13-1.15].

Nonlinear *elasticity-based* models postulate that the nonlinear behavior of concrete can be represented by appropriate changes of the tangent moduli (incremental formulation) or the secant moduli (total stress-strain formulation). They are simple to use and can match certain experimental results with good accuracy. There is no fundamental difference between these and some of the other models described below, as they all result in variable incremental material stiffness matrices, applicable for certain loading ranges (e.g. one matrix for loading and one for unloading). Elasticity-based models are generally derived directly by intuitive or approximate considerations without the use of loading functions, flow rules, etc. In the total or secant formulations, the current state of stress, σ_{ij}, is assumed to be uniquely determined by the state of strain, ϵ_{ij}, and vice versa. This excludes any load path dependencies and residual strains after unloading that concrete certainly exhibits. In incremental formulations, the state of stress is not only a function of the current state of strain but also of the stress path followed to reach that state. Even though such models can retrace cyclic loading paths quite realistically, a number of fundamental problems exist.

As one example, the "equivalent uniaxial strain" model of Darwin and Pecknold [1.16] might be mentioned. This was developed to reproduce two-dimensional states of stress. Incremental stress-strain relations for an orthotropic material take the form,

$$\begin{Bmatrix} d\sigma_1 \\ d\sigma_2 \\ d\tau_{12} \end{Bmatrix} = \frac{1}{1-\nu_1\nu_2}\begin{bmatrix} E_1 & \nu_2 E_1 & 0 \\ \nu_1 E_2 & E_2 & 0 \\ 0 & 0 & (1-\nu_1\nu_2)G \end{bmatrix}\begin{Bmatrix} d\epsilon_1 \\ d\epsilon_2 \\ d\gamma_{12} \end{Bmatrix} \quad (1.5)$$

where $\nu_1 E_2 = \nu_2 E_1$, and subscripts 1 and 2 denote the current principal stress axes. After defining an "equivalent Poisson's ratio",

$$\nu^2 = \nu_1\nu_2 \quad (1.6)$$

and assuming the shear modulus G to be independent of axis orientation,

$$(1-\nu^2)G = \frac{1}{4}(E_1 + E_2 - 2\nu\sqrt{E_1E_2}) \quad (1.7)$$

the introduction of incremental equivalent uniaxial strains, measured in the principal stress directions,

$$\begin{aligned} d\epsilon_{1u} &= \frac{1}{1-\nu^2}\left[d\epsilon_1 + \nu\sqrt{\frac{E_2}{E_1}}d\epsilon_2\right] \\ d\epsilon_{2u} &= \frac{1}{1-\nu^2}\left[d\epsilon_2 + \nu\sqrt{\frac{E_1}{E_2}}d\epsilon_1\right] \end{aligned} \qquad (1.8)$$

leads to the uncoupling of the constitutive equations, Eqs. 1.5,

$$d\sigma_1 = E_1 d\epsilon_{1u}, \quad d\sigma_2 = E_2 d\epsilon_{2u}, \quad d\tau_{12} = G d\gamma_{12} \qquad (1.9)$$

For 3-D models, the work by Bazant and Tsubaki [1.17] and Elwi and Murray [1.18] should be mentioned.

The theory of *plasticity* has a long history of successful applications to metals. In recent years it has been applied also to frictional materials such as concrete, rocks, and soils. There are a few modifications necessary, though.

The classical theory of plasticity is based on three main assumptions - an initial yield surface that defines the beginning of yielding, a hardening rule, and a flow rule. To apply this theory to concrete, also a failure surface is needed to mark the beginning of fracture. The initial yield surface is usually assumed to be smaller but geometrically similar to the failure surface. These surfaces have considerably different shapes than yield surfaces for metals. Both the von Mises and Drucker-Prager yield criteria require tension cut-offs because of the low tensile strength as compared to the compressive strength. Also, the von Mises yield criterion disregards the first stress invariant or hydrostatic stress component of the stress state, whereas concrete exhibits volume dilatancy near failure.

The flow rule of plasticity theory is expressed mathematically as,

$$d\epsilon_{ij}^p = \lambda\frac{\partial f}{\partial \sigma_{ij}} \qquad (1.10)$$

where $\lambda > 0$ is a scalar proportionality factor and $f(\sigma_{ij})$ the yield surface function. Experimental evidence shows that both the associated flow rule and the normality requirement do not apply strictly for frictional materials like concrete. Thus, non-associated flow-rules are gaining popularity [1.19].

Plasticity-based models with strain-hardening can reproduce stress-history dependent behavior, because they allow for unloading, residual strains and reloading, so that cyclic loading problems can be considered. But softening behavior cannot be reproduced. The use of the standard normality rule of classical plasticity theory may predict unreasonably high volume expansions - another reason for the more recent models to include non-associated flow rules and possibly a dilatancy factor, which is a function of the hardening parameter.

The basic idea of *plastic-fracturing* models is that inelastic strains are in part due to plastic deformation and in part due to microcracking [1.20]. The plastic deformations are defined by the flow theory of plasticity, but plastic slip does not lead to strain-softening, i.e., a decrease in stress with increasing strain. But microcracking or fracturing models can do. A combined plastic-fracturing constitutive law can be formulated simultaneously in stress and strain space. Such models are very complicated and still have limited capabilities.

In *endochronic* models, inelastic strains are not obtained from a loading surface, but from the evolution of a measure of irreversible damage, referred to as intrinsic time [1.21, 1.22], which is a nondecreasing scalar variable that depends on strain increments. The inelastic strains relate to the intrinsic time through a series of mapping functions that depend on the current states of stress and strain. Thus, the model is incrementally nonlinear. Endochronic theory, in its original form, has been shown in certain cases to lead to non-unique solutions, and small errors in initial and/or boundary conditions can lead to rapid deterioration in the accuracy of the subsequent computations. To correct these short-comings, loading surfaces and jump-kinematic hardening were introduced, which are quite analogous to the yield surfaces of classical plasticity theory.

Numerous recent developments have evolved from *fracture mechanics* [1.23-1.25]. These are covered elsewhere in this chapter.

1.2.5 Rate-Dependent Models

The stress- or strain-rate dependency of concrete behavior expresses itself in a significant increase in strength and Young's modulus with increasing rate, Fig. 1.17. The strength at the high strain rates associated with impact conditions (approximately $10^2/sec$) can be twice that at a quasi-static rate of approximately $10^{-3}/sec$. Such variations cannot be ignored in realistic analyses.

In the framework of elasto-plasticity, yield and failure surfaces are employed, which describe the union of various stress states, for which the material ceases to behave elastically and non-recoverable deformation takes place. Failure surfaces for concrete are typically cone-like and pressure dependent. Different surfaces have been proposed, with more or less complexity and suitability for computational purposes. The implementation of such models within finite element analysis and flow theory of plasticity requires the definition of a plastic strain tensor increment, and accounting for any changes that the surface may undergo as a result of hardening or softening during the strain increment (so-called consistency condition). Some computational algorithms for the stress increment computation iterate to both satisfy global equilibrium and ensure that the stress and strain states at all integration points correspond to the elasto-plastic constitutive law. By definition, there is no *real time* involved in plastic analysis. All inelastic processes are assumed to have sufficient time to take place, as if occurring at an infinitely slow time rate. When "time steps" are mentioned in the context of plasticity computations, they refer to a "pseudo-time" and should never be confused with real time.

It is precisely this issue that makes modeling of rate-dependent concrete behavior so fundamentally different from modeling its static behavior. Failure surfaces are now rate dependent. They may expand or shrink, depending on what strain rate a material point experiences. For example, in a finite element model of a concrete structure under impact loading, some integration points may be subject to vastly different stress or strain rates than others, which in turn causes the corresponding instantaneous failure surfaces to be completely different.

Clearly, the inclusion of real time plays a crucial role. Inelastic processes cannot develop in fast-rate situations, hence the change of the instantaneous failure surface.

A crude approximation of the rate dependency may be obtained by adjusting the elasto-plastic failure surface parameters so that they correspond to some average rate, or by employing for every integration point an instantaneous failure surface, which depends on the instantaneous strain rate at that point. This means that the failure surfaces are allowed to soften or harden for a given instant in real time, the so-called *rate-hardening* or *softening*, depending on the strain rate.

A more appropriate mathematical framework is found in elasto-viscoplasticity theory, which has been used by various researchers for developing rate-dependent concrete models.

The elasto-viscoplastic model proposed by Perzyna [1.26], differs from the elasto-plastic model insofar that a stress point can be outside the elastic limit surface. The elasto-plastic model recognizes stress states only within and on the yield surface, and most of the computational effort is spent to return stress points to the current yield surface. In an elasto-viscoplastic model, the rate of inelastic flow is defined as some function of the distance between the stress point and the elastic limit surface and the so-called fluidity parameter γ,

$$\dot{\epsilon}_{ij}^{vp} = \gamma < \Phi(F) > \frac{\partial F}{\partial \sigma_{ij}} \tag{1.11}$$

where the dot ($\dot{}$) denotes differentiation with respect to time, and the $< >$ bracket implies that the expression is activated only if $F > 0$, i.e., when the stress point is outside the elastic limit surface. The adoption of an elasto-viscoplastic model requires the definition of, 1) some elastic limit surface (sometimes also called a discontinuity surface), 2) the fluidity parameter, and 3) the functional relationship $< \Phi(F) >$.

To choose an elastic limit surface is relatively easy. Any convenient failure surface description could be used, with proper scaling, since one is dealing with the elastic limit surface, rather than the failure surface. The other two model parameters, the fluidity parameter and the functional relationship for $\Phi(F)$ should ideally be obtained from three-dimensional experiments conducted under different rates of loading. As such experiments are scarce even for uniaxial compression, the models that are in use aim to reproduce one-dimensional responses and are assumed to apply also for other stress histories.

Probably the most complete rate-sensitive model for concrete has been suggested by Nilsson [1.27, 1.28], who introduced an elasto-viscoplastic-plastic-brittle model for concrete at high rates of loading. The closed elastic limit surface is defined by a generalized ellipsoidal surface,

$$F = \left[\left(\frac{\frac{2\sigma_o}{H_r} - \xi_u - \xi_l}{\xi_u - \xi_l} \right)^2 + \left(\frac{\frac{\tau_o}{H_r}}{b(\theta)} \right)^2 \right]^{1/2} - 1 \tag{1.12}$$

where σ_o, τ_o are the octahedral normal and shear stress, and θ the Lode angle of similarity. ξ_u and ξ_l are surface intersections with the space diagonal. The hardening parameter H_r indicates the change of the elastic limit surface as a result of inelastic processes. The hardening parameter is used also to model the rate hardening, where the normalized hardening for the uniaxial compressive strength is obtained by curve fitting of uniaxial experimental data.

For multiaxial cases, a scalar measure of the process rate, the "effective strain rate", ϵ_{ef}, is assumed to be given by the root mean square of the normal and shear octahedral strain rates,

$$\dot{\epsilon}_{ef} = \left(\dot{\epsilon}_o^2 + \frac{1}{4}\dot{\gamma}_o^2\right)^{\frac{1}{2}} \tag{1.13}$$

The fluidity parameter γ is also dependent on the effective strain rate,

$$\gamma(\dot{\epsilon}) = \dot{\epsilon}_{ef}\ e^{-r ln|\frac{\dot{\epsilon}_{ef}}{\dot{\epsilon}^r_{ef}}|} \tag{1.14}$$

where $\dot{\epsilon}^r_{ef}$ is the reference value for the effective strain rate, and r is a material constant.

For the functional relationship $\Phi(F)$, Nilsson adopted the exponential function of a form suggested by Perzyna,

$$\Phi(F) = \left(\frac{F}{F_o}\right)^{\alpha} \tag{1.15}$$

and the appropriate value for α to fit the experiments was found to be 0.40.

Finally, Nilsson introduced the strain-rate dependency of the outer (failure) surface, in the context of instantaneous rate hardening. Two parts of the failure surface, covering the tension/compression and tension/tension regime, are allowed to harden or to soften, depending on the instantaneous effective strain rate.

Other investigators followed similar lines in developing rate-dependent models for plain concrete. Bicanic [1.29, 1.30] proposed a multi-axial elasto-viscoplastic degradation model, in which only the fluidity parameter was elastic strain-rate dependent, and the elastic limit Mohr-Coulomb-like surface (discontinuity surface) was kept constant in the pre-failure regime.

Meyer and Delgado-Saavedra [1.31] expanded the Bicanic model. Two surfaces in stress space describe the material behavior, Fig. 1.22: a discontinuity surface S_D that plays the role of a yield surface, and a bounding failure surface S_F, which is essentially a damage monitoring device to control the failure stress level. The fluidity parameter is considered to be dependent on the elastic strain rate. The model has viscous properties in the plastic region only, i.e. elasto-viscoplastic material behavior, so that the initial yield condition is the same as in plasticity theory. For a stress point within S_D, the material behavior is assumed to be linear elastic, outside it is viscoplastic, and when it reaches S_F, Fig. 1.23, failure takes place with concurrent unloading. During inelastic straining, the S_F surface shrinks, as damage of the material accumulates. Once a stress point reaches S_F, also the S_D surface shrinks as a function of the post-failure dissipated energy density. The five-parameter model of Willam and Warnke [1.32] is used to describe both the discontinuity and strength limit surfaces. The model was calibrated, for fast loading rates only, against the test results reported by Hatano and Tsutsumi [1.33], and it has been applied to the three-dimensional stress analysis of concrete specimens.

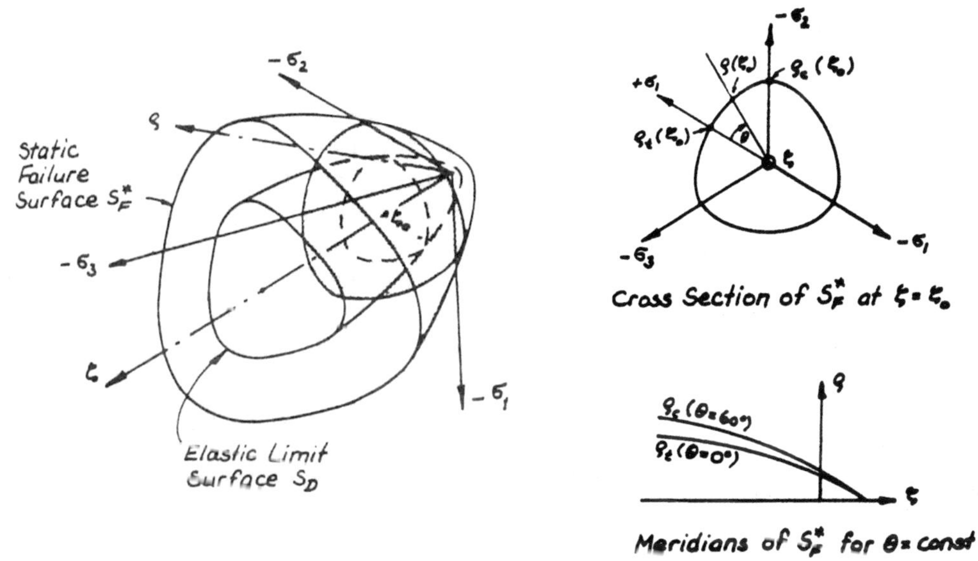

Fig. 1.22 Geometry of static failure surface S_F^* and elastic limit surface S_D [1.31]

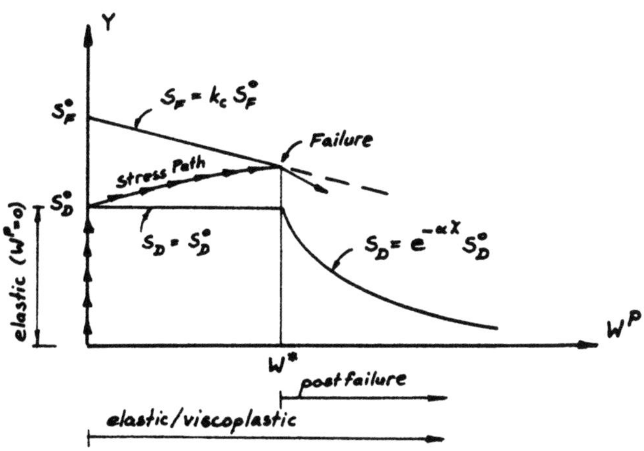

Fig. 1.23 Degradation of surfaces S_D and S_F during viscoplastic straining [1.31]

The rate sensitivity of both Bicanic's and Delgado-Saavedra's model is introduced on two levels: the fluidity parameter is strain-rate dependent, and the degrading strength limit surface ensures the rate-dependence of the failure stress level.

None of the models proposed to date consider the fact that the elastic modulus is affected by the strain rate. Concrete cracking models based on fracture mechanics employ either the fracture energy release rate G_f (fracture energy models) or the fracture toughness K_{Ic} (linear elastic fracture mechanics models). Both parameters are strain-rate dependent, and numerous tests have confirmed that there is an increase of both G_f and K_{Ic} with increased strain rate.

1.2.6 Continuum Damage Mechanics

Damage mechanics has evolved as a subdiscipline of continuum mechanics. Its conception is generally credited to Kachanov in 1958 [1.34], who studied brittle creep rupture under uniaxial tension at elevated temperatures. The state of the material deterioration was characterized by a scalar field variable ψ, referred to as "continuity". For a defect-free material, $\psi = 1$, and $\psi = 0$ characterizes a material with no remaining load carrying capacity. Accordingly, damage D is defined as being complementary to continuity, $D = 1 - \psi$. Rabotnov [1.35] added to this the concept of effective stress, $\overline{\sigma} = \sigma/(1 - D)$. Lemaitre [1.36] then stated the important Strain Equivalence Principle, which says that any strain constitutive equation for a damaged material may be derived in the same way as for a virgin material if the usual stress is replaced by the effective stress.

Before reviewing some of the damage theories, the importance of the representative volume should be mentioned. For any experimental as well as numerical work, this volume should be small enough to avoid smoothing of stress or strain gradients and large enough to represent an average of microprocesses. Lemaitre has shown [1.36] that for concrete this volume should be a 10 cm or 4 in cube and pointed to the localized nature of damage as compared to strain - whereas damage in the form of cracks is tied to areas, strain is associated with volume.

Lemaitre, Mazars and their coworkers have developed a series of damage mechanics models, based on the thermodynamics of irreversible processes. Each equilibrium state is defined by a scalar thermodynamic potential,

$$\rho\psi = f(\epsilon^e, D, \epsilon^p) \tag{1.16}$$

Only the elastic properties of the material are assumed to be affected by damage, i.e.,

$$\rho\psi = \rho\psi^e + \rho\psi^p \tag{1.17}$$

where

$$\rho\psi^e = \frac{1}{2}\mathbf{\Lambda}(\mathbf{D})\epsilon^e\epsilon^e \tag{1.18}$$

$\mathbf{\Lambda}(\mathbf{D})$ is a fourth-order symmetric tensor, representing the secant stiffness properties of the damaged material. A sufficient condition to satisfy the Clausius Duhem inequality is $\dot{D} > 0$. The damage growth is governed by a loading surface described by a function f,

$$f(\epsilon, \Lambda, K_o) = 0 \tag{1.19}$$

In order to assure uniqueness, f is a function of strain, not stress. By definition,

$$\begin{aligned} \dot{D} &= 0 \quad \text{if} \quad f < 0 \quad \text{or} \quad f = 0 \quad \text{and} \quad \dot{f} < 0 \\ \dot{D} &= F(\epsilon) \quad \text{if} \quad f = 0 \quad \text{and} \quad \dot{f} = 0 \end{aligned} \tag{1.20}$$

In Mazars' Scalar Damage Model [1.37], the material is assumed to behave elastically and remain isotropic. The damage growth is governed by the equation,

$$f(\epsilon, \Lambda, K_o) = \tilde{\epsilon} - K(D) \tag{1.21}$$

where

$$\tilde{\epsilon} = \sqrt{< \epsilon_1 >_+^2 + < \epsilon_2 >_+^2 + < \epsilon_3 >_+^2} \tag{1.22}$$

is an equivalent strain, with $< \epsilon_i >_+ = (|\epsilon_i| + \epsilon_i)/2$. ϵ_i are the principal strains, and $K(D)$ is the largest value of $\tilde{\epsilon}$ ever experienced by the material.

The assumption that only positive (extensional) strains are responsible for crack propagation (Eq. 1.21) restricts the model to only type I damage (separation). The dissimilarity between concrete response to tension and compression is taken into account by introducing two damage measures, D_t and D_c, for uniaxial tension and compression, respectively. But this formulation cannot reproduce the stiffness recovery during load reversal, when cracks close and have at most little effect on subsequent compressive behavior.

This shortcoming was eliminated in the Unilateral Damage Model, where the two scalars D_t and D_c grow independently for positive and negative strains, respectively. For complex loading, damage is a combination of D_t and D_c. The material is assumed to behave elastically and to remain isotropic. By decomposing the stress tensor into a positive part σ_+ and a negative part σ_-, the expression for the thermodynamic potential becomes,

$$\rho\psi^e = \rho\psi^e(\sigma_+) + \rho\psi^e(\sigma_-) \tag{1.23}$$

where, again, $\dot{D}_t \geq 0$ and $\dot{D}_c \geq 0$. The growth of the two damage parameters is governed by two separate damage loading surfaces, thereby preserving the differences in tension and compression response, and the damage produced by tension has no effect on the response in compression, and vice versa.

Pijaudier-Cabot [1.38] developed a damage model for high compressive loadings, assuming that the damage produced by hydrostatic pressure preserves the material's isotropy and describing it by a scalar damage function δ. The reduction of the material's stiffness is represented by a second-order tensor, d. Mazars' concept of equivalent strain is generalized by introducing the equivalent deviatoric strain as defined in Eq. 1.22, where again, $< \epsilon_i >_+ = (|\epsilon_i| + \epsilon_i)/2$, but ϵ_i are now the eigenvalues of the deviatoric strain tensor. This generalization becomes important in cases of high compressive loading without positive principal strains. Thus, for compression,

$$\begin{aligned} \dot{d}_j &= 0 \quad \text{if} \quad e_j \geq 0 \\ \dot{d}_j &\neq 0 \quad \text{if} \quad e_j < 0 \end{aligned}$$

Simo and Ju [1.39] pointed out that the definition of equivalent strain as the J_2 invariant of the strain tensor results in non-symmetric elastic-damage tangent moduli. Also the rate-dependency of material behavior is not addressed.

1.2.7 Finite Element Implementation

The finite element implementation of the material models described herein is part of an important new discipline called computational mechanics. This field is concerned with computational aspects of nonlinear solutions of problems in plasticity, fracture and damage mechanics. For practicing structural engineers this field is gaining in importance because of the steady trend towards strength design. Strength or ultimate design is contingent on our ability to compute load levels that cause structural collapse. In skeletal or frame structures, ultimate strength is tied to well-understood redundancy, and limit analysis methods are available to determine collapse mechanisms and the corresponding failure load levels. These methods are applicable only if the members have sufficient ductility to permit the required moment redistributions, i.e. members are assumed to be able to undergo large inelastic deformations (which are tantamount to local failure) without impairing the structure's capacity to carry further load.

In continuum problems, the situation is somewhat similar. Here we are dealing with local regions that may get overstressed (as sampled at individual integration points) and forced into the post-peak regime, thereby triggering stress redistributions. In frame analysis, members are often assumed to exhibit elastic-perfectly plastic moment-curvature behavior, and local failure is neither a mechanical modeling nor a numerical problem, only a design problem. In continuum mechanics, the situation is different. Here the constitutive equations should be capable of tracing the material behavior realistically beyond the peak both in tension (fracture) and compression (crushing or splitting). This requirement places new demands on our numerical solution algorithms.

In frame analysis, a collapse mechanism is characterized by a singular tangential structure stiffness matrix, i.e., the appearance of each plastic hinge deflates the matrix until it turns singular. The corresponding problem in fracture analysis of a finite element discretization of a continuum is a bifurcation problem not unlike structural problems with snap-through effects that require advanced numerical integration algorithms such as the arc-length method.

Much progress has been made in the last decade in the modeling of failure mechanisms, such as cracking and crushing of concrete, bond slip, yielding of reinforcement and interface sliding. However, the difficulties inherent in failure computations slowed down development of robust and reliable solution techniques, that can simulate numerically the response of structures beyond their maximum load-carrying capacity and into the regime of ductile unloading. Since it is the goal of structural design to arrive at ductile failure modes, it would be of great practical value to have available numerical tools to demonstrate the absence of premature brittle failure modes.

Such failure computations involve both soft and hard nonlinearities. Soft nonlinearities are gradual and continuous both in terms of their temporal evolution and spatial distribution. An example is the stress-strain curve of concrete in compression, which reflects a gradual change in stiffness and the internal redundancy of the multiple load paths for stress redistribution. Hard nonlinearities, in contrast, are associated with discontinuous phenomena, also both in time and space. Here, mechanical properties change abruptly, thereby placing a higher demand on the structure to physically redistribute the released forces, and on the numerical algorithm to simulate this redistribution process. Examples are the abrupt change in slope of the stress-strain curve for steel and the brittle fracture of concrete both on the material as well as the element level.

A comprehensive overview of the computational aspects of finite element analysis of reinforced concrete can be found in the proceedings of a workshop held at Columbia University in 1991 [1.15]. The workshop brought together researchers pursuing both the continuum-based approach and the discrete fracture approach. Traditional concepts of plasticity and fracture have been extended far beyond their original scopes in order to capture brittle-ductile failure mechanisms within a computational strategy. For example, localized failure in the form of discontinuous bifurcation is normally attributed to the destabilizing effects of non-symmetric operators, such as in the presence of non-associated plastic flow. The alternative is discrete fracture mechanics, which suffers under severe mesh-size dependence and requires the definition of a fracture process zone based on continuum concepts, which describe progressive decohesion in terms of damage mechanics and/or plasticity. The principal issue is whether progressive failure can be captured reliably by the continuum-based approach with a fixed finite element grid strategy, or whether the transition from a continuum to a failed discontinuum requires a discrete approach with grid adaptation.

The problems that are dealt with in this context are strain-softening, loss of normality, extreme mesh sensitivity, and lack of objectivity and localization due to discontinuous bifurcation.

1.3 Test Methods for Impact

1.3.1 Introduction

Experimental data using a variety of test methods have shown that a rate effect exists which can increase strength and deformation of concrete. However, such experimental results should be interpreted with caution since other effects may also have a significant influence. Inertia effects, local damage, or stress wave reflection can be mentioned which should receive due attention. The test methods are differently prone to such effects, which requires a careful evaluation.

There are several questions concerning loading rate effects with respect to testing. First, the test method must be clear and physically sound so that the quantity of interest can be measured. Second, test methods should be described or developed which are suitable for investigating the reason for the rate effect. An example may clarify two aspects. First, in a bending test, inertia effects must be considered in order to calculate the true fracture energy. Second, it should be investigated whether the increase of fracture energy at high loading rates is caused by an increase of the process zone size or by the forced fracture of aggregate or other causes.

1.3.2 Bending Test

The bending test is frequently used at high rates of loading since the forces to fracture are moderate and can be supplied rather easily. In metal testing, the Charpy impact test on notched specimens, for instance, is widely recommended for the determination of total fracture energy. It is used especially to determine the temperature influences on the ductility of metals. The results are of limited validity, however, since stiffness of the pendulum, size and geometry of the specimen and stiffness of the supports influence the results. Thus, in metal testing the entire setup is standardized and the remaining error may be acceptable.

1.3.2.1 Inertia effect

Let us assume a beam of span l_0 and with two cantilevering overhangs of length l_1 is hit by a central force $P_t(t)$, Fig. 1.24. This beam will deform according to material properties and geometry. If there is an edge crack (notched beam), the deformations are likely to be linear, with their maximum occurring at the beam center. If there is no crack or if the beam is reinforced, the deflection line will be a higher order polynomial, which may be approximated by a sine function. The inertia effects in both cases have been analysed. In case of constant cross-section, the inertia resistance caused by a force $P_t(t)$ is,

$$P_I(t) = \rho A \ddot{u}_0(t) \left[\frac{l_0}{3} + \frac{8}{3} \frac{l_1^3}{l_0^2} \right] \tag{1.24}$$

for linear displacement distribution and

$$P_I(t) = \rho A \ddot{u}_0(t) \left[\frac{l_0}{3} + 2 \frac{\pi^2}{3} \frac{l_1^3}{l_0^2} \right] \tag{1.25}$$

for sinusoidal displacement distribution. A is the cross-sectional area, ρ the mass density of the material, and $\ddot{u}_0$ the acceleration of the beam center.

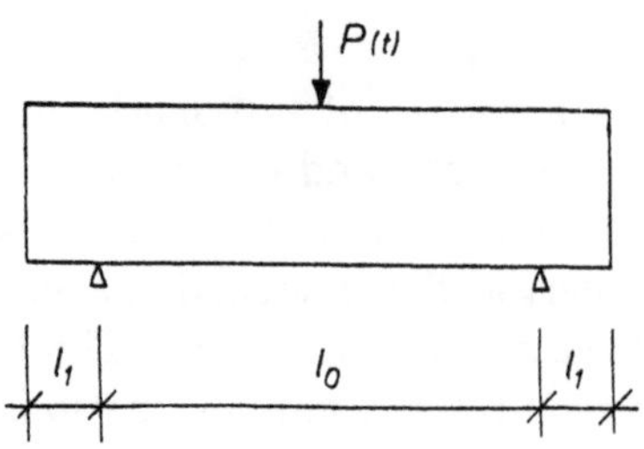

Fig. 1.24 Test beam specimen

If the total load $P_t(t)$ and the acceleration $\ddot{u}_0$ are measured, the effective load which causes bending is,

$$P_b(t) = P_t(t) - P_I(t) \tag{1.26}$$

The deflection time history $\delta(t)$ can be measured separately or obtained by integration of the acceleration signal. $P(t)$ and $\delta(t)$ lead to the fracture energy G_F. The drop-weight-type and the Charpy-bending-type impact test have been analyzed using a two-degree-of-freedom model [1.40, 1.41]. The test beam is represented by a lumped mass with a certain stiffness and the hammer-tup assembly is represented by another lumped mass. The tup-specimen contact zone has an appropriate effective stiffness. The result of the analysis shows an increasing influence of inertia oscillations on the measured result with decreasing stiffness of the specimen and increasing stiffness of the contact area. The analytical peak load is given by

$$P_{t,max} = V_0\sqrt{k_e m_t} \tag{1.27}$$

where V_0 is the striker (tup) velocity, k_e the contact stiffness, and m_t the lumped mass of the hammer-tup combination. The tup force oscillation is calculated by

$$R_t = \sqrt{\frac{m_b}{m_t}\frac{1}{\rho}(1+\rho)^2} \tag{1.28}$$

with m_b = specimen mass, and ρ = ratio between specimen stiffness and contact stiffness. Fig. 1.25 shows the arrangement of the testing equipment, and Fig. 1.26 shows how the amplitude ratio varies with the stiffness ratio. It can be seen how sensitively the recorded tup load reacts to stiff beam-tup contact behavior. The situation improves drastically if, for instance, a rubber or aluminum pad is placed between tup and specimen; however, the consequence would be a reduction in strain rate.

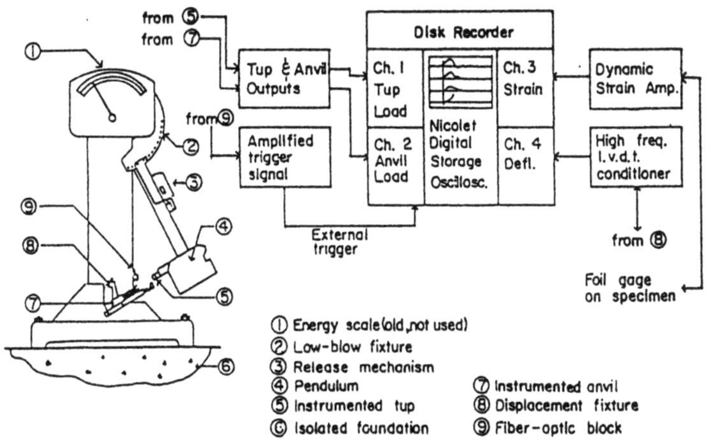

Fig. 1.25 Schematic of modified instrumented Charpy test

The above analysis is also appropriate for bending by a drop-weight [1.42]. It has been used for plain concrete and fiber-reinforced concrete. From the results, the energy can be calculated which is consumed up to maximum load and failure load and, maybe, also up to the load at which the load-displacement curve deviates from a straight line, i. e. some kind of elastic limit.

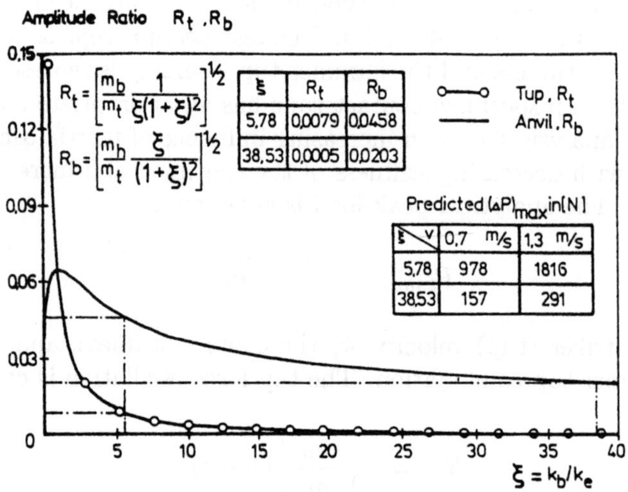

Fig. 1.26 Inertia influence on impact bending response

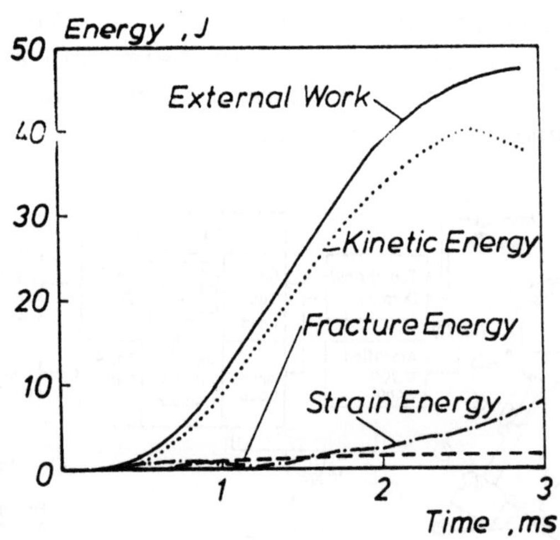

Fig. 1.27 Partition of energy in an impact test

To illustrate the importance of appropriately accounting for kinetic effects, Figs. 1.27 and 1.28 show the partition of energy for two bending tests, first with an impact loading by a falling hammer, second with a displacement-controlled load at smaller velocity. Figure 1.27 [1.43] shows the results of a finite element analysis of the test [1.44]. It can be seen that the kinetic energy increases dramatically during the impact loading, while fracture energy and strain energy are at a comparably low level.

In contrast, Fig. 1.28 shows the result of a test with a maximum strain rate of 0.1 sec^{-1} (Test 21) and 0.005 sec^{-1} (Test 41) [1.45]. At these rather low velocities, the fracture energy dominates, and the kinectic energy is negligible. The examples demonstrate quite clearly that "static" loading leads directly to fracture energy, whereas high strain-rate loading has to be evaluated very carefully if the dynamic fracture energy is to be determined.

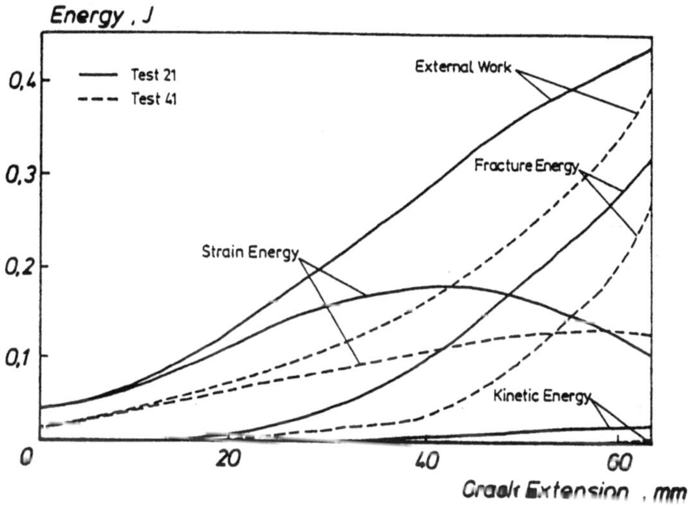

Fig. 1.28 Partition of energy in quasi-static testing

1.3.2.2 Local Damage

During impact of a hammer on a concrete beam, concentrated stresses will occur in the contact area. These can lead to damage of the concrete by which a certain amount of energy is dissipated. This amount depends on the hardness of the hammer, its shape and flatness, on the mechanical properties of the beam, its geometry and flatness, and on the measure of coincidence of the hammer and beam surface. In order to avoid arbitrary influences on the results, the contact surfaces should fit as well and reproducibly as possible and/or the contact surface of the beam should always be covered by a pad of the same material.

The contact force can be approximated by the Hertz formula

$$F = ka^{3/2} \tag{1.29}$$

where a is the penetration of the striker into the material, and k is a constant which is a measure of the stiffness of the impact zone. k depends on the mechanical properties of the specimen and the striker, the loading intensity, and on the geometries involved. F and a are time-dependent. Values for k have been found in experiments, ranging from 4 $kN/mm^{3/2}$ for a rubber pad and 24 $kN/mm^{3/2}$ for a 12 mm ply pad, to 200 $kN/mm^{3/2}$ for a steel plate [1.46]. But k is not really a constant. It varies with stress intensity. Tests of colliding concrete bodies have shown k-values of 50 $kN/mm^{3/2}$ for a spherical/planar surface combination, about 1 for conical/planar, and 80 for truncated conical/planar combinations [1.47]. The concrete quality had a rather small influence on these values. Since k is not constant, the analysis referred to in the preceding section can only be an approximation.

1.3.2.3 Instrumentation

Electrical and optical methods are appropriate for measuring relevant quantities during impact testing. As concrete is concerned, optical methods can be used for the triggering or recording instruments or for crack occurrence measurements. A light beam, which is crossed by the drop weight of an impact testing equipment, can trigger a storage oscilloscope or disc recorder. A bundle of fibres cast into the specimen can detect cracking in the interior of a specimen [1.48, 1.49]. High-speed cameras (>10,000 frames per second) may be used for measuring displacements and crack propagation. However, since concrete is a softening material exhibiting a process zone, it is difficult to locate a crack tip properly; on the other hand, a full-field picture always gives very useful information [1.50].

Electrical devices use strain gauges, LVDTs, cracking gauges, load cells, accelerometers and load-sensitive pads. Strain gauges follow all frequencies which occur at impact testing of concrete. LVDTs should be attached very tightly to the surface in order to prevent spurious oscillations [1.51]. Cracking gauges have been used successfully on bending specimens [1.52], see Fig. 1.29. The so-called KRAK gauges (TTI Division, Hartrum Corporation, Chaska, MN, USA) are mounted with epoxy resin on the specimen surface. Glue and backing material should be brittle in order to follow a crack and not to bridge it.

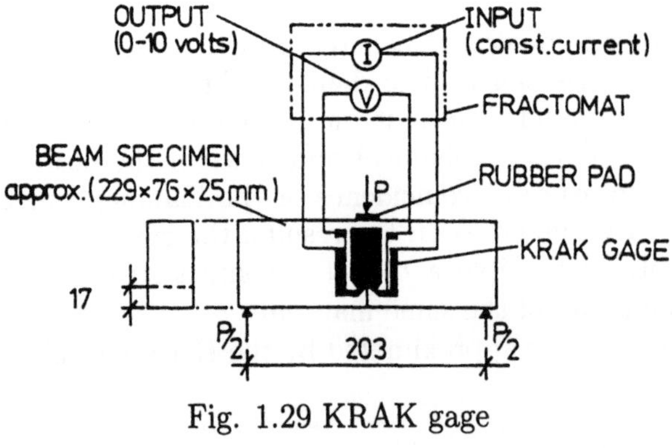

Fig. 1.29 KRAK gage

Accelerometers must be mounted in a clearly defined position which should exclude misalignment. Load cells should be designed in such a way that the resonance frequency is higher than the highest expected frequency of a test. An interesting alternative for load cells may be polymer pressure gauges. These consist of a material (polyvinylidene fluoride) which becomes piezoelectrically and pyroelectrically active when subjected to a large electric field (2 MV/cm at room temperature). Two about 12 μm thin foils of sensitive material are sandwiched between two protective layers of polycarbonate [1.53]. These gauges are mainly used for air pressure measurements, but have also been tested in solids.

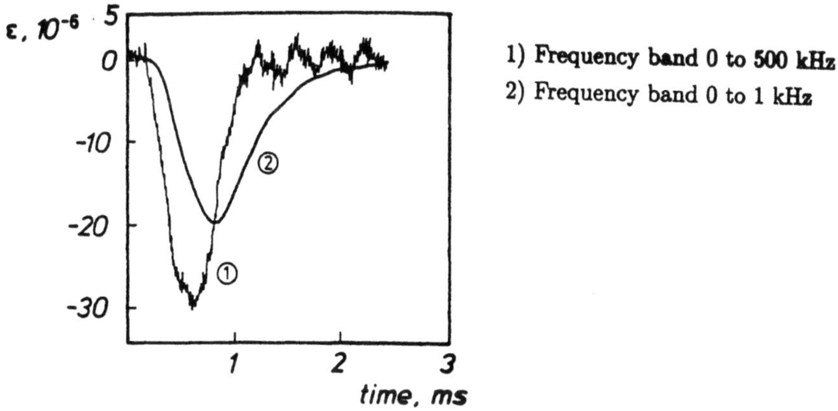

Fig. 1.30 Strain time histories

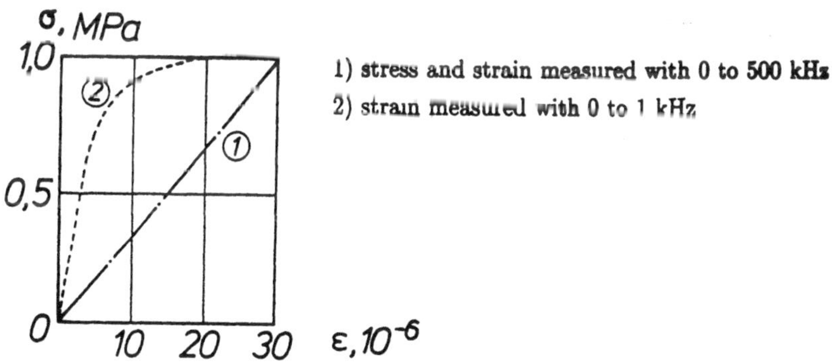

Fig. 1.31 Stress-strain plot

The recording system is an essential part of the measuring chain, having a sufficiently high measuring frequency to follow the event. Fig. 1.30 shows two examples of a strain history record. One was received by a 1 kHz bandwidth oscilloscope, the other by a 500 kHz oscilloscope [1.54]. It can be seen that peak strain and time of occurrence of peak strain are poorly determined by too low a frequency band. When stress and strain are plotted together in a stress-strain plot, Fig. 1.31, it appears that the erroneous strain measurement leads to a large overprediction of Young's modulus and to a nonlinear shape of the curve, whereas the correct measurement leads to the expected linear stress-strain relationship. The concrete has actually been loaded to a fraction of its compressive strength.

As far as concrete testing is concerned, there should be a measuring frequency of at least 10 kHz. Oscilloscopes with a carrying frequency of only 5 kHz are not useful. Various publications have been found where storage oscilloscopes with sampling frequencies larger than 1 MHz have been applied.

1.3.2.4 Various Test Methods

Acoustic emission analysis is one possible method [1.55]. The application of this well-known method to high strain rates (see [1.56], e.g.) requires special arrangements regarding the registration and storage of the acoustic signal.

Impact loading subjects the specimen surface to accelerations as high as 2000 g, which makes it necessary to protect the transducer. Therefore a so-called wave-guide, which carries acoustic waves in a longitudinal direction to the transducer but is weak in the lateral direction, has to be installed. To avoid losses due to wave reflections this wave guide has to be made of a lightweight material (aluminum) with an acoustic impedance of about 17 $MNsec/m^3$. The acoustic impedance of concrete is about 9 $MNsec/m^3$. The general arrangement is shown in Fig. 1.32.

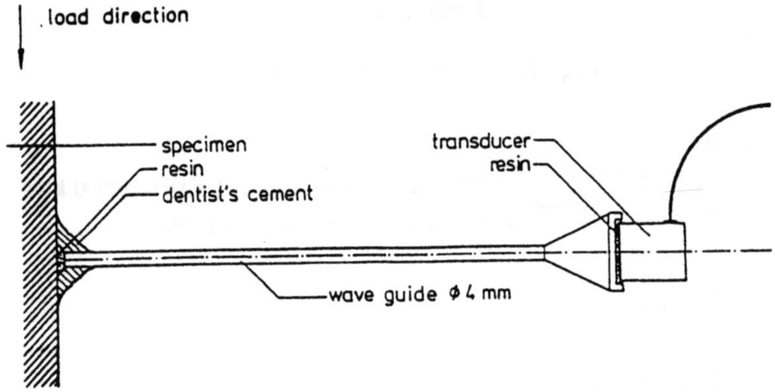

Fig. 1.32 Wave guide to avoid damage of transducer to avoid damage

The analysis of the acoustic signal using traditional equipment is not possible because of the short experiment duration. Instead, the original acoustic emission signal has to be stored either in an analog way on a magnetic tape or in digital form using a transient recorder. After the test the information may be copied to a computer and analyzed numerically.

1.3.3 Uniaxial Loading

Axial loading is often assumed to be the physically soundest test because displacements are distributed uniformly over the cross-section. This is true for a homogeneous isotropic material but becomes questionable when cracks occur. In a plastic hardening material, equilibrium will force the uniaxially loaded specimen into a centric position when it starts to become eccentric. However, a specimen of an elastic softening material tends to become more eccentric because softening areas carry less and less load [1.57]. This means that the descending branch of the stress-deformation curve may be erroneous if too long a specimen is used, which is able to rotate even if the specimen ends do not rotate.

With this in mind, reliable tests can be performed on short specimens which are loaded by a servo-hydraulic actuator, by drop-weights like in the split-Hopkinson bar, or by stress waves.

1.3.3.1 Inertia Effect

Let us consider a tensile specimen which is loaded at the lower end according to Fig. 1.33a. The vertical displacement and acceleration are distributed linearly over the specimen length l. The inertia resistance then is

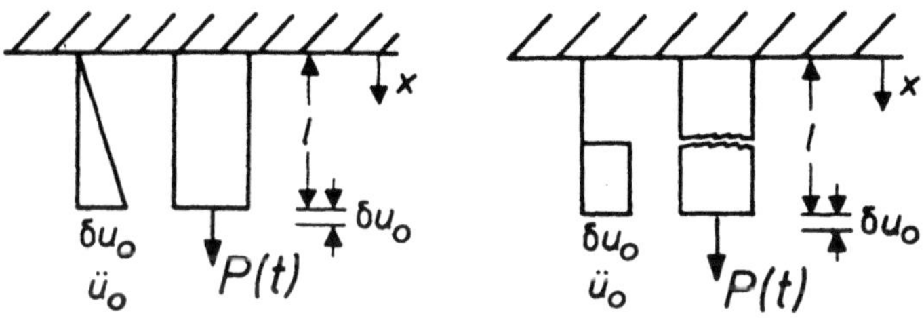

a) uncracked b) with discrete cracking zone

Fig. 1.33 Tensile specimen with displacement and acceleration distribution

$$P_1(t)\delta u_0 = \int_0^l \rho A \ddot{u}_0 \delta u_0 \frac{x^2}{l^2} dx \qquad (1.30)$$

With $\rho = \text{const}$ and $A = \text{const}$,

$$P_1(t) = \rho A \ddot{u}_0 \frac{l}{3} \qquad (1.31)$$

This means that for constant displacement there is no inertia effect, and acceleration and length of the specimen are directly proportional to the inertia effect.

If we assume that a cracking zone starts to develop in the center of the specimen (Fig. 1.33b), the inertia force becomes

$$P_1(t) = \rho A \ddot{u}_0 \frac{l}{2} \qquad (1.32)$$

which is larger than in the previous case.

An example can illustrate how relevant interia forces are in a uniaxial test. If we assume a specimen of 100 mm length and a maximum crack opening of 200 μm at failure, which is reached after 1 $msec$, we get for normal weight concrete with 2400 kg/m^3 a mean acceleration of $\ddot{u}_0 = 400\ m/sec^2$ and an inertia stress

$$\frac{P_1(t)}{A} = (2400)(400)\frac{0.1}{2} = 48,000Pa = 0.048MPa$$

For a tensile strength of 4.8 MPa, the inertia effect is only 1%.

1.3.3.2 Strain and Deformation Measurement

In high-rate loading tests strain gauges can be used, LVDTs and proximity transducers. They should fulfill the requirements with respect to the frequency of the signal to be measured. The frequencies can reach about 10 kHz.

1.3.3.3 Force Measurement

It is common to measure the force by a load cell which is placed in series with the specimen. In order to minimize inertia effects, it is preferable to mount the load cell directly at the non-loaded end of the specimen.

In the split-Hopkinson bar, the transmitter bar is used as a force-measuring device [1.51] if rather long loading pulses are applied. In the original split-Hopkinson pressure bar, where a short rectangular pulse is generated, the stress in the specimen can be calculated from the reflected wave, which travels backwards in the incident bar, see Fig. 1.34. For details of the analysis, the reader is referred to Ref. [1.58].

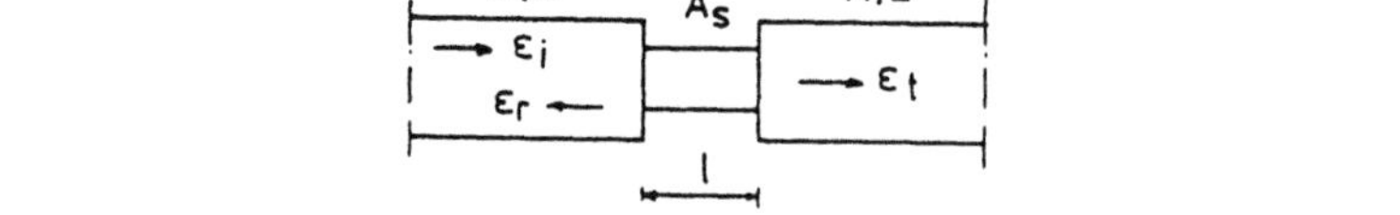

Fig. 1.34 Split Hopkinson bar

1.3.3.4 Loading Devices

There are hydraulic actuators that can reach piston velocities of up to 20 m/s. These are rather expensive devices and only a few laboratories have them at their disposal. On the other hand, these actuators also need a certain displacement to develop the full velocity, and since concrete is rather brittle, this necessary displacement may exceed the fracture displacement of concrete in tension.

Usually, hydraulic equipment is able to reach lower displacement rates, which enable strain rates up to 0.2 sec^{-1} [1.59]. Extra pressure storage can increase the straining rate considerably [1.60]. Higher strain rates can be achieved with electromagnetic loading devices [1.61]. The split-Hopkinson bar mentioned earlier also allows higher loading rates. The same is true for drop-weight and blast loadings.

An interesting method has been proposed in [1.62], where two accelerated masses are used that hit a circular cylinder on both ends simultaneously. The compressive waves are reflected at the remote ends and converted into tensile pulses which meet in the center of the specimen where they cause fracture. Strain rates as high as 20 sec^{-1} were recorded at the center.

1.3.4 Other Types of Loading

There are other types of loading such as the wedge splitting test [1.63], the CT-specimen, the splitting test, the point load test, and maybe others. However, all of these need loading equipment which is basically compressive or tensile. They also require measuring and recording devices which are essentially the same as the ones already described. Therefore it is felt at the moment that these tests do not need further desription.

1.4 Examples of Dynamic Loading with Local Damage

1.4.1 Scope

Local phenomena of impact loading are difficult to assess analytically because dynamic contact between two bodies and material behavior under high bearing stresses are not easy to model. Experiments can help to understand the behavior, and sometimes static loading results can give a hint to capture dynamic response.

Two examples of dynamic loading will be discussed. First, the penetration of a rigid punch into concrete is assessed by a numerical procedure incorporating static stress-displacement curves. Second, the collision of concrete bodies of different shapes is investigated experimentally, leading to a semi-empirical response relation.

1.4.2. Penetration of a Rigid Punch into Concrete

1.4.2.1 Introduction

Impact of rigid bodies on concrete structures occurs quite often during service life as a consequence of handling of material on industrial floors, by flying objects, traffic accidents, and intentional aggression. There are empirical formulas available, which relate the penetration depth to other relevant quantities of the projectile and the target. However, the extrapolation to other conditions than those tested is difficult. Therefore, an attempt is made to develop a formula for hard impact based on physical tests which allow to measure the real behavior.

A hybrid approach will be used that combines static testing with numerical analysis. Static testing leads to a force-penetration curve which is considered to represent a nonlinear spring in the numerical analysis of a dynamic single-degree-of-freedom system. First, the static tests will be described very shortly and, second, the numerical analysis will be presented together with some results.

1.4.2.2 Static Penetration Tests

A steel cylinder has been pushed into a concrete body, and the force-displacement relation has been measured [1.64]. The steel punch diameter and concrete quality were varied. Independent of parameter combination, all force-displacement curves have a similar shape, with a first maximum at small displacements and a second maximum after an intermediate drop of force. The first maximum was accompanied by spalling of concrete around the punch, whereas the second was caused by splitting of the large concrete cylinder, i.e. the second maximum is a structural property which bounds the range of applicability of the results.

The force-displacement curves can be normalized in such a way that they include punch diameter and concrete compressive strength. Fig. 1.35 shows the normalized stress-penetration curve with the two maxima of stress. It appears that a higher concrete quality leads to a lower normalized stress and to relative smaller penetration. There are four characteristic points: 1) the limit of linear elastic response, 2) the first maximum stress, 3) the minimum stress after a sharp decay, 4) the second maximum stress, which is also the limit of applicability. For $f_c = 40\ MPa$, the appropriate normalized stresses are 6, 12.5, 3.5, and 25, the normalized displacements are 0.02, 0.2, 0.5, and 2, respectively. This normalized curve will be used to calculate penetration by impact loading and the appropriate force time history during impact.

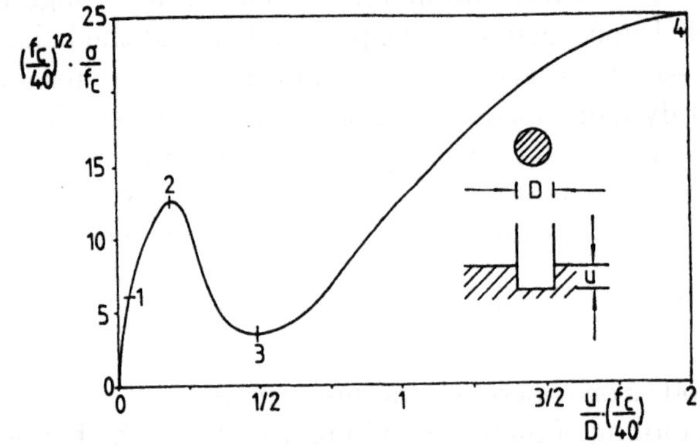

Fig. 1.35 Normalized stress-displacement diagram of static penetration tests

1.4.2.3 Impact Penetration

It is assumed that the projectile is rigid and does not deform during impact ("hard" impact), whereas the concrete exhibits nonlinear behavior. The collision is assumed to be plastic, i.e. the whole kinetic energy of the projectile is converted into deformation energy of the concrete target. When the velocity of the projectile has dropped to zero the final penetration depth is reached.

The deformation energy is given by

$$W_{def} = A \int \sigma du \tag{1.33}$$

where A = cross-sectional area of projectile, σ = average contact stress, u = penetration. With the expressions of Fig. 1.35 and the definitions

$$\xi = \frac{u}{D}\left(\frac{f_c}{40}\right) \quad \text{and} \quad \eta = \frac{\sigma}{f_c}\left(\frac{f_c}{40}\right)^{1/2}$$

where f_c = the mean cylinder compressive strength, and D = punch diameter, it follows that

$$W_{def} = AD\,40\left(\frac{40}{f_c}\right)^{1/2}\int \eta d\xi \tag{1.34}$$

The kinetic energy of the projectile is

$$W_{kin} = \frac{1}{2}mv^2 \tag{1.35}$$

with m = mass and v = initial velocity. Equating Eqs. 1.34 and 1.35 yields with $A = \pi D^2/4$,

$$\int \eta d\xi = \frac{1}{40}\left(\frac{f_c}{40}\right)^{1/2}\frac{2mv^2}{\pi D^3} \tag{1.36}$$

Numerical integration leads to a relationship which can be approximated within two ranges of validity:

$$0 \leq \xi \leq 1 \qquad y = 2.5 - 10(\xi - 0.5)^2 + 6.95625\xi^2 \tag{1.37a}$$

$$1 \leq \xi \leq 2 \qquad y = 6.95625\xi^2 \tag{1.37b}$$

Furthermore, only Eq. 1.37b will be used, since the error only affects small penetrations. Evaluating Eq. 1.36 with 1.37b leads to the normalized penetration,

$$\frac{u}{D} = 24.0583\, f_c^{-3/4}\left(\frac{mv^2}{D^3}\right)^{1/2} \tag{1.38}$$

with the units Pa, kg, m/sec, m. The constant has the unit $Pa^{1/4}$. This formula could be checked by comparison with test results. However, it is preferable to compare it with the widely used empirical NDRC formula [1.65]. This formula, based on numerous tests, has been adopted by the UKAEA [1.66] and reads

$$G = 3.8 \times 10^{-5} N m v^{1.8} f_{ck}^{-0.5} D^{-2.8} \tag{1.39}$$

with N = nose shape factor (= 0.72 for flat nose), m = mass, v = initial velocity, f_{ck} = characteristic cylinder strength of concrete, D = diameter of projectile. From G, the penetration is calculated within three ranges of validity:

$$\frac{u}{D} < 0.22 \qquad G = 0.55\frac{u}{D} - \left(\frac{u}{D}\right)^2 \tag{1.40a}$$

$$0.22 < \frac{u}{D} < 2 \qquad G = \left(\frac{u}{D}\right)^2 + 0.06505 \tag{1.40b}$$

$$\frac{u}{D} > 2 \qquad G = \frac{u}{D} - 0.9395 \tag{1.40c}$$

It appears that Eqs. 1.38 and 1.40 have some common features at least in the range $0 < u/D < 2$. However, the formulas can be compared best by plotting u/D vs $(mv^2/D^3)^{1/2}$. Fig. 1.36 shows the results of the NDRC formula for three concrete qualities together with the results of Eq. 1.38, which are straight lines in this coordinate system.

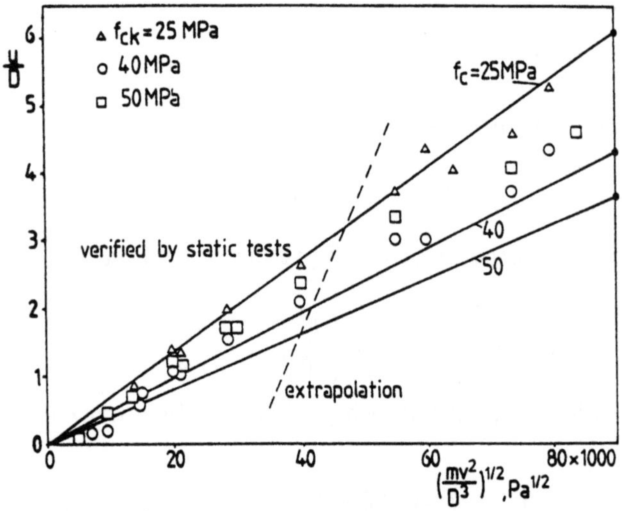

Fig. 1.36 Comparison of theoretical prediction with prediction by NDRC formula

It can be seen that the agreement between the two approaches is better for lower concrete quality, whereas at higher concrete quality the NDRC formula leads to a 10 to 20% larger penetration. However, this is a rather small difference in light of the considerable scatter of impact test results. The limit of validity of Eq. 1.38 is noted by the dashed line because the static penetration tests ended there. On the other hand, the results of the NDRC formula beyond the dashed line may indicate that the formula may be used also for larger values. Real limits can only be verified by further static testing.

1.4.2.4 Force Time History During Impact

The force time history is needed for the correct design of a concrete member which may be hit by a projectile. If an advanced computer code is available together with an adequate material model for concrete, the force time history may be calculated (and the penetration too). However, experience shows that local phenomena of concrete are still difficult to predict by theoretical analysis. Therefore, a hybrid way is chosen, which uses the static results of penetration tests as input for a dynamic single-degree-of-freedom system. The differential equation of motion of a system consisting of a mass m and resistance $R(u)$ reads,

$$m\ddot{u} + R(u) = 0 \tag{1.41}$$

where u = displacement and $R(u)$ = a nonlinear function which has been determined in static tests. Eq. 1.41 can be solved by the central difference method which allows $R(u)$ to follow an arbitrary path. The derivatives of the displacement u with respect to time are represented by the following difference equations,

$$\dot{u}^t = \frac{u^{t+1} - u^{t-1}}{2h} \quad (1.42)$$

$$\ddot{u}^t = \frac{u^{t-1} - 2u^t + u^{t+1}}{h^2} \quad (1.43)$$

where the superscripts $t-1, t, t+1$ indicate respective time steps, and h denotes the length of a time step. The differential equation 1.41 then reads

$$u^{t+1} = -\frac{h^2}{m} R(u^t) + 2u^t - u^{t-1} \quad (1.44)$$

The initial conditions at $t = 0$ are,

$$u^0 = 0 \quad \dot{u}^0 = v \quad \ddot{u}^0 = -\frac{R(u^0)}{m}$$

which lead to

$$u^{0-1} = \frac{R(u^0)}{m}\frac{h^2}{2} - \dot{u}^0 h + u^0 \quad (1.45)$$

Eq. 1.44 has been evaluated for various combinations of variables v, m, f_c, D. The result is the relationship between $F^t = R(u^t)$ and time. This procedure can be used as long as the displacement keeps increasing. At the instant when the velocity becomes zero, maximum penetration is reached. Unloading takes place elastically with a constant spring stiffness, which follows from the very first loading step (initial stiffness).

Fig. 1.37 shows the influence of initial velocity on the stress time history. It appears that the velocity of 20 m/sec leads to a maximum stress of about 500 MPa, which is equal to the stress of the first peak in Fig. 1.35. After the peak unloading starts already.

With a velocity of 80 m/sec a first maximum is reached after 0.1 $msec$, a minimum stress at about 0.3 $msec$ and a second peak after 1.5 $msec$. The second peak almost reaches the maximum attainable stress of Fig. 1.35. Comparing the stress increase at the beginning, it can be seen that the increase is steeper for higher impact velocity. The force is stress times cross-sectional area of the projectile.

Fig. 1.38 shows the influence of projectile diameter on the stress time history for the same combination of mass, concrete quality, and initial velocity. For rather large diameters (0.08 and 0.16 m) the stress reaches only one maximum and then drops to zero. In all other cases, stress reaches subsequently a minimum and a second peak which is higher for smaller diameters. This is a consequence of the nonlinear behavior of concrete because penetration is smaller for a large diameter projectile, i.e. there is little irreversible deformation. For smaller diameters, however, the penetration becomes larger and needs more time and is accompanied by more irreversible deformation. In the case of $D = 0.06$ m there are two maxima of about the same magnitude.

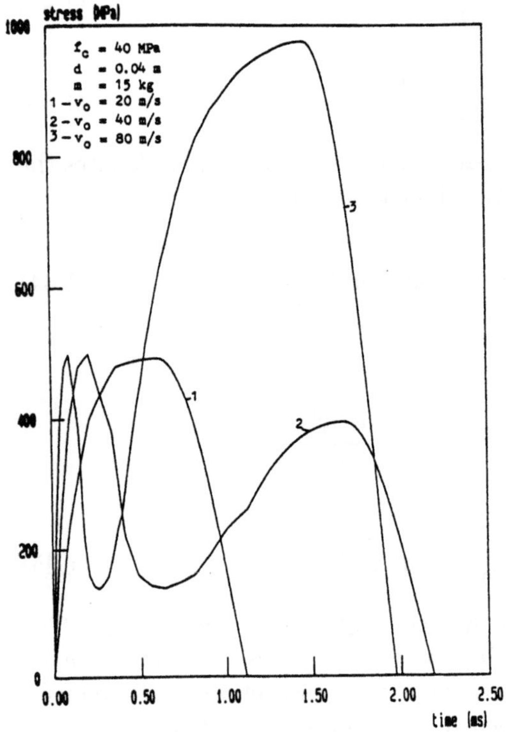

Fig. 1.37 Influence of velocity on stress time history

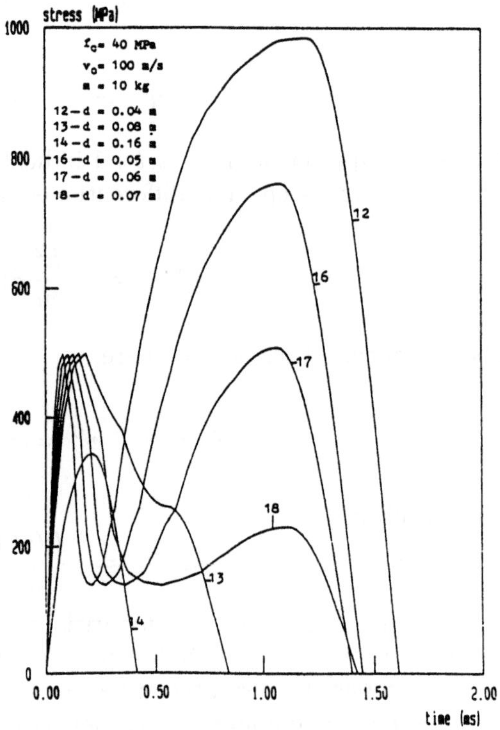

Fig. 1.38 Influence of projectile diameter on stress time history

The mass of the projectile has a strong influence on the response, as can be seen from Fig. 1.39. In this particular example, a mass of 40 kg leads to a single peak stress with a total contact time of less than 2 $msec$. A mass of 80 kg causes one peak and after a first stress drop a stress plateau on a lower level. Masses of 160 and 320 kg lead to a second peak which is higher and occurs at a later instant with larger masses.

The influence of concrete quality on stress time history is shown in Fig. 1.40. Concrete strength is varied between 20 and 60 MPa, whereas velocity, mass, and diameter are kept constant. It can be seen that a higher concrete strength leads to a shorter response with larger peak stress. Depending on the strength, the first peak can be lower or higher than the second one.

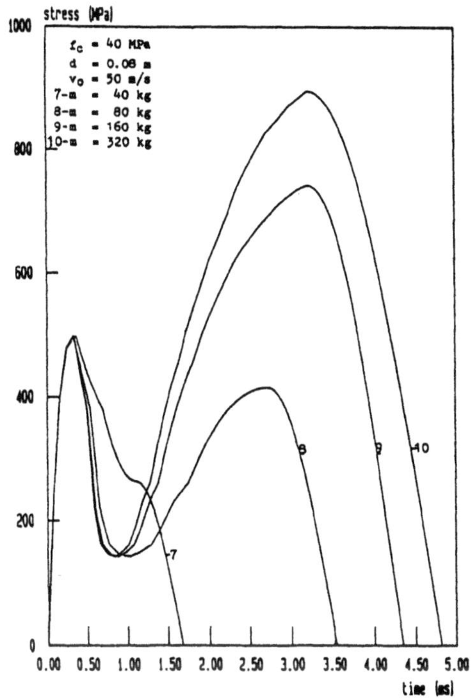

Fig. 1.39 Influence of mass on stress time history

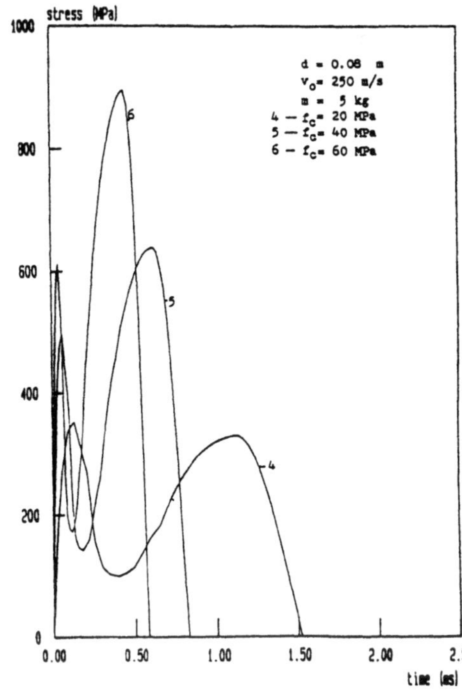

Fig. 1.40 Influence of concrete strength on stress time history

1.4.2.5 Discussion of the Results

The approach which has been used in this study combines static testing with numerical analysis. This has been done with the aim to achieve reliable and reproducible results with a minimum of effort. The method is principally equivalent to the pseudodynamic method [1.67], with the difference that the deformation path in time is monotonic, while in most pseudodynamic tests, an earthquake excitation is simulated. However, the idea of evaluating static results and translating them into a dynamic problem is the same [1.68]. The advantage of the method is that the nonlinear behavior of concrete is considered implicitly together with a complex state of stress, cracking and crushing. Also, partial unloading and redistribution of stresses which occur at the onset of cracking are taken into account automatically. Due to the simplicity of the experiments and numerical methods, many variables can be investigated. As has been shown, there is a large degree of conformity among the various concrete qualities, which reduced the need for tests considerably.

There are disadvantages inherent in the method. Since testing is performed "statically", rate effects do not appear, whereas it is known that concrete is a rate-sensitive material [1.69]. This means that the strength encountered in a static test is less than in a dynamic test, i.e., the resistance of a structure in a static test is generally lower than in a dynamic event. On the other hand, a higher strength leads to larger response, as could be seen in Fig. 1.40. Taking the intermediate curve of Fig. 1.40 as an example, the stress rate to the first peak stress amounts to about 5×10^6 MPa/sec. The dynamic compressive strength increases with a power law [1.69],

$$\frac{f_c}{f_{c0}} = \left(\frac{\dot{\sigma}}{\dot{\sigma}_0}\right)^{\alpha} \tag{1.46}$$

with $\alpha = (5 + 3f_{cm}/4)^{-1}$. For a cube compressive strength of $f_{cm} = 50\ MPa$, α becomes 0.0235. With $\dot{\sigma}_0 = 1\ MPa/sec$, the strength increase becomes 44%. The first peak stress of Fig. 1.35 is given by $\sigma_{max} = 12.5\ f_c^{1/2}$. If strength is increased by a factor β, the maximum stress is only increased by a factor $\beta^{1/2}$, which means that the force encountered is less than the strength increase due to the rate effect. Consequently, the pseudodynamically determined response underestimates the stresses, and overestimates the load duration but does not lead to an unsafe prediction.

A second point of concern is the numerical procedure, which may not be stable and objective. It turned out that consistent results could be achieved with time intervals of the order of a few microseconds.

1.4.3 Response History of Colliding Concrete Bodies

1.4.3.1 Introduction

In several practical situations, the impact between different structural elements may cause severe damage. Examples are heavy objects dropped from offshore production platforms and the collision between concrete armor elements in breakwaters. Here, attention shall be focussed on the response history of concrete-to-concrete impact. Results of concrete-to-concrete impact experiments will be discussed, and a simple model for predicting the dynamic response on the basis of data derived from static tests will be presented. For details see [1.47].

1.4.3.2 Modelling

Elastic Approximation of Impact. When the impact is completely elastic, a symmetric load time history is measured and full restitution is observed. The elastic impact of two elements with effective masses m_1 and m_2 can be described by Hertz' law [1.70, 1.71],

$$P = K_e\delta^{3/2} = \frac{(m_1 m_2)}{(m_1 + m_2)}\frac{d^2\delta}{dt^2} \tag{1.47}$$

where δ = the deformation of the contact zone in mm, P = the load in kN, and K_e = a contact parameter with dimension $kN/mm^{3/2}$. The contact parameter depends on the material properties of the colliding elements and on the specific contact surface geometry.

Eq. 1.47 can be solved for the boundary conditions, $d\delta/dt = v_0$, at $\delta = 0$; and $d\delta/dt = 0$, at $\delta = \delta_{max}$. The following expressions are derived for the maximum load

P_{max}, the maximum deformation δ_{max} at P_{max}, and the contact time t_{max} at P_{max},

$$P_{max} = (1.25\, \alpha\, v^2\, K_e^{2/3})^{3/5} \quad \text{in } kN \tag{1.48a}$$

$$\delta_{max} = (1.25\, \alpha\, \frac{v^2}{K_e})^{2/5} \quad \text{in } mm \tag{1.48b}$$

where $\alpha = (m_1 m_2)/(m_1 + m_2)$, the effective mass, and

$$t_{max} = 1.47 \left[1.25\, \frac{\alpha}{(v^{1/2}\, K_e)} \right]^{2/5} \quad \text{in } msec \tag{1.48c}$$

The elastic impact is shown schematically in Fig. 1.41. At the end of impact, at $t = 2t_{max}$, the velocity is $-v_0$, and no energy is dissipated.

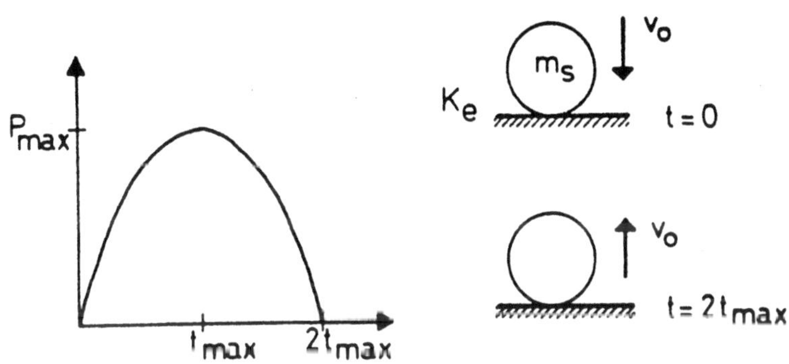

Fig. 1.41 Elastic impact

As an approximation, plastic deformation may be included in the contact parameter K_e when Hertz' model is used for plastic impact, too. Utilizing Jensen and Hoiseth's experiments [1.72], in which considerable crushing of the concrete occurred in the contact surface (more specifically, in the target area), K_e-values ranging from 2 to 44 $kN/mm^{3/2}$ were calculated using Eqs. 1.48a-c. The details of this inverse analysis are given in [1.73]. The analysis showed that an elastic approximation is not very useful when large irreversible deformations occur in the contact zone.

Plastic Model for Impact. A complete description of the response in a plastic contact zone is quite elaborate, but a relatively simple starting point is Meyer's law [1.70],

$$P = K_p\, \delta^n \tag{1.49}$$

where K_p is the contact parameter. For a fully plastic impact between an infinitely rigid sphere and a deformable plane (n = 1), the contact parameter K_p is equal to

$$K_P = \pi\, r^2\, \frac{P_m}{\delta} \approx 2\pi R\, p_m \tag{1.50}$$

where R and r are the radii of the sphere and of the contact zone, respectively, and p_m is the average contact stress. The dimension of the contact parameter is kN/mm. Expressions for the maximum contact load P_{max}, the maximum deformation δ_{max}, and the contact time t_{max}, both at P_{max}, may be derived as follows,

$$P_{max} = v\,(\alpha\,K_p)^{1/2} \quad \text{in } kN \tag{1.51a}$$

$$\delta_{max} = v\left(\frac{\alpha}{K_p}\right)^{1/2} \quad \text{in } mm \tag{1.51b}$$

$$t_{max} = \frac{\pi}{2}\left(\frac{\alpha}{K_p}\right)^{1/2} \quad \text{in } msec \tag{1.51c}$$

The plastic impact is shown schematically in Fig. 1.42: no restitution occurs, the velocity of the striker at the end of impact is zero, and the residual deformation is δ_{max}. The energy dissipated in the impact is equal to $0.5 m_s v_0^2$, where m_s = mass of the striker; and v_0 = the velocity at impact. Utilizing again Jensen and Hoiseth's data, the following K_p-values are calculated. From the $P-v$ relation Eq. 1.51a: $K_p = 10-40\ kN/mm$; from the $\delta - v$ relation Eq. 1.51b: $K_p = 27-69\ kN/mm$; and from the $t-v$ relation Eq. 1.51c: $K_p = 12-22\ kN/mm$. Again, the scatter is relatively large. The differences between the K_e-values (Hertz) and the K_p-values (Meyer) are partly related to dimensional differences.

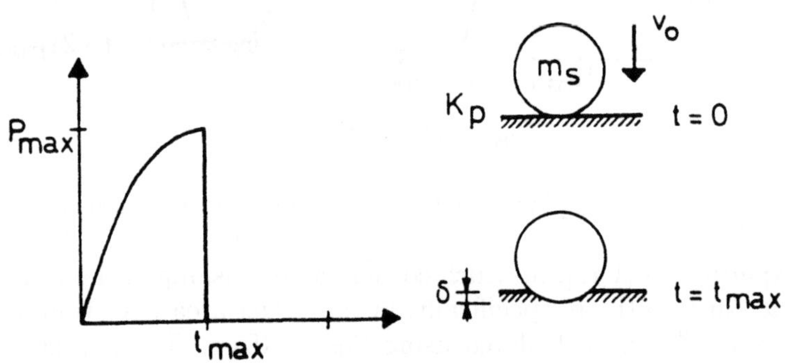

Fig. 1.42 Plastic impact

Elastoplastic Model for Impact. Neither the elastic nor plastic approaches are quite suitable for describing the load-time histories of Fig. 1.43. In the elastic approach, difficulties are encountered as soon as plastic, irreversible deformations occur. With the plastic model the zero restitution is problematic. Therefore, an elastoplastic model seems most suitable, in which the advantages of both approaches are combined.

An elastoplastic model is proposed which consists of the three stages as shown in Fig. 1.44a: the first stage with a constant loading rate; the second with constant load or slightly increasing or decreasing load; and the third with a constant unloading (restitution) rate. The first and third stages can be attributed to elastic material behavior, whereas the second stage is due to "plastic" material behavior. This load time history is a crude simplification of the real behavior, which is influenced by cracking, friction, and lateral confinement of the concrete. However, it will be shown that it is good enough to accurately predict the response history of colliding concrete bodies.

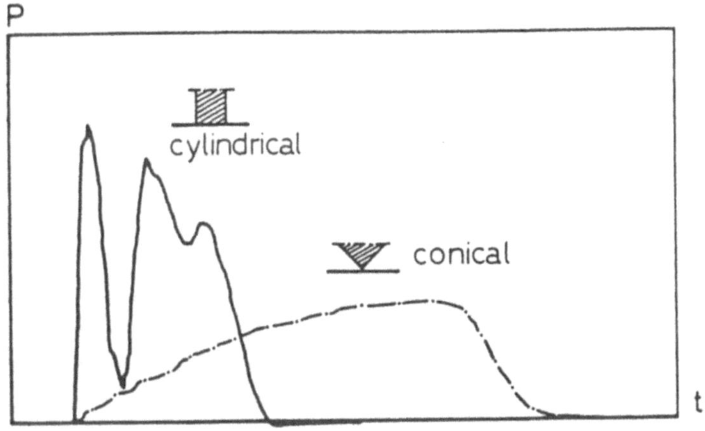

Fig. 1.43 Load time histories for steel and concrete impact [1.72]

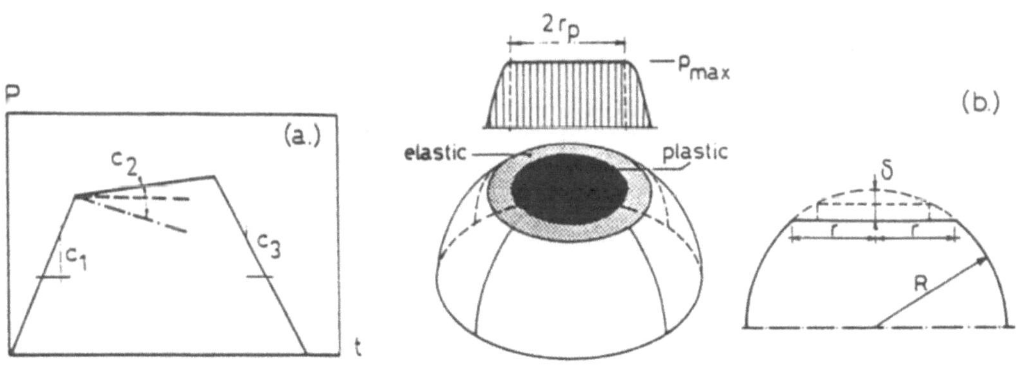

a) Simple elastoplastic load time history model b) Plastic phase [1.47]

Fig. 1.44 Elastoplastic impact

The value of the loading rate c_1 depends on the elastic properties of the materials and the geometry involved. The most important input parameter is the contact parameter K_e in Hertz' law, Eq. 1.47, which determines the maximum load, the contact time, and the displacement according to Eq. 1.48. The load time history for two colliding elastic bodies can be closely approximated by a sine function [1.71]. However, if nonlinear effects, e.g., local cracking and crushing of the concrete, affect the contact zone behavior in a very early stage, the sine function may be further approximated by a straight line, i.e., a constant loading rate c_1, Fig. 1.44a.

The first stage is terminated as soon as "plastic" material behavior prevails, or in the case of concrete, when cracking and crushing play a dominant role. In the contact zone, partially elastic and partially plastic deformations occur as shown in Fig. 1.44b. Under the assumption of a parabolic stress distribution in the elastic zones, the contact force is

$$P = A_{pl}p_0 + A_{el}\beta p_0 \tag{1.52a}$$

The average stress in the contact zone, $p_m = P/\pi r^2$ becomes

$$\frac{p_m}{p_0} = \left(\frac{r_p}{r}\right)^2 (1-\beta) + \beta \tag{1.52b}$$

where r_p = the radius of the plastic zone, and p_0 = the yield level of the critical stress. A_{pl} and A_{el} represent the plastic and elastic deformed areas, respectively. The value of β depends on the stress distribution and varies between 1/2 and 2/3. Depending on the size of the plastic zone, i.e., the ratio r_p/r, the average contact stress will vary between β and one. The relationship between the average contact stress and the critical stress is shown in Fig. 1.45 for $\beta = 1/2$. The increase of the average contact stress is less than proportional to the increase of the contact surface, which explains the relatively high values for K_p obtained from the inverse analysis of Jensen and Hoiseth's data from the $\delta - v$ relationship (see previous section).

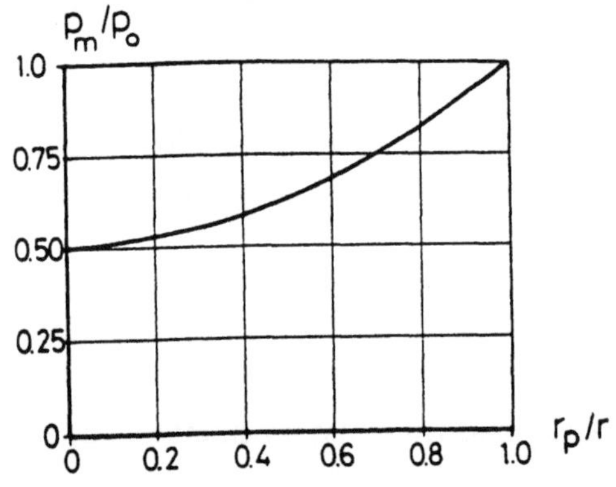

Fig. 1.45 Average contact stress as function of plastic zone size

The maximum load is a function of the geometry and the materials involved. The loading rate c_2 depends on the material behavior. If a stable elastic-plastic stress distribution is achieved, $c_2 = 0$. If the material is strain-hardening and/or the plastic contact area increases, the load may increase in time, and $c_2 > 0$. On the other hand, when softening material is involved, or when the plastic contact area decreases due to cracking of outer layers, $c_2 < 0$.

Finally, as shown in Fig. 1.44a, elastic restitution occurs after a maximum load is reached. In principle, this elastic restitution can be calculated following Hertz' law, or alternatively from the elastic unloading of the material surrounding the plastically deformed contact zone, in which case the unloading stiffness can be related to the elastic behavior of the material. In the proposed model, a constant rate of unloading (restitution) c_3 is assumed, which is a simplification of real behavior.

In conclusion, the input parameters needed for the elastoplastic model are the contact parameter K_e for the elastic phase, the critical stress and the critical size of the contact surface (from which the maximum load can be derived), and the unloading stiffness. When these parameters are known, the complete load time history can be calculated for any impact with given impulse. In principle, all these parameters can be determined from dynamic experiments. The size of the contact surface is determined from static experiments. The critical stress depends on the concrete quality, defined by the uniaxial cube compressive strength, taking into account loading rate effects as well as the influence of triaxial stress. These aspects will be considered when presenting the results of the dynamic experiments.

1.4.3.3 Experiments

In order to obtain some insight in the response of two colliding concrete elements, a number of dynamic tests were carried out. In addition, several static experiments were done, enabling a comparison of the dynamic and static strength of the material as a function of the contact surface geometry. Details can be found in [1.47].

1.4.3.4 Conclusions

Results of load tests on an extremely small bearing area on concrete have been evaluated, which were obtained in static tests with constant displacement rate. The results were used to predict the penetration of an impacting projectile. Equating the kinetic energy of the projectile and the deformation energy as determined in the static test, good agreement was achieved between the theoretical prediction and an empirical relationship used in practice. This agreement gave confidence in the use of static results also for predicting the force time history during impact using the central difference method. It may be concluded that,

- penetration depth due to impact can be determined from static testing;
- force time histories during impact can be predicted from static tests by solving the differential equation of motion numerically; and
- generally, the use of experimental techniques, together with numerical methods, prove to be a powerful tool to solve complicated problems where nonlinear material behavior is involved.

It is recommended to use the hybrid technique more widely for the solution of local phenomena in inhomogeneous materials with nonlinear stress-strain behavior, such as concrete.

1.5 Models for Strain Rate Effects

1.5.1 Introduction

This section deals mainly with a few theoretical approaches to the explanation of the stress or strain rate influence on concrete strength. These theories are based on thermodynamics or fracture mechanics or both. Their predictions are compared with experimental results as far as possible. It is attempted to show the influence of concrete composition and temperature on the stress rate dependency. Most of these theories treat concrete on a meso-scale considering flaws and individual particles, but arrive at a prediction of the strength, i.e. an average property. Some of the theories are concerned with a discrete crack in a homogeneous and isotropic material, thus leading to a dynamic stress intensity factor or crack opening resistance.

1.5.2 Fracture Mechanics Approach

Linear elastic fracture mechanics can be applied to concrete in two ways: first, on a macro-scale, taking concrete as an elastic, homogeneous, isotropic material, and second, on a meso-scale, considering concrete as a regularly flawed material. In the first case the size of the concrete element and the crack should exceed a certain minimum value which depends on the concrete grade. The order of this dimension is meter. The second model can be a fictitious material with equidistant flaws, which are physically due to pores in the hydrated cement paste and the shrinkage and thermal cracks around the large aggregate particles. It can also be a numerical model of concrete with aggregate, matrix and bond between the two constituents.

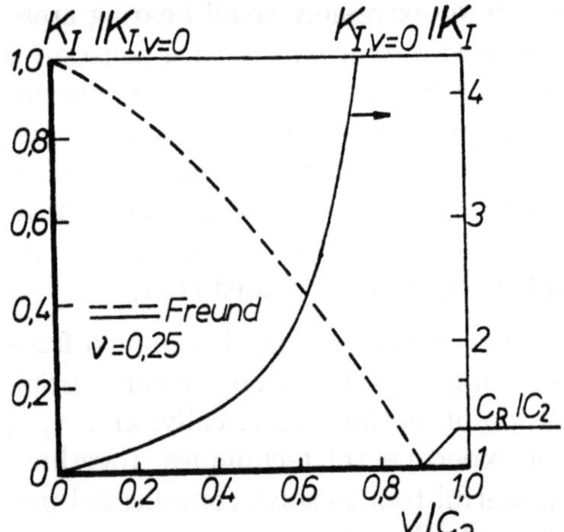

Fig. 1.46 Semi-infinite crack propagating in a finite strip

Macro-level. As linear fracture mechanics with one discrete crack is concerned, there are analytical solutions for the stress intensity factor under dynamic loading [1.74]. One loading configuration shall be recalled. This is the example with a semi-infinite crack in a finite strip. At time $t = 0$, the faces of the strip are moved by an instantaneous displacement of magnitude δ_0, which is kept constant. The crack propagation velocity v is also assumed constant. Fig. 1.46 gives the theoretical solution [1.75] in terms of normalized dynamic stress intensity factor versus crack velocity v/c_2, with c_2 the propagation velocity of a shear wave. This normalized factor is equal to unity at $v = 0$ and shows an almost linear decay with increasing velocity until $v = c_R$, the Rayleigh wave speed.

This theoretical result may be discussed by regarding the stress waves, which cause a stress intensity at the crack tip. At low crack velocities, the information from the stressed body and the wake region of the crack is transferred to the crack tip more rapidly than the crack velocity. At higher crack velocities, the crack faces may not move fast enough to provide the strains at the crack tip necessary for a high stress intensity factor. If the limiting value of the energy transport along the crack faces is reached, i.e. the Rayleigh wave speed, the stress intensity factor drops to zero.

Since $K_I \sim \delta_0$, the result means that a running crack causes less stress intensity at the same displacement than a static crack. If fracture is initiated at $K_I = K_{IC}$, and K_{IC} is a material property, this would mean that the loading capacity is larger the greater the crack velocity is. The dashed line in Fig. 1.46 gives an idea of such an interpretation which cannot be more than a rough indication.

Kipp et al. [1.76] extended the theory of constant stress to arbitrary stress loading by an appropriate use of the stress history. From the special loading case of a constant strain rate ϵ_0 and thus a constant stress rate σ_0 in an elastic material, the following relationship for the stress intensity factor for a penny-shaped crack is derived,

$$K_I(t) = \frac{4\alpha}{3\sqrt{\pi}}\dot{\sigma}_0 c_s^{1/2} t^{3/2} \tag{1.53}$$

where α is a geometric coefficient equal to 1.12 for the penny-shaped crack, c_s the shear wave velocity, and t = loading time. If K_{IC} is regarded as a fracture criterion, a relationship between strain rate $\dot{\epsilon}_0$ and fracture stress σ_c can be established,

$$\sigma_c \sim \left[\frac{9\pi E K_{IC}^2}{16\alpha^2 c_s}\right]^{1/3} \dot{\epsilon}_0^{1/3} \tag{1.54}$$

This cube root law holds for high strain rates and/or sufficiently large cracks. By use of the known relationship between stress intensity factor and stress, $K_{IC} \sim \sigma_c$, it can be shown that a relation of the form,

$$a \sim \left[\frac{c_s K_{IC}}{E\dot{\epsilon}_0}\right]^{-2/3} \tag{1.55}$$

exists, which links crack length a and strain rate $\dot{\epsilon}_0$. For average concrete properties,

$$a \sim 0.1\dot{\epsilon}_0^{-2/3} \tag{1.56}$$

i.e., a strain rate of 1 sec^{-1} requires a crack length of 0.1 m, a strain rate of 100 sec^{-1} one of 5 mm in order to make the theory applicable.

Meso-level. Linear fracture mechanics has been applied to concrete at a fictitious level by Weerheijm [1.77]. Concrete is schematized as a material containing penny-shaped cracks of single size and equal distance. Fig. 1.47 shows a representative concrete element with flaws of diameter $2a$ and distance $2b$.

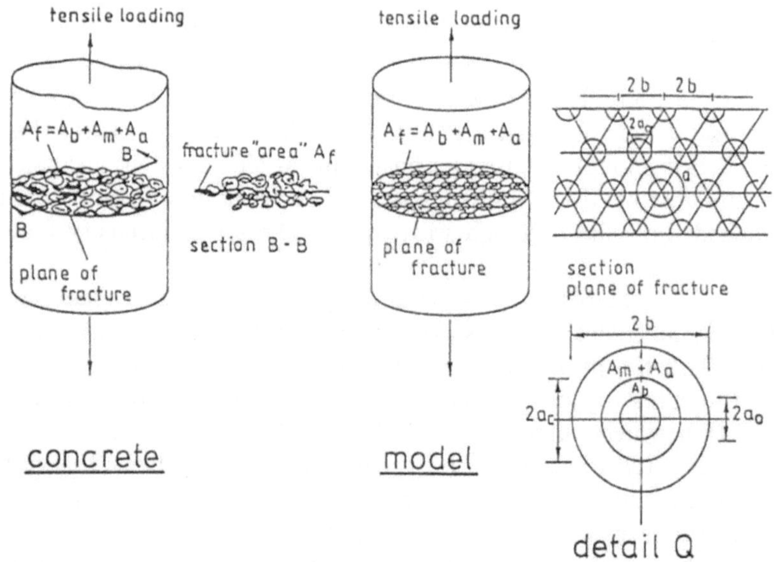

Fig. 1.47 Voids in concrete represented by penny-shaped cracks

First, the ratio a/b is calculated from the total pore volume of the concrete, n, which includes gel pores, capillary pores and initial shrinkage cracks. Assuming spherical pores of radius a in a fictitious sphere with radius b, the ratio a/b becomes $n^{1/3}$. A concrete porosity of 10% leads to $a/b = 0.46$, and $n = 20\%$ leads to 0.58. This is of course a rough simplification. The absolute values of a and b are determined from the critical stress intensity factor, which is determined in macroscopic experiments, and a uniaxial tensile stress of 0.6 times the static tensile strength f_t. This means, this stress level is assumed to be a critical stress level at which unstable crack propagation starts. From these two conditions an expression follows for the starting value of a. From

$$K_{IC} = \sigma\sqrt{\pi a}\, f(geom, a/b) \tag{1.57}$$

it follows, with $\sigma = 0.6\ f_t$, that

$$a = \frac{1}{\pi}\left[\frac{K_{IC}}{0.6 f_t \cdot f(a/b)}\right]^2 \tag{1.58}$$

When the initial values of a and b are known, the dynamic aspect is treated by considering the kinetic energy during crack propagation. The total energy consists of three parts, i.e.,

1) the surface free energy at the crack faces;
2) the irrecoverable energy due to plastic deformation and friction;
3) the kinetic energy, E_{kin}.

The first two parts are equal to the fracture toughness, G_{IC}, which is a material property. The sum of all three parts is equal to the externally supplied energy G_I. As Weerheijm has shown, E_{kin} and G_I depend on stress, stress rate, initial crack length, crack velocity, Young's modulus, and Poisson's ratio. From the condition,

$$E_{kin} = \int_{a_1}^{a_2} (G_I - G_{IC}) da \tag{1.59}$$

where a_1 and a_2 are two states of cracking, a relation follows between stress and fracture time. For a constant stress rate, the tensile strength can be calculated.

Since the mathematical formalism is complicated, only one result of this study is shown in Fig. 1.48 for a certain concrete mix: in the region between static testing and $\dot{\sigma} = 10^{11}\ N/m^2sec$, a slight continuous increase of strength and thereafter, a steep increase within one order of magnitude of stress rate is noted. According to this model this increase is completely due to the fact that external energy is converted to kinetic energy. The attractive feature of this model is that the complete range of stress rates is covered by one approach.

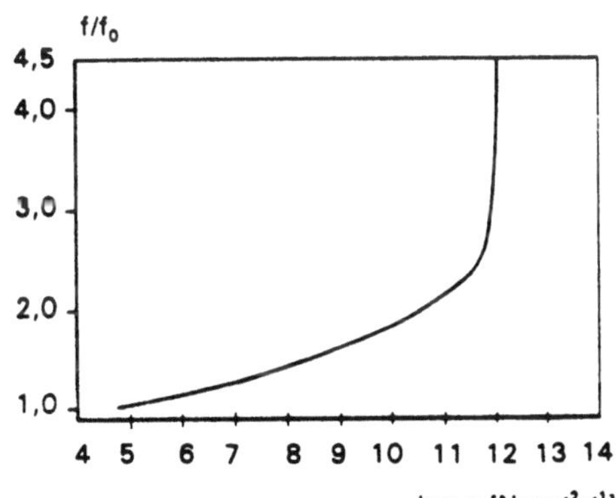

Fig. 1.48 Prediction of tensile strength as function of stress rate [1.77]
$f_o = 5\ N/mm^2$, $n = 6.4\%$, $K_{IC}^2 = 0.3 \times 10^{12}\ N^2/m^3$, $E = 35,000\ N/mm^2$

1.5.3 Thermodynamic Approach

To treat concrete by thermodynamics means to consider it on an atomic level. Atoms are in a state of continuous motion, while attracting and repulsing forces are acting on them. Each atom is situated on a certain energy level. Due to continuous motion there is always a chance that an atom overcomes the inherent energy barrier and moves to another place in the system. If external energy is added to a system of atoms, the atoms may overcome the energy barrier (activation energy) more easily. Energy can be supplied by mechanical loading, heating, or concentration gradients. The greater these external influences are the more likely such place changes are to occur. Place changes of atoms can be detected in an average way by deformations, cracks or chemical reactions.

Mihashi and Wittmann [1.78] used this approach to predict the loading rate influence on the strength of concrete. They combined the thermodynamic approach to some extent with fracture mechanics. They state that the fracture of concrete may be caused by a series of local failure processes in the hydration products of cement and interfaces between cement and aggregate. As soon as a failure criterion is satisfied in one part of the phase, a crack is initiated. Extension of cracks and coalescence of cracks cause fracture. The concrete system consists of a group of elements linked in series, Fig. 1.49. Each element contains a circular crack, the length of which depends on the pore sizes of hardened cement paste (for the prediction of the rate influence, the absolute value of the crack length is not important). The distribution of the material defects and the characteristic properties of each element are statistically equal over the whole material.

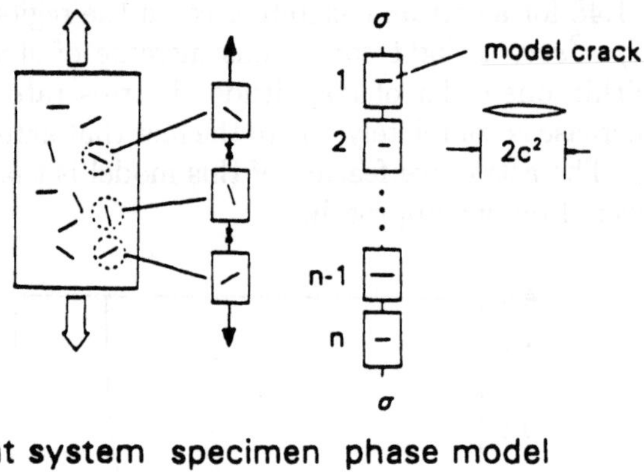

Fig. 1.49 Model of hardened cement with linked elements [1.78]

To this material schematization the rate theory is applied. The rate of crack initiation is a function of activation energy, stress and temperature. The rate of crack initiation is expressed by

$$r = \frac{kT}{h} exp\left(-\frac{U_0}{kT}\right)(q\sigma)^{\frac{1}{n_b kT}} \tag{1.60}$$

where k = Boltzmann constant; h = Planck constant; T = absolute temperature; U_0 = activation energy; q = local stress concentration factor; and n_b = a material constant. Eq. 1.60 is a simple relationship between crack initiation rate r and stress σ, if all other parameters are taken constant,

$$r \sim \sigma^\alpha \tag{1.61}$$

The authors calculate the mean probability of fracture during a time interval and end up with a relationship between stress rate and tensile strength, which can be simplified to,

$$\frac{f}{f_0} = \left(\frac{\dot{\sigma}}{\dot{\sigma}_0}\right)^{\alpha} \tag{1.62}$$

where f and f_0 are the tensile strengths under impact and static loading, respectively, and $\dot{\sigma}$ and $\dot{\sigma}_0$ are the corresponding stress rates. The coefficient α depends on the material, temperature and humidity. The larger α the more sensitive a material to stress rate, Fig. 1.50.

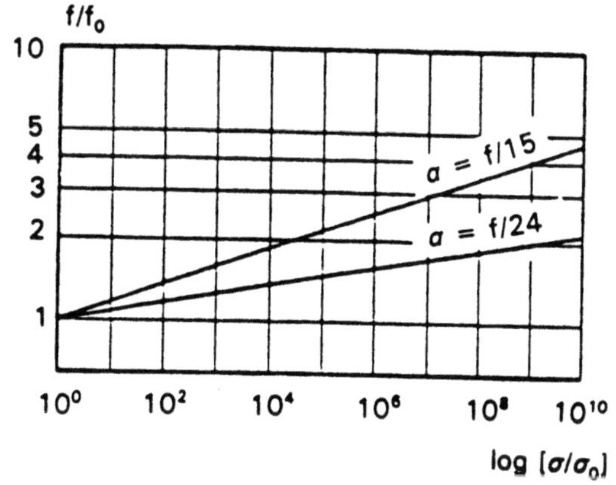

Fig. 1.50 Strength vs stress rate according to Mihashi and Wittmann [1,78]

Lindholm et al. [1.79] applied the rate theory to rock in order to predict rate and temperature influence on strength under multiaxial loading. They assumed the activation energy to be a linear function of stress,

$$U(\sigma) = U_0 - a(\sigma - \sigma_0) \tag{1.63}$$

where U_0 = the total activation energy of the process, a = the activation volume, σ_0 = a constant, and σ = applied stress. The rate equation is used in its simplest form,

$$\dot{\epsilon} = \dot{\epsilon}_0 \, exp\left(-\frac{U(\sigma)}{RT}\right) \tag{1.64}$$

$\dot{\epsilon}$ = strain rate, $\dot{\epsilon}_0$ = a constant, R = gas constant. This relation has been used earlier by Zhurkov [1.61]. Eqs. 1.63 and 1.64 lead to

$$\sigma_{max} = f = \frac{U_0}{a} + \sigma_0 - \frac{RT}{a} \, ln\frac{\dot{\epsilon}_0}{\dot{\epsilon}} \tag{1.65}$$

In this relation, $(U_0/a + \sigma_0)$ is the limiting stress when $T = 0$ or when $\dot{\epsilon} = \dot{\epsilon}_0$. According to Eq. 1.65 the strength decreases linearly with temperature and increases linearly with the logarithm of the imposed strain rate. A schematic plot of Eq. 1.65 is given in Fig. 1.51. For temperatures between 0 and 275^oK, the relationship between strength and strain rate follows a linear-logarithmic line. At higher temperatures these lines are truncated by a lower threshold value of σ_0. Comparing Eqs. 1.62 and 1.65, the difference between theoretical predictions is obvious, and it shall depend on experimental data to determine which relationship is the more appropriate for concrete. (It should be noted that Lindholm presented his theory with regard to compressive stress but did not explicitly restrict its applicability to compressive stress).

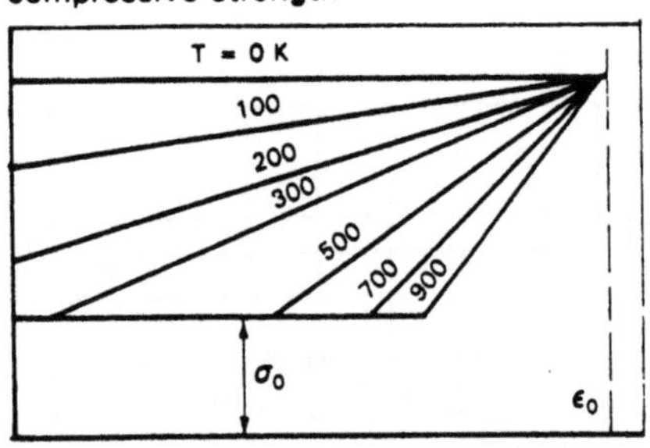

Fig. 1.51 Influence of temperature and strain rate on compressive strength, after Lindholm et al [1.79]

The rate theory was checked for predicting the stress rate and low temperature influences on the fracture energy of concrete. Starting from the same point as Lindholm et al., the energy is considered during cracking deformation. The deformation rate $\dot{\delta}$ is given as a function of mechanical energy Q_m and activation energy Q_a, according to [1.80],

$$\dot{\delta} \;=\; c_1 exp\left(-\frac{Q_a}{kT}\right) exp\left(\frac{Q_m}{kT}\right) \tag{1.66}$$

where c_1 is a constant. At fracture the mechanical energy is equal to the fracture energy G_F, which is the area under a complete stress-deformation curve from a tensile experiment. To adjust the macroscopically measured quantities for the microscopic level to which Eq. 1.66 refers, the number of gel particles, N_g, and the appropriate fracture area A should be considered in Eq. 1.66, which then appears in the form [1.81],

$$\dot{\delta} \;=\; c_1\; exp\left(-\frac{Q_a}{N_g kT}\right) exp\left(\frac{G_F A}{N_g kT}\right) \tag{1.67}$$

Taking the logarithm, this relation reads,

$$G_F \;=\; \frac{Q_a}{A} - (lnc_1 - ln\dot{\delta})\frac{N_g kT}{A} \tag{1.68}$$

which gives the fracture energy G_F as a function of deformation rate and temperature. At $T = 0$, the limiting value of this energy is Q_a/A. The limiting value of the deformation rate is c_1. Of course, Eqs. 1.65 and 1.68 correspond to each other.

Another thermodynamic approach is based on the damage evolution during loading, which is treated as an internal state variable which influences the Helmholtz-free energy function. Shah [1.82] has worked out this theory and came to the interesting result that, first, the strength increases almost linearly with increasing stress rate, and second, tensile strength is more sensitive than compressive strength to stress rate.

1.5.4 Other Models

Curbach [1.83] showed that crack propagation velocity has a very important influence on the tensile strength of concrete. By use of a finite element code with appropriate formulation of inertia effects it was demonstrated that strain rates beyond 1 sec^{-1} lead to an increase of tensile strength. At $\dot{\epsilon} = 100\ sec^{-1}$, the predicted strength is six times the static strength, and 60% of the increase is due to crack velocity influence.

Considering the inhomogeneous structure of concrete, there is a secondary effect which depends on inertia: it is the crack path which is forced to go through aggregate particles rather than around them. Since strength and fracture energy of the aggregate particles are larger than those of the matrix, this effect increases the total fracture energy of concrete and the tensile strength as well. This has been shown by Zielinski and Reinhardt [1.84] and has been confirmed by Brühwiler [1.85].

1.5.5 Experimental Results

Strength. Experimental results confirm the theoretical predictions, as shown by Fig. 1.52. It can be seen that some models are appropriate for small to intermediate stress rates, whereas other models predict the high stress rates best. The first category is based on rate theory and free energy considerations, whereas the second uses inertia effects as a basis.

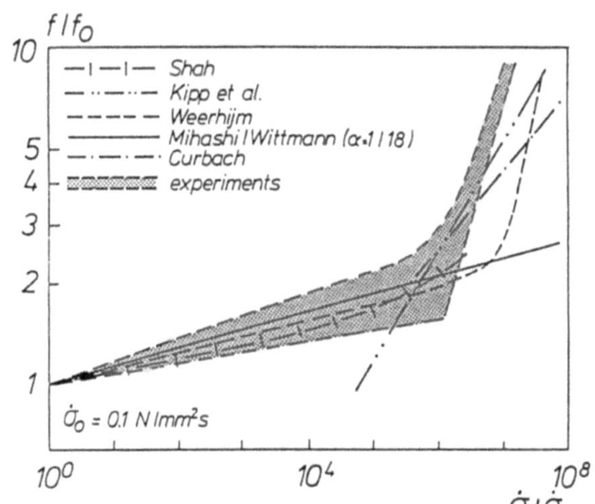

Fig. 1.52 Comparison of model predictions and experiments with respect to tensile strength of concrete

The left-hand part of the diagram is best described by the power law,

$$f/f_0 \quad = \quad (\dot{\sigma}/\dot{\sigma}_0)^{\alpha} \qquad (1.69)$$

The exponent α depends on composition of concrete, humidity and temperature. It has been determined to fit numerous results. For tensile loading and $\dot{\sigma}_0 = 0.1$ $MPasec^{-1}$, it was found to be $\alpha = (10 + f_{cm}/2)^{-1}$. Beyond $\dot{\sigma} = 10^6$ $MPasec^{-1}$, the cube root law from linear fracture mechanics applies,

$$f_t/f_{t0} \quad = \quad c\,\dot{\sigma}^{1/3} \qquad (1.70)$$

with $c = 7\alpha - 2$.

Since compressive strength is indirectly governed by tensile strength, it turned out that again a power function applies, however, with a different fitting constant,

$$f_c/f_{c0} \quad = \quad (\dot{\sigma}/\dot{\sigma}_0)^{\beta} \qquad (1.71)$$

with $\beta = (5 + 3f_{cm}/4)^{-1}$ and $\sigma_0 = 1$ $MPasec^{-1}$. The cube root law's validity has not yet been clearly demonstrated.

Young's modulus. Concrete can be treated as a linear elastic material up to a stress which is about 60% of the tensile strength. In that range, the attainable strain is rather small (about 50×10^{-6}), and there are only micro-flaws in the material, which extend over only a small area around a running crack. This contributes to the resistance due to inertia effects. From experiments the following relation is suggested, which does not seem to depend on concrete grade,

$$E/E_0 \quad = \quad (\dot{\sigma}/\dot{\sigma}_0)^{0.016} \qquad (1.72)$$

with $\dot{\sigma}_0 = 0.1$ MPa/sec for tensile loading. For compression loading, the exponent is 0.025 and $\dot{\sigma}_0 = 1$ MPa/sec.

It should be noted that Eq. 1.72 is only valid for $\dot{\sigma}/\dot{\sigma}_0 > 1$. At slow loading rates or even at constant stress, creep comes into play, which may cause an elastic modulus apparently much smaller than predicted by Eq. 1.72. However, creep and elastic strain should be treated appropriately in order to prevent confusion.

Stress-free crack opening displacement. As the stress-free crack opening is concerned, there are contradicting results in the literature. A recent careful study [1.77] has shown that w_{max} decreases with loading rate. It is proposed to use,

$$w_{max}/w_{max,0} \quad = \quad (\dot{\sigma}/\dot{\sigma}_0)^{-0.042} \qquad (1.73)$$

This means that the size of a crack process zone decreases with increasing loading rate or, in other words, the material becomes more brittle.

Fracture energy. Fracture energy G_F is the area under the $\sigma - w$ curve. If we assume w_0 = const, then G_F depends on tensile strength and the shape of the descending branch. If this remains similar, then G_F increases similarly to f_t. If the shape of the descending branch changes, then G_F shows a different relation to displacement rate than f_t does.

The scarce experimental results were evaluated and fitted to a single power law which reads,

$$G_F/G_{F,0} = (\dot{w}/\dot{w}_0)^\gamma \tag{1.74}$$

with $\gamma = (10 + f_{cm}/2)^{-1}$ and $\dot{w}_0 = 1\ \mu m/sec$. For usual concretes, γ varies from 0.025 to 0.05.

Crack propagation velocity. The crack velocity in concrete has a considerable influence on the behavior of concrete under high strain rates, as it has been shown by Weerheijm [1.77]. The crack velocity in linear elastic materials is limited by the Rayleigh-wave velocity, but almost none of the materials have shown this velocity.

Not many results are available on the crack velocity in concrete. Curbach [1.83] has carried out tests using crack gauges consisting of silver paint marks which were destroyed by the propagating crack. The tests indicate that the maximum value of crack propagation velocity of an unstable crack is independent of the strain rate. A value of about

$$v_{crack} \approx 500\ m/sec$$

has been found, which is less than a quarter of the theoretical maximum value of about 2100 m/sec, the approximate Rayleigh-wave velocity for concrete.

Crack propagation velocities of about 200 m/sec have been reported by Shah et al.[1.86], while Muria Vila et al.[1.87] report results from compressive tests on concrete, where velocities up to 1200 m/sec have been measured. It should be noted, however, that the crack propagation measurement device used by Shah was more sensitive than the one used by Muria Vila.

1.5 References

[1.1] Frenaij, J.W.: Time-dependent shear transfer in cracked reinforced concrete. Doctoral Thesis, Delft University of Technology, Delft 1989.

[1.2] Gambarova, P.G. and M. DiPrisco: Interface Behaviour, in: CEB Bulletin No. 210, Lausanne 1991.

[1.3] Kupfer, H., H. K. Hilsdorf, and H. Rüsch: Behavior of concrete under biaxial stresses, ACI Journal, August 1969, 656-666.

[1.4] Schickert, G. and J. Danssmann: Behavior of concrete stressed by high hydrostatic compression, Int. Conf. on Concrete under Multiaxial Conditions, Toulouse, May 22-24, 1984.

[1.5] Balaguru, P. N. and S. P. Shah: Fiber-Reinforced Cement Composites, McGraw-Hill Book Co., New York 1992.

[1.6] Neville, A. M.: Properties of Concrete, 3rd Ed., Longman Scientific and Technical, 1987.

[1.7] Sinha, B. P., K. H. Gerstle, and L. G. Tulin: Stress-strain relations for concrete under cyclic loadings, ACI Journal, February 1964.

[1.8] Bennet, E. W. and N. K. Raju: Cumulative fatigue damage of plain concrete in compression, Structures Solid Mech. and Engin. Design (1971), 1089-1102.

[1.9] Paskova, T.: Low-Cycle Fatigue Behavior of Concrete With and Without Fiber Reinforcement, Ph. D. Dissertation, Columbia University, New York 1994.

[1.10] Meyer, C.: An Energy-Based Damage Model for Inelastic Dynamic Analysis of Reinforced Concrete Frames, in: Nonlinear Seismic Analysis and Design of Reinforced Concrete Buildings (Eds. P. Fajfar and H. Krawinkler), Elsevier Applied Science, London 1992.

[1.11] Su, E. C. M. and T. C. Hsu: Biaxial compression fatigue and discontinuity of concrete, ACI Materials Journal, May-June 1988.

[1.12] Stankowski, T. and K. H. Gerstle: Simple formulation of concrete behavior under multiaxial load histories, ACI Journal, Mar-Apr 1985.

[1.13] Nilson, A. H. (Ed.): Finite Element Analysis of Reinforced Concrete, Special Publication, ASCE, New York 1982.

[1.14] Meyer, C. and H. Okamura (Eds.): Finite Element Analysis of Reinforced Concrete Structures, Special Publication, ASCE, New York 1986.

[1.15] Isenberg, J. (Ed.): Finite Element Analysis of Reinforced Concrete II, Special Publication, ASCE, New York 1993.

[1.16] Darwin, D. and D. A. Pecknold: Analysis of RC shear panels under cyclic loading, ASCE J. Struct. Div., February 1976.

[1.17] Bazant Z. P. and T. Tsubaki: Total strain theory and path-dependence of concrete, ASCE J. Eng. Mech. Div., Dec. 1980.

[1.18] Elwi, A. A. and D. W. Murray: A 3D hypoelastic concrete constitutive relationship, ASCE J. Eng. Mech. Div., Aug. 1979.

[1.19] de Borst, R.: Non-Linear Analysis of Frictional Materials, Ph. D. Dissertation, Delft University of Technology, Delft, The Netherlands, 1986.

[1.20] Bazant, Z. P. and S. S. Kim: Plastic-fracturing theory for concrete, ASCE J. Eng. Mech. Div., June 1979.

[1.21] Valanis, K. C.: A Theory of viscoplasticity without a yield surface, Archiwum Mechaniki Stossovaneij (Archives of Mechanics, Warsaw) 23 (1971), 517-551.

[1.22] Bazant, Z. P. and P. Bhat: Endochronic theory of inelasticity and failure of concrete, ASCE J. Eng. Mech. Div., 1976, 701-722.

[1.23] Bazant, Z. P., ed.: Fracture Mechanics of Concrete Structures, Elsevier Applied Science, London and New York 1992.

[1.24] Wittmann, F. H.: Fracture Mechanics of Concrete, Elsevier Applied Science, 1983.

[1.25] Carpinteri, A. and A. R. Ingraffea: Fracture Mechanics of Concrete, Martinus Nijhoff Publishers, 1984.

[1.26] Perzyna, P.: Fundamental problems in viscoplasticity, Advances in Applied Mechanics 9, 1966, 243-377.

[1.27] Nilsson, L.: Finite element analysis of impact on concrete structures, in: Finite Elements in Nonlinear Mechanics, (Eds. P. Bergan et al), Tapir, Trondheim 1978.

[1.28] Nilsson, L. and R. Glemberg: A constitutive model for concrete in high rate of loading conditions, Proc., IABSE Colloquium, Delft (1981), 159-174.

[1.29] Bicanic, N.: Nonlinear Finite Element Transient Analysis of Concrete Structures, Ph. D. Thesis, C/Ph/50/78, University College of Swansea, Wales, 1978.

[1.30] Bicanic, N. and O. C. Zienkiewicz: Constitutive model for concrete under dynamic loading, Earthquake Engineering and Structural Dynamics 11 (1983), 689-711.

[1.31] Meyer, C. and H. E. Delgado-Saavedra: Modeling large inelastic cyclic response of concrete, in: Finite Element Analysis of Reinforced Concrete Structures (Eds. C. Meyer and H. Okamura), Special Publication, ASCE, 1986.

[1.32] Willam, K. J. and E. P. Warnke: Constitutive model for triaxial behavior of concrete, Proc., IABSE Seminar on Concrete Structures Subjected to Triaxial Stresses, Paper III-1, Bergamo, Italy, 1974.

[1.33] Hatano, T. and H. Tsutsumi: Dynamical compressive deformation and failure of concrete under dynamic load, Proc., Second World Conference on Earthquake Engineering, Tokyo, Vol. 3, 1960, 1963-1978.

[1.34] Kachanov, L. M.: Time of rupture process under creep conditions, Izvestia Akademii Nauk USSR 8 (1958), 26-31 (in Russian).

[1.35] Rabotnov, Y. N.: Creep rupture, Proc., XII Int. Congress of Applied Mechanics, Stanford University, Springer Verlag, 1968.

[1.36] Lemaitre, J.: A Course on Damage Mechanics, Springer Verlag, Berlin 1992.

[1.37] Mazars, J.: Application de la mecanique de l'endommangement au comportement non lineaire et a la rupture du beton de structure, These de Doctorate d'Etat, L. M. T., Universite Paris, France, 1984.

[1.38] Pijaudier-Cabot, G.: Caracterisation et modelisation du comportement du beton par un essai multiaxial au tomatique, These de 3eme cycle, L. M. T., Universite Paris, France, 1985.

[1.39] Simo, J. C. and J. W. Ju: Strain- and stress-based continuum damage models - I. Formulation, Int. J. Solids and Structures 23, no. 7 (1987), 821-840.

[1.40] Suaris, W. and S.P. Shah: Inertial effects in instrumented impact testing of cementitious composites, ASTM J. Cement, Concrete and Aggregates 3, no. 2, (1981), 77-83.

[1.41] Gopalaratnam, V.S., S.P. Shah, R. John: A modified instrumented Charpy test for cement-based composites. Experimental Mechanics 24, no. 3, (1984), 102-111.

[1.42] Suaris, W. and S.P. Shah: Properties of concrete subjected to impact, ASCE J. Struct. Eng. 109, no. 7 (1983), 1727-1741.

[1.43] Kobayashi, A.S., N.M. Hawkins, J.J. Du: An impact damage model of concrete. University of Washington, Seattle (1989).

[1.44] Bentur, A., S. Mindess, N. Banthia: The behaviour of concrete under impact loading: Experimental procedures and method of analysis, Materials and Structures 19, no, 113 (1986), 371-378.

[1.45] Yon, J.H., N.M. Hawkins, A.S. Kobayashi: On the strain rate sensitivity of concrete mechanical properties. University of Washington, Seattle, paper (5-15-89) (1989).

[1.46] Hughes, G. and A.W. Beeby: Investigation of the effect of impact loading on concrete beams, The Structural Engineer 60 B, no. 3 (1982), 45-52.

[1.47] van Mier, J.M., A.F. Pruijssers, H.W. Reinhardt, T. Monnier: Load-time response of colliding concrete bodies, ASCE J. Struct. Eng. 117, no. 2 (1991), 354-374.

[1.48] Rossi, P. and F. Le Maour: Ouvrages d'art: le contrôle des fissurations. La Revue des Laboratoires d'essais, no. 6, (1986), 25-27.

[1.49] Rossi, P.: Fissuration du béton: Du materiau à la structure. Application de la méchanique linéaire de la rupture. Doctoral thesis ENPC, Paris, Dec. 1986.

[1.50] Mindess, S., N.P. Banthia, A. Ritter, J.P. Skalny: Crack developments in cementitious materials under impact loading, in: MRS Symp. Proc. Voc. (Eds. S. Mindess and S.P. Shah), 64, Pittsburgh, PA 1986, 217-233.

[1.51] Reinhardt, H.W., H.A. Körmeling, A.J. Zielinski: The split Hopkinson bar, a versatile tool for the impact testing of concrete, Materials and Structures 19, no. 109 (1986), 55-63.

[1.52] John, R. and S.P. Shah: Fracture of concrete subjected to impact loading, ASTM Cement, Concrete and Aggregates 8 (1986), 24-32.

[1.53] Bur, A.J. and S.C. Roth: A polymer pressure gage for dynamic pressure measurements. Proc. 2nd Symp. Interaction of non-nuclear munitions with structures. Panama City Beach, Florida, April 1985, 291-295.

[1.54] Frank, T.: Beeinflussung der Prüfergebnisse durch die Messeinrichtung bei der Stossbeanspruchung von Beton. Materialprüfung 26, no. 4 (1984), 96-100.

[1.55] Curbach, M., P. Maliszkiewicz, J. Eibl: Acoustic emission measurement of concrete under high loading rates. Transactions of the 10th International Conference on Structural Mechanics in Reactor Technology SMiRT, 1989.

[1.56] Diederichs, U., U. Schneider, M. Terrien: Formation and propagation of cracks and acoustic emission, in: Fracture Mechanics of Concrete (Ed. F.H. Wittmann), Elsevier, Amsterdam 1983.

[1.57] Hillerborg, A.: Stability problems in fracture mechanics testing, in: Fracture of Concrete and Rock: Recent Developments (Eds. S.P. Shah, S.E. Swartz, B. Barr). Elsevier Appl. Sci. London, New York 1989, 369-378.

[1.58] Zukas, J.A. et al.: Impact Dynamics, John Wiley & Sons, New York 1982.

[1.59] Paulmann, K. and J. Steinert: Beton bei sehr kurzer Belastungsgeschichte, Beton 32, no. 6 (1982), 225-228.

[1.60] Takeda, J.: A loading apparatus for high speed testing of building materials and structures, Proc. 2nd Japan Congress on Testing Materials, Jap. Soc. Testing Mat., Kyoto, Japan (1959), 236-238.

[1.61] Zhurkov, S.N.: Kinetic concept of the strength of solids, Int. J. Fracture Mechanics 1 (1965), 311-323.

[1.62] Gran, J.K.: A new technique for studying the dynamic tensile response of concrete, Proc. 2nd Symp. Interaction of non-neclear munitions with structures, Panama City Beach, Florida (1986), 206-210.

[1.63] Brühwiler, E. and F.H. Wittmann: Effect of rate of loading on fracture energy and strain softening, Contribution to RILEM Committee 89-FMT, Stevin report no. 25-87-16 (1987), Delft University of Technology, April 1987.

[1.64] Lieberum, K.H. and H.W. Reinhardt: Strength of concrete on an extremely small bearing area, ACI Structural Journal 86, no. 1 (1989), 67-76.

[1.65] Kennedy, R.P.: A review of procedures for the analysis and design of concrete structures to resist missile impact effects, Nuclear Eng. Design 37 (1976), 183-203.

[1.66] Barr, P.: Guidelines for the design and assessment of concrete structures subjected to impact, Safety and Reliability Directorate, SRD R439, Issue 2, UKAEA, Calcheth, March 1988.

[1.67] Mahin, S. and P.B. Shing: Pseudodynamic method of seismic testing, ASCE J. Struct. Eng. 111, no. 7 (1985), 1482-1503.

[1.68] Block, K.: Der harte Querstoss - Impact - auf Balken aus Stahl, Holz und Stahlbeton. Dissertation, Universtität Dortmund, 1983.

[1.69] Reinhardt, H.W.: Simple relations for the strain rate influence on concrete, Darmstadt Concrete 2 (1987), 203-211.

[1.70] Goldsmith, W.: Impact: The Theory and Physical Behaviour of Colliding Solids. Edward Arnold Publishers Ltd., London, U.K. 1960.

[1.71] Hughes, G. and D.M. Speirs: An investigation of the beam impact problem, Technical Report 546 (1982), Cement and Concrete Assoc.

[1.72] Jensen, J.J. and K. Hoiseth: Impact of dropped objects on lightweight concrete, Nordic Concr. Res., 2 (1983), 102-113.

[1.73] Pruijssers, A.F., J.G.M. van Mier, A. Mol: Impact behaviour of concrete breakwater elements, TUD/TNO Research Report (1987), Delft, The Netherlands.

[1.74] Sih, G.C. ed.: Dynamic Crack Propagation, Noorhoff Intern. Publ., Leyden 1972.

[1.75] Freund, L.B.: Crack propagation in an elastic solid subjected to general loading - I. Constant rate of extension. J. Mech. Phys. Solids 20 (1972), 129-140.

[1.76] Kipp, M.E., D.E. Grady, E.P. Chen: Strain-rate dependent fracture initiation. Int. J. Fracture 16 (1980), 471-478.

[1.77] Weerheijm, J.: Concrete under impact tensile loading and latual compression. Doctoral Thesis, Delft University of Technology, Delft 1992.

[1.78] Mihashi, H. and F.H. Wittmann: Stochastic approach to study the influence of rate of loading on strength of concrete. Heron 25, no. 3, Delft, The Netherlands 1980.

[1.79] Lindholm, U.S., L.M. Yeakley, A. Nagy: The dynamic strength and fracture properties of Dresser basalt, Int. J. Rock Mech. Min. Sci. & Geomech. Abstr. 11 (1974), 181-191.

[1.80] Krausz, A.S. and H. Eyring: Deformation Kinetics, J. Wiley & Sons, New York 1975.

[1.81] Körmeling, H.A.: Impact of steel fibre concrete at -170^oC, Stevin Report 5-84-13, Delft University of Technology, Delft, December 1984.

[1.82] Shah, S.P.: Constitutive relations of concrete subjected to a varying strain rate. Symp. Proc. The interaction of non-nuclear munitions with structures. US Air Force Academy, Colorado, May 10-13, 1983, 81-84.

[1.83] Curbach, M.: Festigkeitssteigerung von Beton bei hohen Belastungsgeschwindigkeiten. Schriftenreihe des Instituts für Massivbau und Baustofftechnologie, Heft 1, Karlsruhe 1987.

[1.84] Zielinski, A.J. and H.W. Reinhardt: Stress-strain behaviour of concrete and mortar at high rates of tensile loading, Cement and Concrete Research 12, (1981), 309-319.

[1.85] Brühwiler, E.: Versuche über den Einfluss von Druckvorlasten auf das Verhalten von Staumauerbeton unter Zug bei hohen Dehngeschwindigkeiten. Ecole Polytechnique Fédérale de Lausanne, Lab. des Matériaux de construction, Dec. 1987.

[1.86] Shah, S.P. and R. John: Rate-sensitivity of mode I and mode II fracture of concrete, Mat. Res. Soc. Symp. Proc. Vol. 64, eds. S. Mindess and S.P. Shah, Pittsburgh, 1986, 21-37.

[1.87] Muria Vila, D. and P. Hamelin: Comportement au choc des bétons et mortiers à matrices hydrauliques, in: Combining materials: Design, production and properties (Ed. J.C. Maso), 1st Int. RILEM Congress Vol.2, Chapman and Hall, London New York 1987, 725-732.

[1.88] Carrasquillo, R.L., A.H. Nilson, and F.O. Slate: Properties of high-strength concrete subject to short-term loads, ACI Journal, May-June 1981.

FLEXURAL MEMBERS AND BEAM-COLUMN JOINTS

C. Meyer
Columbia University, New York, NY, USA

F.C. Filippou
University of California Berkeley, Berkeley, CA, USA

P. Gergely†
Cornell University, Ithaca, NY, USA

2.1 Introduction

2.1.1 General Comments on Structural Analysis

The purpose of structural analysis is to predict or explain in a rational way the response of structures to specified loads. The task of analyzing reinforced concrete frames subjected to earthquake loads is among the most difficult ones known, because the loading is highly random and the load-deformation relationships for the structure and its individual components very nonlinear, with marked stiffness and strength degradation. As a result, the choice of an appropriate analysis method is not an easy one, but it has far-reaching consequences.

When analyzing static linear elastic response of structures, approximate methods are quite acceptable, because inelastic moment redistribution can "correct" for the errors inherent in the underlying assumptions. Whenever reference is made to an exact analysis, the use of quotation marks is in order as a reminder that our mathematical tools and analysis methods have their limits. This applies particularly to concrete structures. As long as a structure possesses sufficient ductility, discrepancies between analysis results and actual behavior are relatively unimportant.

In the case of dynamic loading, the differences between analytical and actual response can become unacceptable. The analysis results may be at times barely better than guesswork, especially if a static analysis for so-called "equivalent" lateral loads is involved. Yet the use of such static analysis results for design purposes is rather standard practice and typically justified because of the unpredictable nature of the loading, which does not seem to warrant a more exact analysis. The logical way for dealing with the uncertainty of the loading is to treat it as a random process and to use random vibration analysis techniques to predict structural response in a statistical sense (see Appendix). If numerical simulation is used, such as a Monte Carlo technique, then each artificially generated earthquake ground motion history will permit the determination of the corresponding response. The differences between the results for different sample runs may then be taken as a measure of accuracy that would be acceptable for the selected analysis method. For example, if the standard deviation for maximum interstory drifts is 10%, calculations to five significant figures are obviously not warranted.

When nonlinear dynamic response is involved, the picture changes again. Linear static analysis results may bear no resemblance with reality anymore. Not even the randomness of the loading can any longer be used to justify grossly simplified analyses, because these are likely to provide the designer with no or completely unreliable information on strength and ductility demands.

As far as strength is concerned, the situation is not as hopeless as it may seem, because with capacity design we have a tool to control strengths. But neither ductility demands nor interstory drifts can be predicted with sufficient accuracy using overly simplified analysis methods. Ductility demands computed for critical members on the basis of a linear elastic static analysis can easily be off by an order of magnitude. Thus, if a reinforced concrete frame is designed to safely withstand a strong earthquake, and inelastic response is relied upon as a source of energy dissipation, then a realistic nonlinear time history analysis may be the only means to obtain the information needed by the designer.

Any method of analysis that claims to be "realistic" must take into account the nonlinear response characteristics of the structure and the randomness of the load input. Traditionally, a number of arguments have been raised against such detailed analyses:

1. High computing cost. This factor has become all but irrelevant, as most PCs have now the computing power to perform such analyses. Only if extremely long computation times are involved does this argument carry any weight, in which case the computation time is of concern, not the cost.
2. Limitations of existing mathematical models to simulate nonlinear material and structural response. This is a valid and serious argument. In fact, all models that are available at present are basically empirical in nature and therefore valid only within the range of parameters for which they have been calibrated against experimental results. This problem will be addressed in detail later.
3. The randomness of load input. This problem can be addressed by treating the earthquake as a random process. Monte Carlo-type numerical simulations have a number of advantages over other, more sophisticated quasi-nonlinear random

vibration solution methods.

4. Difficulties associated with performing such analyses and verifying the correctness of the results. This is also a serious problem. Analysts who want to perform nonlinear dynamic analyses must have a sufficient educational background in material behavior, dynamic analysis and numerical solution techniques as well as practical experience. Even then, such analyses can be very costly in terms of man-hours. Not until computer codes with robust material models and numerical algorithms are available will this user problem be reduced to that of a more conventional computer analysis.

2.1.2 Mechanics-Based Models

The most difficult part of nonlinear dynamic analysis of RC frames is the establishment of a mathematical model that can faithfully reproduce the member behavior. All approaches have in common that they start out with basic assumptions and then calibrate any built-in parameters against experimental results. The various methods differ fundamentally by the level on which this calibration takes place.

One category of models is often referred to as "mechanics-based" or"rational". Here the theories of elasticity, plasticity, fracture or damage mechanics are used to establish constitutive relationships for material behavior that need to be confirmed experimentally on the material level. Such material models are readily incorporated into finite element computer codes, and the structural response can be computed, assuming that certain computational difficulties are overcome. Such an approach is sometimes referred to as microscopic modeling. In reference to reinforced concrete it implies constitutive relationships for both concrete and steel and for the bond behavior on their interface.

A model which treats reinforced concrete as a single composite material with its own composite stress-strain relationship is defined as macroscopic. This approach permits one to represent an entire panel with a single macro- or superelement, provided there are no excessive stress gradients that require further mesh refinements. The derivation of appropriate stress-strain relationships to be prescribed at the integration points of such an element is not straightforward at all. Because we cannot benefit anymore from the "rationality" of classical mechanics for a material point, empirically derived expressions are used, which need to be confirmed experimentally. Yet, we know enough about the mechanics of reinforced concrete to assure that such empirical methods be sufficiently "rational".

Frame elements can be considered a third category. These are of significant practical importance and differ from the other macroelements in that a set of simplifying assumptions is made regarding the kinematics of member deformations (e.g. plane sections remain plane). It is this category of models that is possibly most difficult to calibrate because the basic engineering mechanics aspects are so much integrated that they are virtually unidentifiable to all but the most expert eye. This makes this category of models most susceptible to inappropriate use, because lack of physical insight is easily covered up by force-fitting of experimental data during the calibration phase, especially if complex phenomena such as strength and stiffness degradation are to be reproduced.

There is no single "best" approach to model such complex behavior, and the nomenclature used here implies no inherent value system. All so-called "rational" models contain empirical elements, and all "empirical" models are more or less rational. It is also worth mentioning at this point that the "accuracy" of numerical models is always evaluated by comparing their results with experimental data. Yet, we should not forget that such data vary widely from experimentalist to experimentalist and from laboratory to laboratory, not mentioning the inherent randomness of concrete material properties. Thus, any validation should utilize only highly reliable and (within limits) repeatable experimental data, preferably with error bounds for specific confidence levels. Unfortunately, few data published in the open literature contain this information.

2.1.3 Comments on Modelling of Damage

In conventional design for static loads, the basic design requirement can be stated in the general form,

$$Design\ Strength \quad \geq \quad Load\ Effect$$

When designing for dynamic loads, for example, those resulting from a strong earthquake, such a design requirement is often supplemented with a corresponding ductility requirement,

$$\mu_{capacity} \quad \geq \quad \mu_{demand} \tag{2.1}$$

Use of a concrete member's or structure's ductility property in this simple format is incorrect, however, because it implies that both the ductility capacity and demand are independent of the number of load cycles. It is known that this is not the case. In fact, the decrease of both strength and ductility capacity with the number of load applications is generally designated as fatigue. The number of load applications adds a second dimension to the problem. A more appropriate design requirement of the type of Eq. 2.1 would therefore take on the form,

$$\mu_{capacity}(N) \quad \geq \quad \mu_{demand}(N) \tag{2.2}$$

where N is the number of load applications. Eq. 2.2 is still an oversimplification. It ignores the fact that in reality a structure is subjected to loads of varying amplitudes. In addition, the order in which a series of load impulses of different amplitudes are applied, has also an effect. For example, the damage resulting from one strong load impulse followed by a series of smaller loads can be substantially different from the damage resulting from the same strong load impulse preceded by the series of smaller loads.

The concept of damage is familiar to everybody. Yet, if we have to define it in precise scientific terms, it appears that our notion of damage is in general rather vague. In the context of building response to destructive earthquakes, a large number of damage definitions have been proposed [2.1]. Most of these are of a more or less empirical nature and often prone to subjective influences. To avoid this problem, it is advantageous to tie the damage definition to the degree of physical deterioration with clearly defined consequences regarding the material's capacity to resist further load. This is the approach of damage mechanics discussed in Sect. 1.2.6. Similarly, the notion of failure needs to be stripped of an occasionally arbitrary definition by relating it to a specific level of damage at which the material or member ceases to resist further load. For example, we may measure damage as the amount of energy dissipated after N load cycles, $E(N)$, normalized with the total energy dissipation capacity per unit volume, E_{tot},

$$D = \frac{E(N)}{E_{tot}} \tag{2.3}$$

This definition can be substituted for the cycle ratio (N/N_f) used in conventional fatigue analysis and is in general applicable both on a material and member level.

On the material level we are dealing with the stress-strain response of plain concrete. Failure is logically defined as the point at which the slope of the stress-strain curve ceases to remain positive. The energy dissipation capacity E_{tot} is clearly a material property that can be measured in the laboratory as a function of stress or strain level under constant-amplitude loading [2.2].

On the member level, the moment-curvature or load-deformation response is involved. Failure is now defined as the point at which the moment-curvature (or moment-rotation) curve ceases to have a positive slope, in the absence of an axial force. E_{tot} on the member level can also be measured in the laboratory as a function of load or deformation amplitude, but it obviously is a function of many other factors, such as reinforcement details.

Even though (or rather because) the above damage definition is very basic, it cannot easily be extended to the structure level, because on this level the loading has a spatial distribution component that is absent on the material and to some extent on the element level. For this reason, all so-called *global damage* definitions have an element of arbitrariness and lack the rational component that would be needed to evaluate a structure's residual capacity to resist further load. Therefore we shall not attempt to define damage on the structural (or global) level. It is not missed either, at least not in the context of nonlinear dynamic analysis of degrading structures.

The rationality or mechanics basis for the above damage definition can be illustrated by juxtaposing it to that of the well-known model of Park and Ang [2.3],

$$D = \frac{\delta_{max}}{\delta_u} + \frac{\beta}{Q_y \delta_u} \int dE \tag{2.4}$$

where δ_{max} = maximum deformation experienced so far, δ_u = ultimate deformation under monotonic loading, Q_y = calculated yield strength, dE = dissipated energy increment, and

$$\beta = (-0.447 + 0.73\frac{l}{d} + 0.24n_o + 0.314P_t)0.7^{\rho_w} \tag{2.5}$$

with l/d = shear span ratio, n_o = normalized axial force, ρ_w = confinement ratio, P_t = longitudinal steel ratio. The concatenation in Eq. 2.5 of the four independent influence parameters deemed most significant has clearly an element of arbitrariness. Likewise, there is no physical evidence whatsoever to suggest the linear superposition of the two components that make up the damage index of Eq. 2.4. Yet, the fact that the lack of a physical basis was compensated for by a superbly executed regression analysis of numerous experimental data does not seem to diminish the popularity of this model, which has been incorporated in the widely used IDARC program.

It may be appropriate to note that a general definition of damage such as Eq. 2.3, by referring to the remaining or residual energy dissipation capacity, lends itself to non-mechanistic sources of damage. For example, if a concrete specimen had been exposed to N freeze-thaw cycles or a given history of chemical attack, the inflicted damage D may be correlated to the remaining energy dissipation capacity, $E_{tot} - E(N)$, which can be measured.

A serious drawback of the damage definition of Eq. 2.3 is the lack of an experimental data base for E_{tot}. S-N curves have been generated for some selected cases, but for the low-cycle fatigue range that is of interest for earthquake response analysis, very few data exist.

On the member level, a considerable body of knowledge exists, thanks to extensive experimental investigations, that were carried out primarily at the University of California at Berkeley, the University of Illinois at Urbana, and the University of Michigan. However, because of the large number of influence factors it is essential to organize the available data in a systematic way.

At Columbia University, a comprehensive research project is currently under way to study damage accumulation and the energy dissipation capacity of concrete with and without fiber reinforcement [2.2,2.4,2.5]. Typical damage accumulation curves were shown in Fig. 1.20 of Chapter 1. The evaluation of the data is still ongoing, and functional relationships for damage accumulation at this point are only preliminary. It needs to be pointed out that, once such formulations are available, they lend themselves to remaining life predictions and reliability estimates. For example, for plain concrete, we may write,

$$\begin{aligned} D &= (1.7 - S)\,\overline{N} \quad \text{for } \overline{N} \leq 0.6 \\ D &= \overline{N}^{1.6S} \quad\quad\quad\ \text{for } \overline{N} > 0.6 \end{aligned} \tag{2.6}$$

where $\overline{N} = N/N_f$ is the cycle ratio, and S = stress ratio.

For variable-amplitude loading, Eq. 2.6 may be used to convert the damage produced by N_1 cycles with some stress level S_1 to an equivalent number of cycles N_2^* at a different stress level S_2, by equating the corresponding damage levels D_1 and D_2^*,

$$D_1 = \left(\frac{N_1}{N_{f1}}\right)^{1.6S_1} = \left(\frac{N_2^*}{N_{f2}}\right)^{1.6S_2} = D_2^*$$

from which

$$N_2^* = \left(\frac{N_1}{N_{f1}}\right)^{\frac{S_1}{S_2}} N_{f2} \qquad \text{for } \overline{N} > 0.6$$

The combined damage produced by N_1 cycles of load level S_1 and N_2 cycles of load level S_2 is then,

$$D = \left(\frac{N_2^* + N_2}{N_{f2}}\right)^{1.6 S_2} \tag{2.7}$$

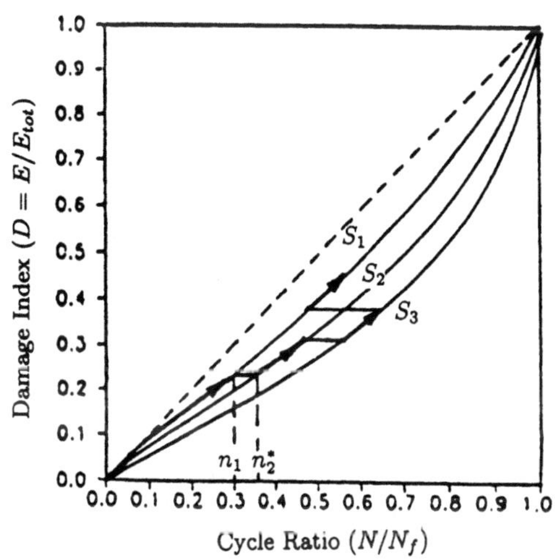

Fig. 2.1 Damage accumulation for variable load amplitudes

Similarly, the evolution of damage can be followed through an arbitrary sequence of loads of variable amplitude, Fig. 2.1, and at each stage the remaining number of cycles can be determined that the material can resist for a given load level.

For material characterization, linear elastic analyses and selected nonlinear analyses require only basic information as input, such as Young's modulus, Poisson's ratio, and the yield strength of steel. For degrading reinforced concrete, also damage accumulation information of the kind of Eq. 2.6 is needed.

On the member level, corresponding information is as yet not available. Because of the large number of influence factors, it is not even clear how such data could be organized, once they are generated. Theoretically, it should be possible to determine the total energy dissipation capacity E_{tot} of a member, as a function of the important influence factors and then to formulate a damage accumulation law equivalent to Eq. 2.6. The most significant parameters aside from the load amplitude will be the cross sectional dimensions, concrete strength, and reinforcing details, both for flexural, shear and confinement reinforcement. The establishment of the required data base is clearly a formidable undertaking, but indispensable for a mechanics-based alternative to the empirical hysteretic load-deformation relationships with ever more complex

Takeda-type rules. Only with such a damage accumulation law that quantifies the residual energy dissipation capacity of a RC member will it be possible to predict its response and probability of failure.

This analysis challenge is further complicated by the existence of different potential failure modes. If a member were known to fail in flexure, then a model could be constructed as described above. But shear and bond failure modes cannot be ignored. Moreover, each failure mechanism has in general its own damage accumulation rate, so that it is a challenge indeed to predict a priori which failure mode will eventually prevail.

2.1.4 Concentrated Plastic Hinge vs Spread Plasticity Models

The simulation of hysteretic behavior of structures under severe seismic excitations is a challenging problem. Indeed, in addition to the problem of modelling the structural system and the interaction between its different components, there is the difficulty of representing the hysteretic behavior of the critical regions that undergo several cycles of inelastic deformation. Past experimental work has been devoted mostly to the understanding of the hysteretic behavior of critical regions, such as the ends of girders and columns, under the combined action of bending moments, shear and axial forces. Very few experiments have addressed entire lateral force resisting systems, because of the prohibitive cost of such studies. The past approach towards improving our understanding of the seismic behavior of structural systems has thus been based on the inherent assumption that analytical models for the hysteretic behavior of critical regions can be derived from the experimental evidence and then used to simulate the seismic behavior of entire structures. Unfortunately, the capabilities of computer hardware have usually limited such attempts to the simplest possible member models, which assume that all inelastic behavior is concentrated in plastic hinges of zero length. In conjunction with this approach, models of different degrees of complexity have been proposed to describe the hysteretic moment-rotation relationship of such hinges. Their main shortcomings are as follows:

1. the neglect of the spread of inelastic deformations as a function of loading history;
2. the neglect of the presence of distributed gravity loads;
3. the inability to account in a rational manner for the interaction between bending moment, shear and axial force;
4. the limited success to account for strength loss and member softening;
5. the inability of proposed hysteresis laws to simulate the evolution of damage and its effect on the subsequent hysteretic behavior of a plastic region.

The continuing rapid increase in the computing power of engineering workstations calls for corresponding advances in software development to address the shortcomings listed above. With computing power or its lack becoming increasingly a non-issue, it is appropriate to develop more rational models that can simulate the hysteretic behavior of structural members and are formulated in a clear, efficient and consistent manner to assure their robust numerical behavior in the nonlinear analysis of structures with many degrees of freedom.

2.2 Concentrated Plastic Hinge Models

When RC frames are subjected to strong seismic excitation, the inelastic behavior occurs primarily near the member ends. Thus, the first approach to model this behavior was to introduce concentrated hinges with nonlinear rotational springs at the member ends. Such models may contain several springs that are connected in series or in parallel.

The earliest parallel component element (Fig. 2.2a) was introduced by Clough and Johnston [2.6] and allowed for a bilinear moment-rotation relationship. The element consists of two parallel components, an elastic-perfectly plastic one to represent yielding and a perfectly elastic one to represent strain-hardening. The stiffness of the member is the sum of the two component stiffnesses. Takizawa and Aoyama [2.7] generalized this model for multilinear monotonic behavior, including cracking. The series model was formally introduced by Giberson [2.8]. Its original form (Fig. 2.2b) consists of a linear elastic element with one equivalent nonlinear rotational spring attached to each end. The inelastic deformations of the member are lumped into these end springs. This model is more versatile than the original Clough model, since it can describe more complex hysteretic behavior by selecting appropriate moment-rotation relationships for the end springs. This makes the model attractive for the phenomenological representation of the hysteretic behavior of RC members.

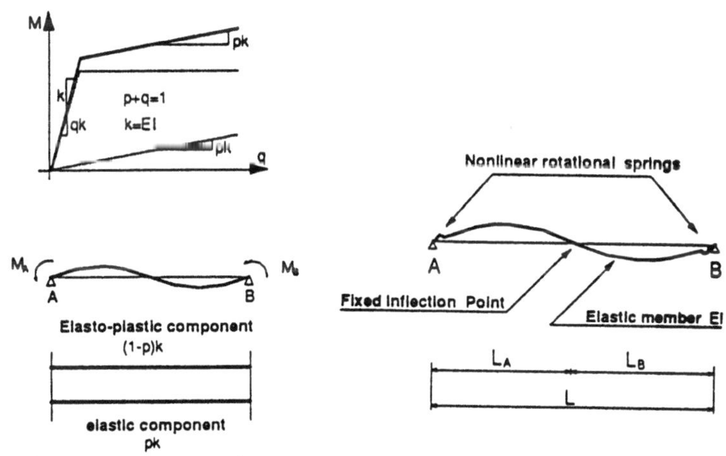

a) Parallel model [2.6] b) Series model [2.8]

Fig. 2.2 Simple lumped plasticity models

Several constitutive models have been proposed to date for the nonlinear springs (Fig. 2.3). These include cyclic stiffness degradation in flexure and shear [2.9, 2.10], pinching under load reversal [2.11], and fixed-end rotations at the beam-column joint interface due to bar pullout [2.12, 2.13]. Axial-flexural coupling is usually neglected. Nonlinear rate-sensitive representations have also been generalized from endochronic theory [2.14] to provide continuous expressions for the hysteretic spring behavior.

An extensive discussion of the mathematical functions that are appropriate for such models is given in [2.15]. A critical issue for these models is the selection of parameters to represent the hysteretic behavior observed in experiments, which involves two basic problems:

a) the model parameters depend not only on the section characteristics, but also on the load and deformation histories, thus limiting the generality of this approach;
b) a consistent and rational method for the selection of model parameters requires special algorithms to assure a least square fit between analytical results and experimental data.

In [2.16], a formal system identification method was used to select the parameters for the moment-rotation relationship.

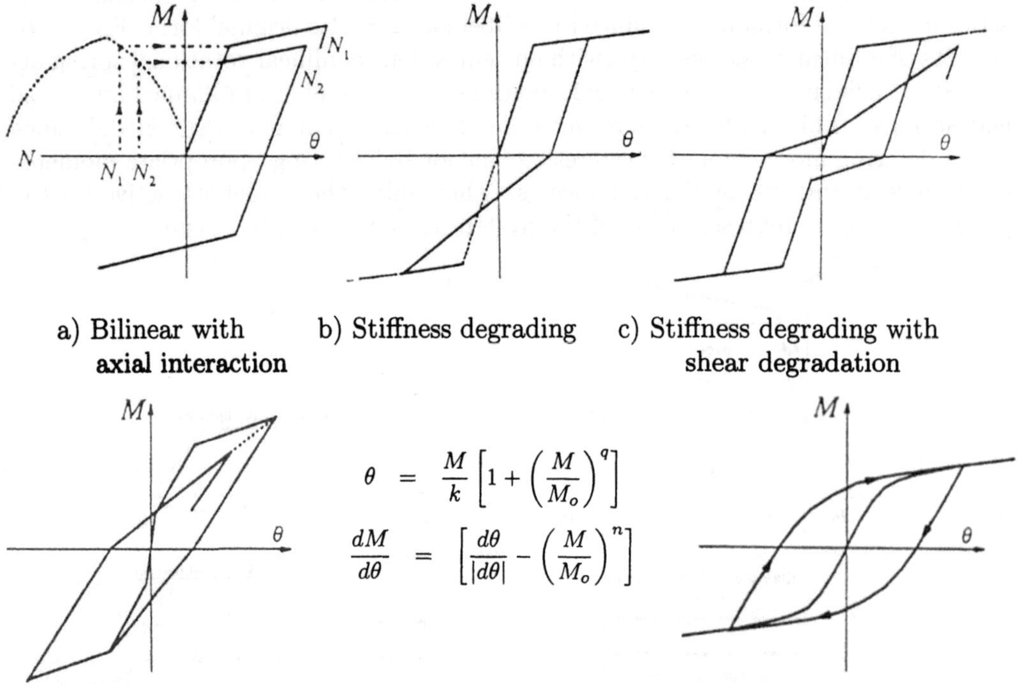

a) Bilinear with axial interaction
b) Stiffness degrading
c) Stiffness degrading with shear degradation

$$\theta = \frac{M}{k}\left[1 + \left(\frac{M}{M_o}\right)^{q}\right]$$

$$\frac{dM}{d\theta} = \left[\frac{d\theta}{|d\theta|} - \left(\frac{M}{M_o}\right)^{n}\right]$$

d) Takeda model [2.10]
e) Ramberg-Osgood model
f) Rate models based on endochronic theory

Fig. 2.3 Constitutive models for nonlinear springs

The dependence of flexural strength on the axial load under uniaxial and biaxial loading conditions has been modelled explicitly for beam-columns and structural walls. In most lumped plasticity models the axial force-bending moment interaction is described by a yield surface for the stress resultants and an associated flow rule according to classical plasticity theory [2.17]. The response is assumed to be linear for stress states that fall within the yield surface, in which case the flexural and axial stiffnesses of the member are uncoupled and independent of the end loads. The introduction of multiple yield and loading surfaces and corresponding hardening rules makes multilinear constitutive representations for the springs possible that include cracking and cyclic stiffness degradation [2.18].

A lumped plasticity model is a simplification of actual behavior with gradual spread of inelastic deformations into the member. As a rule of thumb, plasticity typically extends about one distance d from the end face into the span, where d is the member depth. In the case of a flexural structural wall, this distance can be considerable; yielding of the edge member reinforcement may very well extend over one or two stories. The accuracy of lumped plasticity models was assessed in several correlation studies, particularly those dealing with large resisting elements of wall-frame structures [2.19, 2.20]. The models' basic advantage is their simplicity that reduces storage requirements and computational costs and maintains the numerical stability of the computations. Most lumped models, however, oversimply important aspects of the hysteretic behavior and are therefore of limited accuracy. One such limitation derives from restrictive a priori assumptions for the determination of the spring parameters. Parametric and theoretical studies of girders under monotonic loading demonstrate a strong dependence between model parameters and the imposed loading pattern and level of inelastic deformation [2.21]. Neither factor is likely to remain constant during the dynamic response. The problem is further complicated by fluctuations of the axial force in columns. Because of this history dependency, damage predictions at the global and particularly at the local level may be grossly inaccurate. Such information can only be obtained with more refined models that can describe the hysteretic behavior of the section as a function of axial load. Another limitation of most lumped plasticity models is their inability to adequately describe the deformation softening behavior of RC members. Such behavior can be observed as the reduction in lateral resistance of an axially loaded cantilever column under monotonically increasing lateral top displacement. Again, more advanced models are needed in this case.

The generalization of plastic theory concepts [2.17] to RC column stress and strain resultants limits the applicability of these models to well-detailed members with large inelastic deformation capacities of their critical regions. For a reinforced concrete column section, the yield surface of the stress resultants is actually a function of a reference strain that couples the corresponding displacement components. This contradicts classical plasticity theory, which does not account for deformation softening and assumes the section deformability to be unlimited.

To overcome some of the limitations of classical plasticity theory when describing the interaction between axial force and bending moments, Lai et al [2.22] proposed a fiber hinge model that consists of a linear elastic element extending over the entire member length and one inelastic element at each end (Fig. 2.4). Each inelastic element is made up of four inelastic springs at the section corners that represent the longitudinal reinforcing steel, and a central concrete spring that is effective in compression only. This five-spring representation of the end sections is capable of simulating the axial force-biaxial bending moment interaction in RC members in a more rational way than is possible by classical plasticity theory.

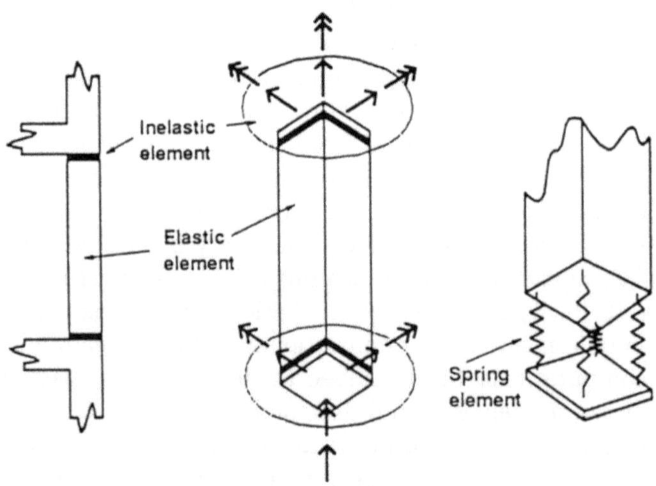

a) Member in frame b) Member model c) Inelastic element

Fig. 2.4 Degrading inelastic element for reinforced concrete [2.22]

2.3 Spread Plasticity Models

2.3.1 General Overview

Distributed nonlinearity models allow a more accurate description of the inelastic behavior of RC members than concentrated plastic hinge models do, because they recognize that nonlinearity can take place anywhere in an element, and the element behavior is derived by weighted integration of individual section responses. Since in practice numerical integration is used, the behavior of only selected sections at the integration points need to be monitored. Either the element deformations or forces are considered the primary unknowns and determined by suitable interpolation functions from the global element displacements or forces, respectively. Discrete cracks are smeared over finite lengths rather than treated explicitly. The constitutive behavior of a section is either formulated in accordance with classical plasticity theory in terms of stress and strain resultants, or derived explicitly by discretizing the section into "fibers" (see Sect. 2.4). The assumption that plane sections remain plane is common to all of these models, so that strains can be considered to vary linearly over a section.

The early beam-column models have neglected the coupling between axial force and bending moment. They typically consist of two cantilever elements connected at the fixed point of contraflexure, Fig. 2.5, as introduced by Otani [2.12]. To derive the cantilever stiffness, independent hysteresis rules are used for the end moment-free end displacement and for the end moment-free end rotation relationships. To overcome some of the numerical difficulties caused by the element formulation, such as the lack of symmetry of the stiffness matrix, Otani lumped the inelastic deformations in two equivalent springs at the member ends, thereby compromising the model's generality. The global model behavior is derived by integrating the curvatures over the two cantilevers. The main limitation of this and similar models is the assumption that the point of contraflexure is fixed.

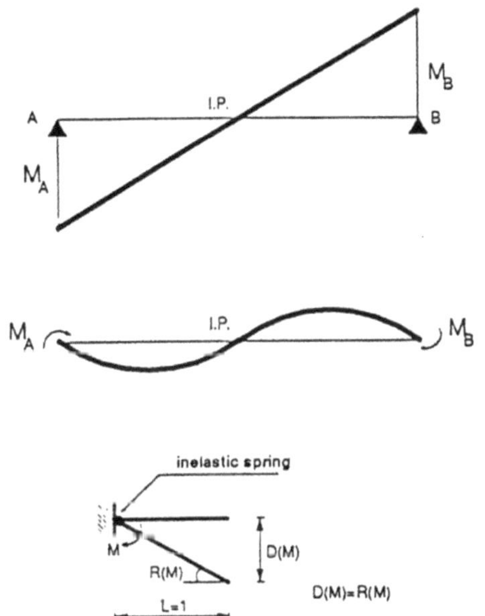

a) Moment distribution

b) Element deformation

c) Equivalent inelastic rotational springs

Fig. 2.5 Otani's model [2.12]

Takayanagi and Schnobrich [2.18] divided the element into a finite number of segments, each represented by a nonlinear rotational spring, Fig. 2.6. The properties of a segment depend on the bending moment at its midpoint and are assumed to be constant over the segment length. Static condensation is used to reduce this multi-spring model to a single beam-column element. Even though the nonlinear behavior is eventually lumped at the end springs, this element belongs to the family of distributed nonlinearity models, because it accounts for inelastic deformations that take place anywhere along the member. This multiple spring model was used to study the seismic response of coupled shear walls, which experience significant variations in axial forces. To account for the interaction between axial force and bending moment, a three-dimensional limit surface was introduced for the rotational springs.

The first elements with distributed nonlinearity were formulated using the classical stiffness method and cubic Hermitian polynomials to approximate the deformations along the element. Torsional response is generally assumed to obey linear elasticity and to be uncoupled from axial and flexural behavior, so that it can be omitted in the following discussion. For the sake of brevity, only the case of uniaxial bending shall be considered. The extension to the biaxial bending case is straightforward.

The general three-dimensional element with axial and flexural degrees of freedom in the local reference system is depicted in Fig. 2.7 and without rigid body modes in Fig. 2.8. Nodal displacements are grouped in vector $\overline{\mathbf{q}}$ for the element with rigid body modes and in vector $\mathbf{q}$ for the element without rigid body modes:

$$\overline{\mathbf{q}} = \{\overline{q}_1 \ \ \overline{q}_2 \ \ \overline{q}_5 \ \ \overline{q}_6 \ \ \overline{q}_7 \ \ \overline{q}_{10}\}^T \tag{2.8}$$

$$\mathbf{q} = \{\overline{q}_1 \ \ \overline{q}_2 \ \ \overline{q}_5\}^T \tag{2.9}$$

Limiting ourselves in the following to the uniaxial bending case, the transverse displacement $v(x)$ and the axial displacement $u(x)$ are approximated by

$$\overline{\mathbf{d}}(x) = \begin{Bmatrix} u(x) \\ v(x) \end{Bmatrix} = \mathbf{a_d}(x)\,\overline{\mathbf{q}} \tag{2.10}$$

where x denotes the member axis, $\mathbf{a_d}(x)$ is a matrix that contains the cubic interpolation functions for the transverse displacements and the linear interpolation functions for the axial displacements,

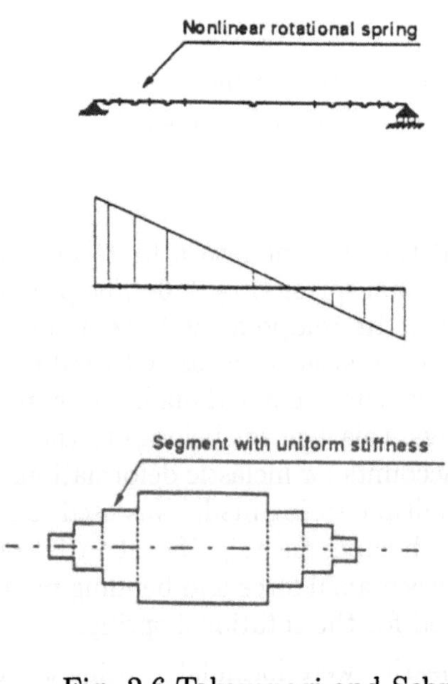

a) Element model

b) Bending moment diagram

c) Section stiffness distribution

Fig. 2.6 Takayanagi and Schnobrich multiple spring model [2.18]

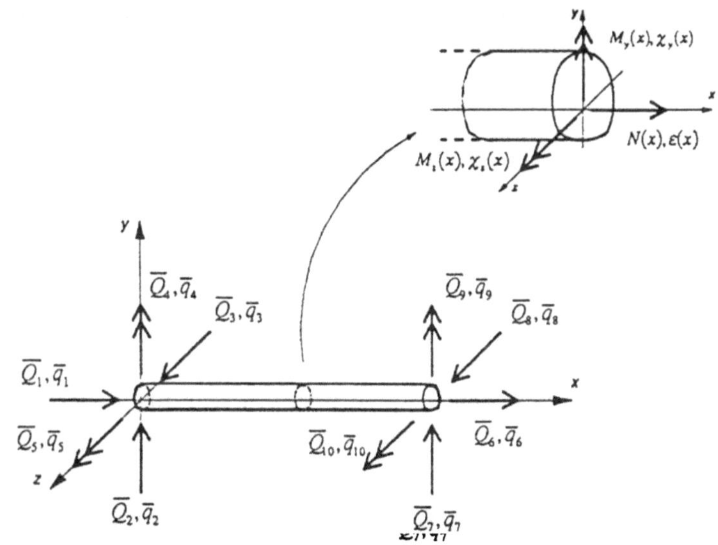

Fig. 2.7 Local reference system of beam element with rigid body modes

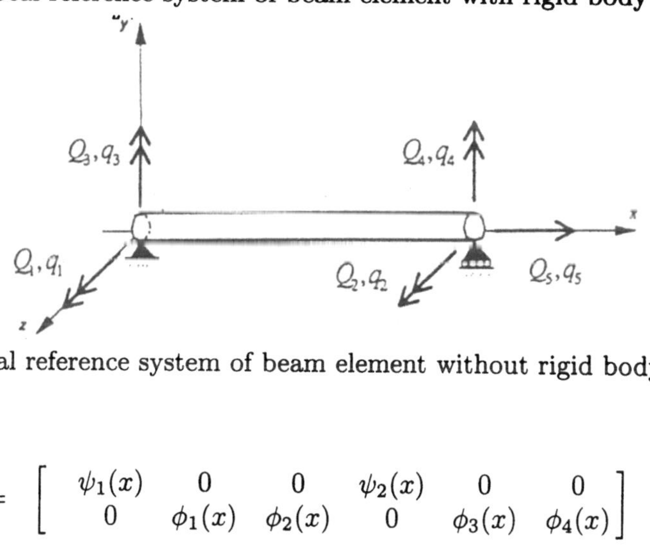

Fig. 2.8 Local reference system of beam element without rigid body modes

$$\mathbf{a}_\mathbf{d}(x) = \begin{bmatrix} \psi_1(x) & 0 & 0 & \psi_2(x) & 0 & 0 \\ 0 & \phi_1(x) & \phi_2(x) & 0 & \phi_3(x) & \phi_4(x) \end{bmatrix} \tag{2.11}$$

with

$$\psi_1(x) = 1 - \frac{x}{L} \qquad \psi_2(x) = \frac{x}{L}$$

$$\phi_1(x) = 2\frac{x^3}{L^3} - 3\frac{x^2}{L^2} + 1 \qquad \phi_2(x) = \frac{x^3}{L^2} - 2\frac{x^2}{L} + x$$

$$\phi_3(x) = -2\frac{x^3}{L^3} + 3\frac{x^2}{L^2} \qquad \phi_4(x) = \frac{x^3}{L^2} - \frac{x^2}{L}$$

In the derivation of the beam and beam-column element stiffness matrices, the axial strain, $\epsilon(x)$, and the curvature about the z-axis, $\chi_z(x)$, are the generalized deformations. Under the assumption that plane sections remain plane the section deformations $\mathbf{d}(x)$ are related to the nodal displacements by

$$\mathbf{d}(x) = \left\{ \begin{array}{c} \epsilon(x) \\ \chi_z(x) \end{array} \right\} = \left\{ \begin{array}{c} u'(x) \\ v''(x) \end{array} \right\} = \overline{\mathbf{a}}(x)\, \overline{\mathbf{q}} \tag{2.12}$$

where $\overline{\mathbf{a}}(x)$ is derived from the displacement interpolation functions,

$$\overline{\mathbf{a}}(x) = \left[\begin{array}{cccccc} \psi_1'(x) & 0 & 0 & \psi_2'(x) & 0 & 0 \\ 0 & \phi_1''(x) & \phi_2''(x) & 0 & \phi_3''(x) & \phi_4''(x) \end{array} \right] \tag{2.13}$$

Using the principle of virtual displacements or the principle of minimum potential energy, the element stiffness matrix $\overline{\mathbf{K}}$ is found as the integral over the section stiffness $\mathbf{k}(x)$,

$$\overline{\mathbf{K}} = \int_0^L \overline{\mathbf{a}}^{\mathbf{T}}(x)\, \mathbf{k}(x)\, \overline{\mathbf{a}}(x)\, dx \tag{2.14}$$

where $\mathbf{k}(x)$ relates the section forces $\mathbf{D}(x)$ to the corresponding deformations $\mathbf{d}(x)$,

$$\mathbf{D}(x) = \mathbf{k}(x)\, \mathbf{d}(x) \tag{2.15}$$

The section forces are the generalized stresses, i.e., the axial force $N(x)$ and the bending moment $M_z(x)$ at section x,

$$\mathbf{D}(x) = \left\{ \begin{array}{c} N(x) \\ M_z(x) \end{array} \right\} \tag{2.16}$$

The application of the virtual displacement principle yields the element resisting forces $\overline{\mathbf{Q}}_{\mathbf{R}}$ as the integral of the section resisting forces $\mathbf{D}_{\mathbf{R}}(x)$,

$$\overline{\mathbf{Q}}_{\mathbf{R}} = \int_0^L \overline{\mathbf{a}}^{\mathbf{T}}(x)\, \mathbf{D}_{\mathbf{R}}(x)\, dx \tag{2.17}$$

Models based on this classical finite element displacement approach have been proposed, among others, by Hellesland and Scordelis [2.23] and Mari and Scordelis [2.24]. The formulation has been extended by Bazant and Bhat [2.25] to include the effect of shear by means of multiaxial constitutive laws based on endochronic theory. In this model the section is subdivided into horizontal layers, such that each layer is allowed to crack at a different angle that is derived from the interaction of normal and shear stress in that layer.

The main shortcoming of stiffness-based models is their inability to describe the behavior of a member near its ultimate resistance and after the onset of strain softening, since they suffer from numerical instability problems.

Since the curvature distribution in a member that has yielded at the ends is not well represented by cubic Hermitian interpolation functions, computational economy with improved representation of internal deformations is achieved by approximating both, the section deformations, which are the basic unknowns of the problem, and the section flexibilities. Menegotto and Pinto [2.26] interpolate both variables based on the values at a few monitored sections and include the axial force-bending moment interaction. The section flexibilities are assumed to vary linearly between monitored sections, which is equivalent to a hyperbolic stiffness variation. This improvement in accuracy makes the approach computationally attractive, since fewer sections need to be monitored and, hence, the number of variables that need to be computed and stored is smaller than in stiffness models of comparable level of discretization.

Further improvement in element accuracy is achieved by introducing variable displacement interpolation functions. A major limitation of the classical displacement approach is the assumption of cubic interpolation functions, which result in linear curvature distributions along the element. These are adequate for linear or nearly linear response. However, when the member undergoes significant yielding at the ends, the curvature distribution becomes highly nonlinear in the inelastic region. This requires a very fine discretization in the inelastic region of stiffness-based elements. Mahasuverachai [2.27] was the first to propose flexibility-dependent shape functions that are continuously updated during the analysis, as inelastic deformations spread into the member, and to approximate deformation increments rather than total deformations. The section deformation increments are written as,

$$\mathbf{\Delta d}(x) \quad = \quad \mathbf{f}(x)\,\mathbf{b}(x)\,\mathbf{F}^{-1}\,\mathbf{\Delta q} \quad = \quad \mathbf{a}(x)\,\mathbf{\Delta q} \tag{2.18}$$

where Δ denotes the increment of the corresponding vector. This new formulation was applied to the development of pipeline elements, where the source of nonlinearity is geometric rather than material.

Recent efforts to develop more robust and reliable reinforced concrete frame elements have shown two parallel trends. First, deviating from the original classical stiffness method, researchers have focussed attention on flexibility-dependent shape functions and, more recently, on flexibility formulations that permit a more accurate description of the force distribution within the element. Secondly, the elements are subdivided into longitudinal fibers, which has two inherent advantages: a) the section behavior can be derived from the uniaxial stress-strain behavior of the fibers; three-dimensional effects such as concrete confinement by transverse reinforcement can be incorporated into the uniaxial stress-strain relation; and b) the interaction between bending moment and axial force can be described in a rational way.

The flexibility approach is based on force interpolation functions within an element. Typically, the element is analyzed without including the rigid body modes. In this case the end rotations relative to the chord and the axial differential displacement are the element generalized deformations, or simply, element deformations. The element forces and deformations without the rigid body modes are shown in Fig. 2.8. Under the assumption of small deformations and small displacements the element deformations $\mathbf{q}$ are related to the element displacements $\overline{\mathbf{q}}$ of Fig. 2.7 by a simple compatibility matrix. In the uniaxial bending case the vector of element forces without rigid body modes is

$$\mathbf{Q} = \{Q_1 \; Q_2 \; Q_5\}^T \tag{2.19}$$

It is common to assume the bending moment distribution within the element to be linear and the axial force to be constant. In vector notation,

$$\mathbf{D}(x) = \mathbf{b}(x)\, \mathbf{Q} \tag{2.20}$$

where $\mathbf{b}(x)$ is a matrix containing the force interpolation functions,

$$\mathbf{b}(x) = \begin{bmatrix} 0 & 0 & 1 \\ (\frac{x}{L} - 1) & (\frac{x}{L}) & 0 \end{bmatrix} \tag{2.21}$$

The application of the virtual force principle yields the element flexibility,

$$\mathbf{F} = \int_0^L \mathbf{b^T}(x)\, \mathbf{f}(x)\, \mathbf{b}(x)\, dx \tag{2.22}$$

where $\mathbf{f}(x)$ is the section flexibility matrix such that,

$$\mathbf{d}(x) = \mathbf{f}(x)\, \mathbf{D}(x) \tag{2.23}$$

The advantage of this formulation derives from the realization that, irrespective of the state of the element, the force interpolation functions of Eq. 2.21 satisfy element equilibrium in a strict sense, as long as no element loads are applied. In other words, whatever material nonlinearities take place at the section level and even as the element starts softening when deformed beyond its ultimate resistance, the assumed internal force distributions are exact.

A critical issue for flexibility-based elements is the implementation in an existing finite element program. Such programs are typically based on the direct stiffness method of analysis, where the global system of equilibrium equations is solved for the unknown structural displacements. After the element displacements are extracted from the structural displacements the phase of element state determination starts. During this phase the resisting forces and the stiffness matrix need to be determined for the given element displacements. If the element formulation is flexibility-based, the element state determination requires a special procedure, since the element resisting forces cannot be derived by integrating the section resisting forces according to Eq. 2.17. An interesting state determination procedure for a flexibility-based finite element is proposed by Ciampi and Carlesimo [2.28] and is discussed at length in Spacone [2.29]. The section moment-curvature relation of this model is based on the endochronic theory presented by Brancaleoni et al [2.30]. The element state is determined from the section deformation residuals that result from the numerical integration of the section constitutive relation.

2.3.2 Columbia Frame Element Model

As an example of an element with spread plasticity, the frame element developed at Columbia University [2.1,2.31,2.32] shall be mentioned. Starting with the stress-strain relationships for concrete and steel, Fig. 2.9, we must realize that confinement reinforcement enhances both the strength and particularly the ductility of concrete. This effect is expressed in the empirical relationships,

$$f_{cu} = (1 + 10\rho") \, f'_c, \qquad \epsilon_{cm} = (2 + 600\rho") \, \epsilon_{cu} \qquad (2.24)$$

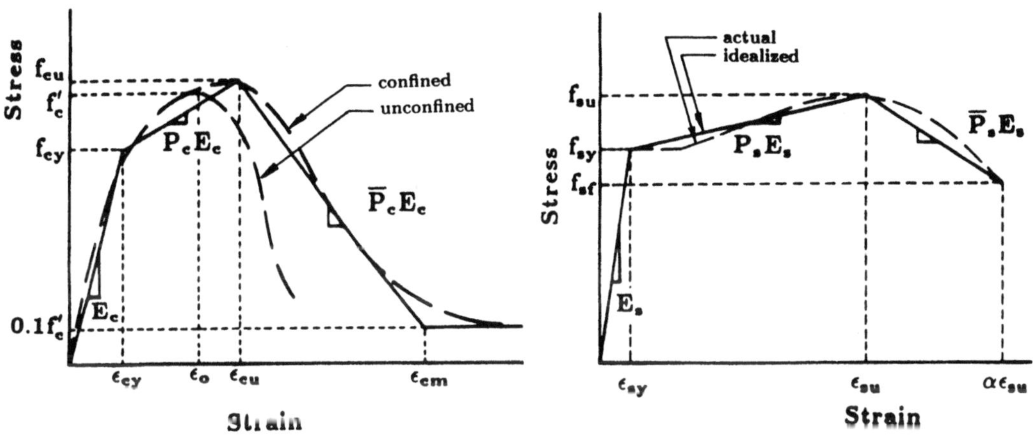

Fig. 2.9 Stress-strain curves for concrete and steel

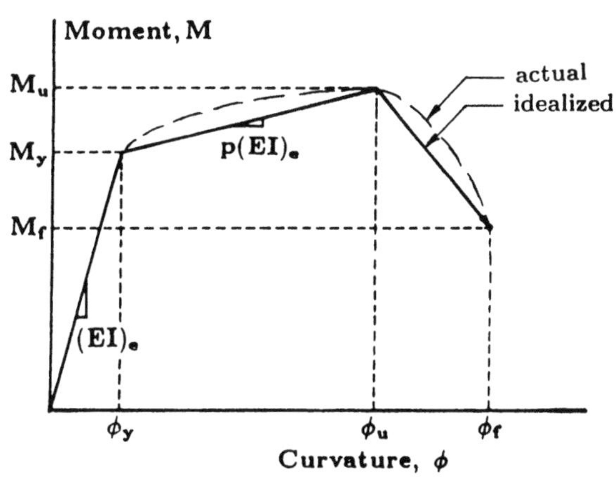

Fig. 2.10 Primary moment-curvature curve

where ρ" is the volumetric confinement steel ratio. Figure 2.9 also indicates a small residual compressive strength for concrete, regardless how large the strain.

With the stress-strain laws for steel and concrete specified and the cross-sectional dimensions of the member known, it is relatively straightforward to compute the bending moment associated with any specific curvature [2.1]. Repeating this calculation for different curvature levels results in the so-called primary or skeleton moment-curvature curve for monotonically increasing load, Fig. 2.10. Highly stressed member sections are assumed to be precracked by service-level loads. Therefore, no slope discontinuity at the cracking moment is expected, and with the yield moment readily determined, it is appropriate to approximate the $M - \phi$ curve with the three linear branches shown.

Once the moment-curvature relationship for a section is known and the bending moment along a member given, Fig. 2.11, the flexibility coefficients are readily obtained through integration. The assumption of spread plasticity permits this explicit integration in closed form, since the finite lengths of the plastic regions are known. The element stiffness matrix is then found by inverting the flexibility matrix.

The presence ot an axial force can be included during each of the preceding steps. It affects the moment-curvature relationship and in particular the member's ductility. This effect is important in highly stressed columns, where the axial force can exhaust a large portion of the concrete's strain capacity even within the elastic range.

Modeling the inelastic response of frame elements to load reversals is a rather difficult task. Figure 2.12 shows a typical load-deformation curve that has been obtained experimentally [2.33]. As can be seen, the member stiffness decreases gradually, once the yield capacity of the member has been exceeded. And if the member is deformed beyond a certain critical level, its strength also decreases. Below that threshold value, the hysteresis loops appear to be stable, i.e., the same load is required to reach a given deformation level in several successive load cycles.

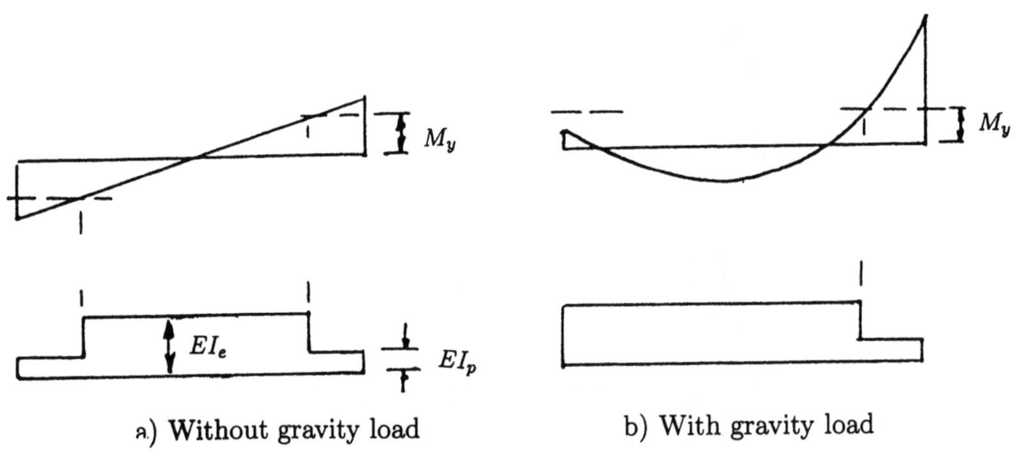

a) Without gravity load b) With gravity load

Fig. 2.11 Frame member with variable stiffness

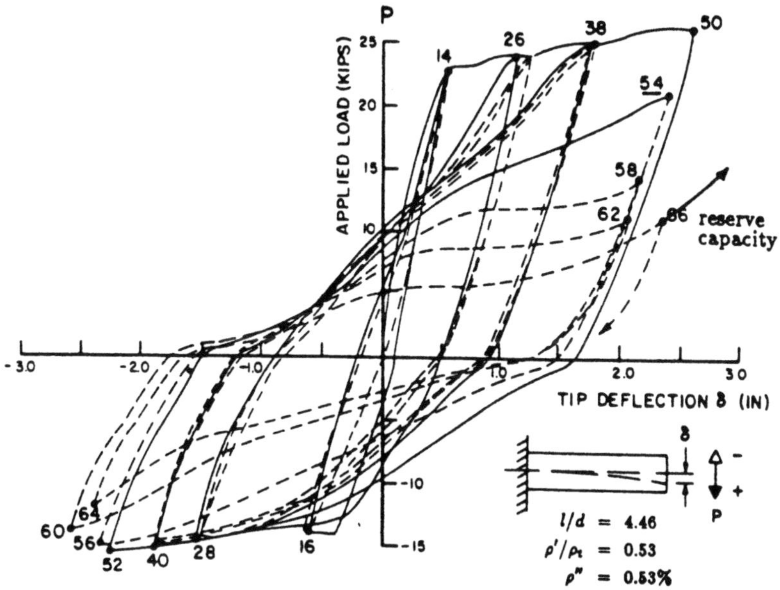

Fig. 2.12 Stable hysteresis loops [2.33]

This is deceptive, however. The number of load cycles for each load (or deformation) level in this particular experiment was too small to exhibit any low-cycle fatigue related strength degradation. The experiment illustrated in Fig. 2.13 shows that such strength degradation can be significant [2.34]. And it is clear from damage mechanics considerations that strains can cause significant damage and softening at levels well below those at which the steel starts to yield.

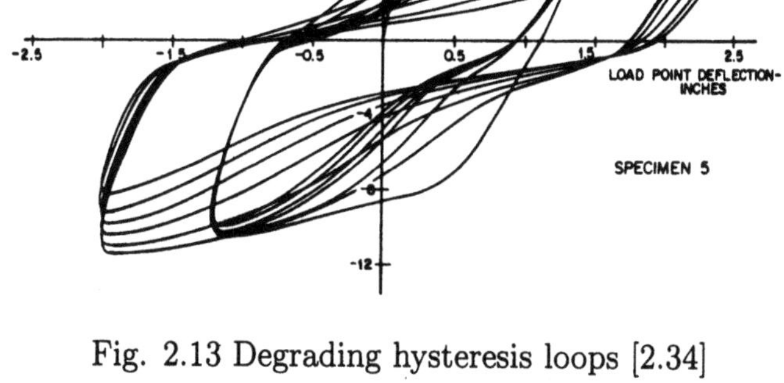

Fig. 2.13 Degrading hysteresis loops [2.34]

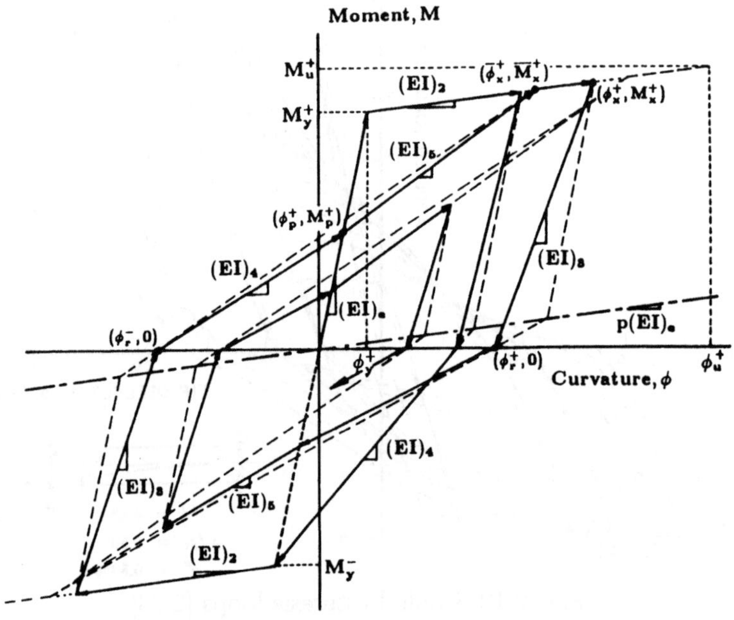

Fig. 2.14 Idealized hysteresis behavior

Almost all frame element models that set out to reproduce load-deformation curves such as those shown in Figs. 2.12 and 2.13 use empirical rules. The first comprehensive set of such rules was listed by Takeda et al [2.10]. Therefore, these empirical models are often referred to as Takeda models. Individual researchers typically add their own modifications to the original set. Our own rules [2.1] are illustrated in Fig. 2.14. There are basically five different branch types in the moment-curvature relationship:

1. *Elastic loading and unloading* - applicable if the moment has never exceeded the yield moment M_y in either direction; moments and curvatures are related through the initial stiffness EI.
2. *Inelastic loading* - applicable if the yield moment is exceeded and curvatures (or moments) still increase; here the initial stiffness is reduced by the strain-hardening factor p.
3. *Inelastic unloading* - after the yield moment has been exceeded; experimental data indicate that the slope of this branch is less than the initial elastic slope; the decrease in stiffness is proportional to the extent of the maximum prior inelastic excursion.
4. *Inelastic reloading during closing of cracks* - in reversed load cycles, previously open cracks tend to close, with a subsequent increase in stiffness and a characteristic pinched shape of the $M-\phi$ curve. The extent of pinching depends on the amount of shear cracking, i.e. it is directly related to the member's shear

span.

5. *Inelastic reloading after closing of cracks* - once the moment exceeds an empirically determined crack-closing moment, the $M - \phi$ curve experiences a sharp increase in stiffness.

In addition to stiffness degradation, RC members experience strength deterioration, especially if loaded beyond the yield moment. A correlation between the onset of such strength reduction and spalling of the concrete cover has been observed, but tests have shown that the damaging process that leads to loss in strength, starts much sooner and depends very much on the number of load cycles. Chung [2.32] therefore introduced a strength drop index, Fig. 2.15,

$$S_d = \frac{\Delta M}{\Delta M_f} = \left(\frac{\phi - \phi_y}{\phi_f - \phi_y}\right)^{\omega} \tag{2.25}$$

where ϕ_y and ϕ_f are yield and failure curvatures, respectively, and ω is a constant that has been determined empirically to be 1.5. ΔM_f is the moment capacity reduction of the member if it is subjected to a single load cycle up to failure curvature ϕ_f, and ΔM is the strength drop due to a single cycle up to curvature ϕ. A failure moment curve can be derived, again empirically, which together with the strength drop of Eq. 2.25 defines the number of load cycles to failure.

Numerical results achieved with this model will be presented in Chapter 6.

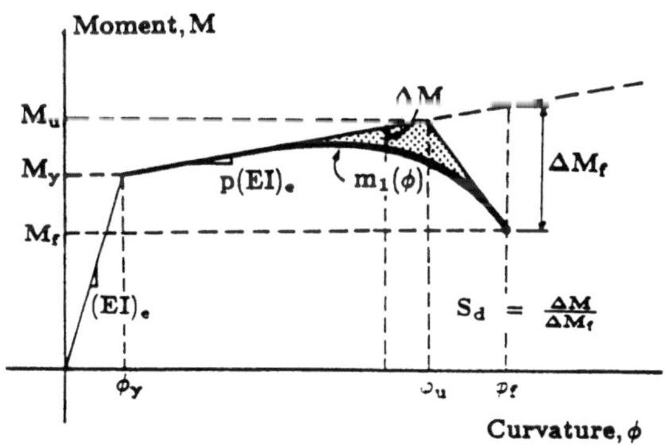

Fig. 2.15 Strength degradation curve

2.4 Fiber Models

The most promising models for nonlinear analysis of RC members appear to be flexibility-based fiber elements. In these models an element is subdivided into longitudinal fibers, as shown in Fig. 2.16. Each fiber is geometrically defined by its location in the local yz reference system and area A. The constitutive relation of the section is not specified explicitly, but found by integrating the fiber responses with their own uniaxial stress-strain relations, as shown in Fig. 2.16. The models proposed to date are limited to small displacements and deformations and assume that plane sections remain plane. Two new tasks appear in the formulation of such elements: a) the element state determination, which involves the determination of the resisting forces for given element displacements, and, b) the determination of the section flexibility $\mathbf{f}(x)$, needed to calculate the element flexibility $\mathbf{F}$ according to Eq. 2.22.

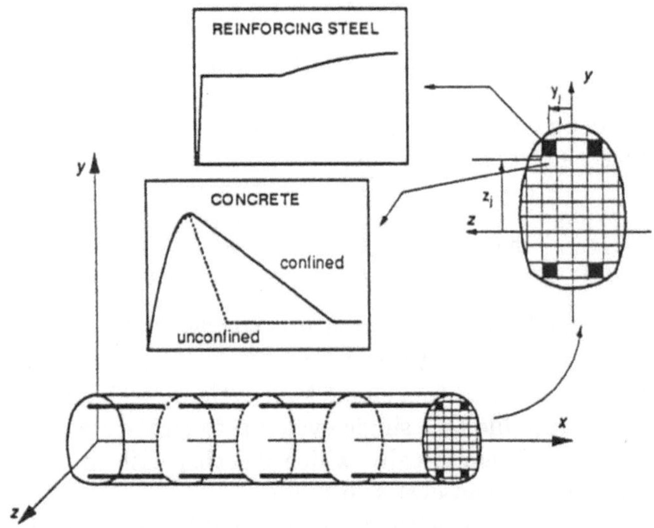

Fig. 2.16 Fiber element

The basic assumption in a flexibility-based model is the selection of the internal force distribution in the element, which is expressed in Eq. 2.20 by the force interpolation functions $\mathbf{b}(x)$. In a consistent state determination process the section forces are determined from the element forces according to Eq. 2.20, followed by the computation of the corresponding fiber stresses based on equilibrium. The fiber strains and flexibilities are determined from the fiber stress-strain relations, and the section deformations as well as the section flexibility are computed by applying the virtual force principle. The determination of fiber stresses from the section forces is, however, a statically indeterminate problem for a section with more than two fibers. The fiber stresses cannot be determined from the axial force and bending moment at the section, since there are only two equilibrium equations in the uniaxial bending case for three or more unknown stresses. One possible solution is to assume a stress distribution within the section, but the problem is then only postponed to the fiber state determination phase of the algorithm, since fiber stress-strain relations are typically expressed as explicit functions of strain. The typical solution is to linearize the section constitutive relation and compute the section deformations from the new section forces and the section flexibility from the previous step. Fiber stresses σ_{if} and stiffnesses $\mathbf{E}_{\text{if}}$ are then determined from the fiber stress-strain relations. The section resisting forces are computed from the fiber stress distribution applying the virtual displacement principle. If ℓ_{if} is the compatibility matrix relating fiber strain ϵ_{if} to section deformations $\mathbf{d}(x)$ according to

$$\epsilon_{\text{if}} = \mathbf{l}_{\text{if}} \, \mathbf{d}(x) \tag{2.26}$$

where, following the hypothesis that plane sections remain plane, $\mathbf{l}_{\text{if}}$ is

$$\ell_{\text{if}} = [-y_{if} \;\; 1] \tag{2.27}$$

then the section resisting forces are

$$\mathbf{D_R}(x) = \sum_{if=1}^{nf(x)} \ell_{\text{if}}^{\mathbf{T}} \, A_{if} \, \sigma_{\text{if}} \tag{2.28}$$

The section stiffness matrix $\mathbf{k}(x)$ is then assembled from the fiber stiffnesses following the virtual displacement principle,

$$\mathbf{k}(x) = \sum_{if=1}^{nf(x)} \ell_{\text{if}}^{\mathbf{T}} \, (EA)_{if} \, \ell_{\text{if}} \tag{2.29}$$

This section stiffness matrix is then inverted to yield the section flexibility $\mathbf{f}(x) = \mathbf{k}^{-1}(x)$. The new element flexibility $\mathbf{F}$ is computed from Eq. 2.22 and is then inverted to obtain the element stiffness. The remaining problem is the determination of the element resisting forces from the section resisting forces along the element. At present, this is the main challenge in the development of flexibility-based fiber elements.

The first flexibility-based fiber element was proposed by Kaba and Mahin [2.35]. It follows the outline of the flexibility approach presented above, using the force interpolation functions $\mathbf{b}(x)$ of Eq. 2.21 to determine the element flexibility matrix. Only uniaxial bending is taken into account in this model. In the state determination phase of the nonlinear analysis the section deformations are computed from the element deformations with the flexibility-dependent deformation shape functions of Eq. 2.18. Due to the nonlinear behavior of the section, $\mathbf{f}(x)$, $\mathbf{F}$ and consequently $\mathbf{a}(x)$ change during the element deformation history. The section deformations are then used to determine the fiber strains based on the plane section assumption, and the corresponding fiber stresses and stiffnesses follow from the fiber stress-strain relation. Subsequently, the section stiffness $\mathbf{k}(x)$ and the corresponding resisting forces $\mathbf{D_R}(x)$ are determined by applying the principle of virtual work at the section. The section stiffness is inverted to yield the section flexibility $\mathbf{f}(x)$. Finally, the element flexibility matrix $\mathbf{F}$ is determined with Eq. 2.22, and the element resisting force increments $\mathbf{\Delta Q_R}$ are derived, using the virtual displacement principle,

$$\mathbf{\Delta Q_R} = \int_0^L \mathbf{a}^{\mathbf{T}}(x) \, \mathbf{\Delta D_R}(x) \, dx = \mathbf{F}^{-1} \int_0^L \mathbf{b}^{\mathbf{T}}(x) \, \mathbf{f}(x) \, \mathbf{\Delta D_R}(x) \, dx \tag{2.30}$$

The integrals over the element length are evaluated by subdividing the element into equally spaced slices and assuming linear flexibility distribution between the slices. This model yields promising results but exhibits convergence problems and is unable to describe element softening. The element formulation is actually based on a mixed approach, since it uses both deformation and force interpolation functions. Unfortunately, the element lacks theoretical clarity and has several theoretical inconsistencies that cause the numerical problems. The first inconsistency is the result of determining the element flexibility matrix from compatibility considerations and the application of the virtual force principle, while determining the element resisting forces from equilibrium considerations and the application of the virtual displacement principle. The second inconsistency appears in the state determination process which violates equilibrium within the element, since the distribution of the section resisting forces $\mathbf{D_R}(x)$ does not satisfy the equilibrium condition of Eq. 2.20 and 2.21. Consequently, the resulting bending moment distribution is not linear and the axial force distribution not uniform, as required by the force interpolation functions $\mathbf{b}(x)$.

Zeris and Mahin [2.36,2.37] improved the original Kaba-Mahin model and extended the formulation to the biaxial case, mainly by improving the element state determination. Once the main program determines the nodal displacement increments $\mathbf{\Delta\overline{q}}$, the element updating sequence consists of the following steps: a) Eq. 2.18 is applied at the end sections of the element to determine the section deformation increments $\mathbf{\Delta d}(0)$ and $\mathbf{\Delta d}(L)$; b) the corresponding bending moments and axial forces at the end sections are established by means of a modified event-to-event advancement method; and c) the deformations at interior sections of the element are updated with an iterative procedure so as to produce section resisting forces that conform with the assumed force distribution in the element.

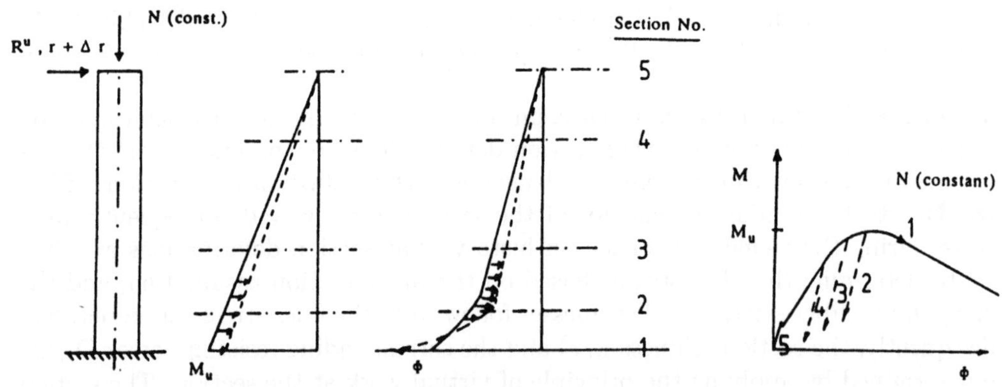

a) Member b) Moments c) Curvatures d) Moment-curvature relation

Fig. 2.17 Behavior of softening cantilever model

An interesting analysis of the softening behavior of a cantilever beam is discussed by Zeris and Mahin [2.36], Fig. 2.17. When a cantilever is displaced beyond the point of ultimate resistance, section 1 at the fixed base loses load carrying capacity and starts softening. Sections 2 through 5 along the cantilever unload elastically in order to satisfy internal equilibrium. Stiffness-based elements cannot trace such behavior. They assume a linear curvature distribution, which deviates significantly from the actual distribution during element softening, as the sharp jump in the curvature value near the fixed end attests, Fig. 2.17. In this case the column has to be subdivided into several elements, but convergence problems are still encountered. Early flexibility elements, such as that described by Kaba and Mahin [2.35], are also unable to correctly trace the softening behavior of the member, because equilibrium is not enforced along the element.

The element proposed by Zeris and Mahin has been used to analyze a two-story space frame, Fig. 2.18, that had been tested by Clough and Gidwani [2.38] on a shaking table. A base excitation based on the Taft record, scaled to 0.57g, was applied simultaneously in both principal directions of the structure. Because of the limited amount of beam yielding expected, the beams were modeled as simple two-component elements. Columns were discretized by five unequally spaced sections. The motions computed for the center of the roof mass are shown in Fig. 2.18c, and Fig. 2.18d shows the base shear histories computed for the two principal directions of the frame. Response maxima in the transverse y direction initially lag behind those in the longitudinal x axis. Subsequently, torsional response is induced as a result of non-symmetric inelastic distributions of the lateral stiffness and strength among the four columns due to axial load fluctuations and biaxial bending.

Even though the element proposed by Zeris and Mahin shows satisfactory performance, the element state determination procedure is not very clear and is derived from ad hoc corrections of the Kaba Mahin model rather than from a general theory. The most recent fiber beam-column element, proposed by Taucer et al [2.39] fills this theoretical gap in two ways: the element formulation is cast in a general mixed method framework, and a consistent element state determination process is proposed that satisfies both internal equilibrium and compatibility.

The derivation follows the two-field mixed method, which uses the integral form of equilibrium and force-deformation relations to derive the matrix relationship between element forces and element deformations. In order to make that relationship linear, the section force-deformation relation is linearized about the present state. An iterative algorithm is then used to satisfy the nonlinear section force-deformation relation within the required tolerance.

In the two-field mixed method [2.40] independent shape functions are used to approximate the force and deformation fields along the element. Denoting with Δ increments of the corresponding quantities, the incremental force field is written as,

$$\mathbf{\Delta D}(x) \quad = \quad \mathbf{b}(x)\, \mathbf{\Delta Q} \tag{2.31}$$

where $\mathbf{b}(x)$ is written as in Eq. 2.21 for the uniaxial case. In general, the deformation field is selected independently of the force field, but in this application the flexibility-dependent interpolation functions $\mathbf{a}(x)$ of Eq. 2.18 are used.

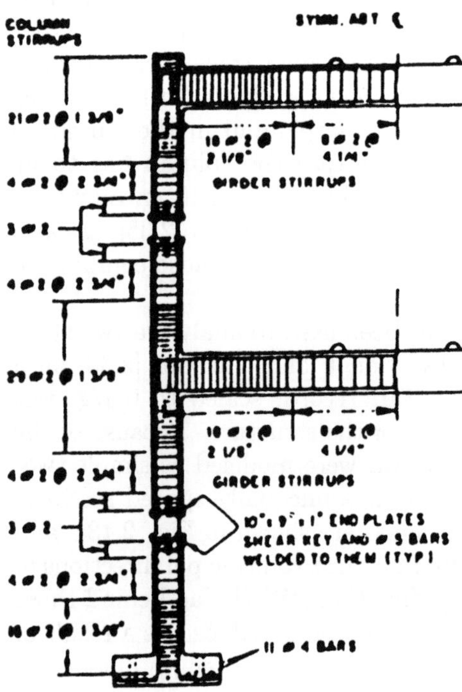

a) Front elevation

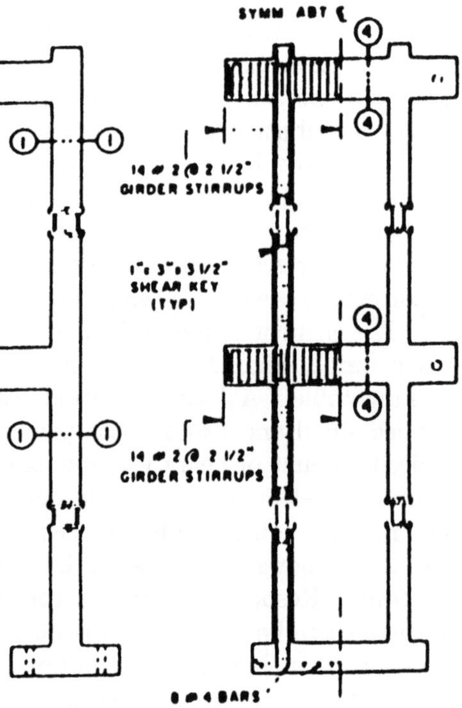

b) Side elevation

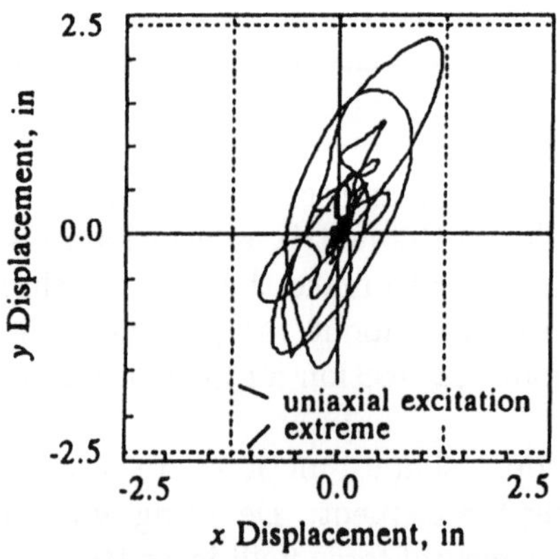

c) Predicted motion of roof centroid

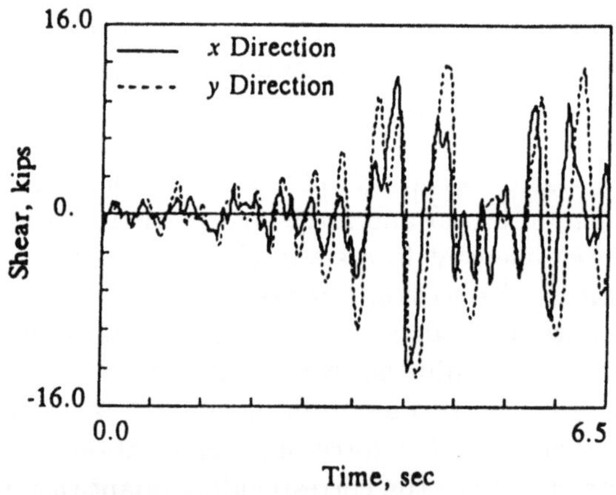

d) Predicted base shear histories

Fig. 2.18 Two-story RC frame

In the mixed method formulation the integral forms of equilibrium and force-deformation relations are expressed first. These are then combined to obtain the relation between element force and deformation increments. The weighted integral form of the linearized section force-deformation relation is,

$$\int_0^L \delta\mathbf{D}^{\mathbf{T}}(x)\ [\mathbf{\Delta d}(x) - \mathbf{f}(x)\ \mathbf{\Delta D}(x)]\ dx \ = \ 0 \qquad (2.32)$$

The section force-deformation relation appears in the flexibility form $\mathbf{\Delta d}(x) = \mathbf{f}(x)\mathbf{\Delta D}(x)$ so that the resulting element flexibility matrix is symmetric [2.40]. Using Eqs. 2.18 and 2.20 and realizing that it must hold for an arbitrary virtual element force increment $\mathbf{\Delta Q}$, Eq. 2.32 is written as,

$$\mathbf{T}\ \mathbf{\Delta q} - \mathbf{F}\mathbf{\Delta Q} \ = \ 0 \qquad (2.33)$$

where $\mathbf{F}$ is the element flexibility matrix defined in Eq. 2.22, and $\mathbf{T}$ is a matrix that only depends on the force and displacement interpolation functions,

$$\mathbf{T} \ = \ \int_0^L \mathbf{b}^{\mathbf{T}}(x)\ \mathbf{a}(x)\ dx \qquad (2.34)$$

Eq. 2.33 is the matrix expression of the integral form of the element force-deformation relation. Next, the equilibrium equation of the beam element is imposed. In the classical two-field mixed method, the integral form of the equilibrium equation is derived from the virtual displacement principle,

$$\int_0^L \delta\mathbf{d}^{\mathbf{T}}(x)\ [\mathbf{D}(x) + \mathbf{\Delta D}(x)]\ dx \ = \ \delta\mathbf{q}^{\mathbf{T}}\ \mathbf{P} \qquad (2.35)$$

where $\mathbf{P}$ is the vector of applied loads. First, Eqs. 2.18 and 2.20 are substituted into Eq. 2.35. Then, observing that the resulting expression must hold for arbitrary $\delta\mathbf{q}$, the following matrix equation is obtained,

$$\mathbf{T}^{\mathbf{T}}\ \mathbf{Q} \ + \ \mathbf{T}^{\mathbf{T}}\ \mathbf{\Delta Q} \ = \ \mathbf{P} \qquad (2.36)$$

This is the matrix expression of the integral form of the element equilibrium. Rearranging and combining Eqs. 2.33 and 2.36 yield

$$\begin{bmatrix} -\mathbf{F} & \mathbf{T} \\ \mathbf{T}^{\mathbf{T}} & \mathbf{0} \end{bmatrix} \begin{Bmatrix} \mathbf{\Delta Q} \\ \mathbf{\Delta q} \end{Bmatrix} \ = \ \begin{Bmatrix} \mathbf{0} \\ \mathbf{P} - \mathbf{T}^{\mathbf{T}}\mathbf{Q} \end{Bmatrix} \qquad (2.37)$$

Solving the first equation of Eq. 2.37 for $\mathbf{\Delta Q}$ and substituting the result in the second equation leads to,

$$\mathbf{T}^{\mathbf{T}}\ \mathbf{F}^{-1}\ \mathbf{T}\ \mathbf{\Delta q} \ = \ \mathbf{P} - \mathbf{T}^{\mathbf{T}}\ \mathbf{Q} \qquad (2.38)$$

If the flexibility-dependent deformation interpolation functions $\mathbf{a}(x)$ of Eq. 2.18 are used, it can be shown that $\mathbf{T}$ reduces to the identity matrix $\mathbf{I}$. Eq. 2.38 simplifies in this case to

$$\mathbf{F}^{-1}\,\Delta\mathbf{q} \;=\; \mathbf{P}-\mathbf{Q} \tag{2.39}$$

which is the linear relation between element stiffness $\mathbf{K} = \mathbf{F}^{-1}$ and unbalanced element forces $\mathbf{P}-\mathbf{Q}$. The element stiffness $\mathbf{K}$ is written in the form $\mathbf{F}^{-1}$ to indicate that it is obtained by inverting the element flexibility. Even though the classical flexibility method results in the same system of linearized equations as those in Eq. 2.39, the derivation in [2.39] was based on the two-field mixed method for the following reasons: a) the mixed method formulation yields directly the expression for the flexibility dependent deformation shape functions $\mathbf{a}(x)$ of Eq. 2.18; b) it reveals the consistent implementation of the state determination process; and c) it is more general in scope, allowing alternative deformation interpolation functions to be explored in future studies.

The proposed element state determination algorithm is based on the computation of section and element residual deformations. Stiffness-based nonlinear algorithms are based on the repeated application of unbalanced forces. It thus appears natural to use "unbalanced", or, preferably, residual deformations for flexibility-based algorithms. The step-by-step procedure can be summarized as follows:

1. from the element nodal deformation increments $\Delta\mathbf{q}$, compute the element force increments $\Delta\mathbf{Q}$, using the last stiffness matrix $\mathbf{K}$;
2. use Eq. 2.31 to compute section force increments $\Delta\mathbf{D}(x)$ and update the section force vector $\mathbf{D}(x)$;
3. compute section deformation increments $\Delta\mathbf{d}(x) = \mathbf{f}(x)\Delta\mathbf{D}(x)$, where $\mathbf{f}(x)$ is the last section flexibility matrix;
4. assuming that plane sections remain plane, evaluate the strain distribution in each section;
5. compute stresses and tangent moduli for all section fibers;
6. based on section equilibrium compute the section stiffness matrix $\mathbf{k}(x)$ and resisting force vector $\mathbf{D_R}(x)$; invert $\mathbf{k}(x)$ to obtain the section flexibility $\mathbf{f}(x)$;
7. determine the unbalanced section forces $\mathbf{D_U}(x) = \mathbf{D}(x) - \mathbf{D_R}(x)$;
8. compute the linearized section deformation vector $\mathbf{r}(x) = \mathbf{f}(x)\mathbf{D_U}(x)$;
9. apply the principle of virtual forces to find the element residual deformations $\mathbf{s} = \int_0^L \mathbf{b}^{\mathbf{T}}(x)\mathbf{r}(x)dx$ and the element flexibility matrix $\mathbf{F}$ according to Eq. 2.22;
10. invert $\mathbf{F}$ to obtain the element stiffness matrix $\mathbf{K}$;
11. since the application of the element residual deformations to the element end deformations would violate node compatibility, a force increment $\Delta\mathbf{Q} = \mathbf{K}\cdot(-\mathbf{s})$ is applied to the element. This is the new element force increment. A new element iteration loop starts.

Steps 2) through 11) are repeated until a prescribed convergence criterion is satisfied. Typically, unbalanced section forces $\mathbf{D_U}(x)$ or, better, the energy measure defined by the product $\mathbf{s}^{\mathbf{T}}\mathbf{K}\mathbf{s}$, are monitored until they become sufficiently small.

It can be verified that during the element iteration, internal equilibrium, expressed by Eq. 2.20, and internal compatibility, expressed in Eq. 2.18, are always satisfied. Element and section constitutive relations are satisfied, within a given tolerance, when the element algorithm converges. Taucer et al [2.39] also propose an exact procedure for the application of distributed loads inside the element, which can be regarded as the generalization of the procedure for applying element loads in the context of the flexibility method.

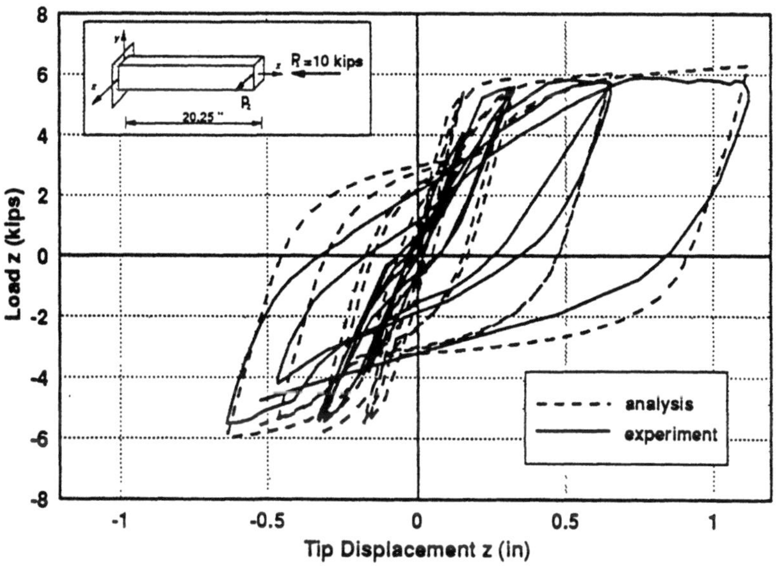

Fig. 2.19 Load-displacement response of cantilever tip for constant axial load and cyclic uniaxial bending

A number of comparisons between analytical and experimental results were made to study the element performance. Results of three analyses of a reinforced concrete cantilever beam, discretized with a single beam-column element, shall be presented here. The cantilever beam has been tested by Low and Moehle [2.41] under a compressive axial load and biaxial or uniaxial flexure. Fig. 2.19 illustrates the uniaxial bending case, in which a constant axial force and a cyclic transverse force along the weak axis z are applied at the free end. Numerical and experimental results agree well. But the fiber model behavior is stiffer, especially for low levels of lateral force, at which the experimental results exhibit "pinching".

The same cantilever was studied under biaxial bending conditions, with two cyclic transverse loads applied at the free end. Fig. 2.20 shows the tip response in the strong direction y. Analytical and experimental results correlate very well both for small and large displacements. When the concrete is fully cracked at the built-in section, bar pullout effects appear in the experimental data, but their contribution to the tip displacement is small.

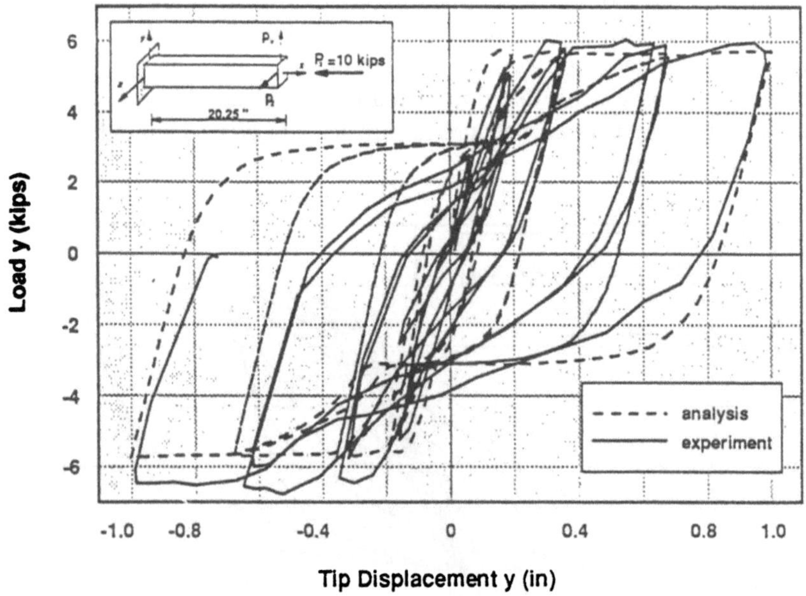

Fig. 2.20 Load-displacement response of cantilever tip for constant axial load and cyclic biaxial bending

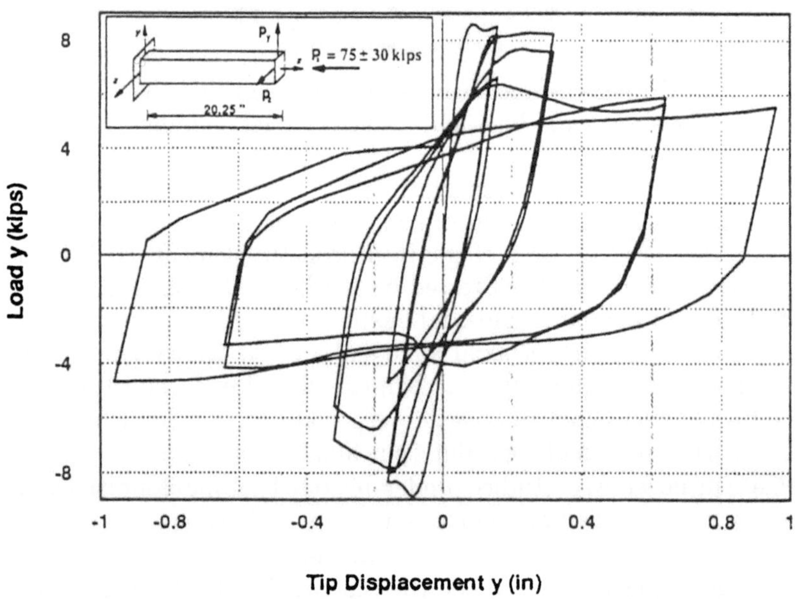

Fig. 2.21 Load-displacement response of cantilever tip for cyclic axial load and cyclic biaxial bending

Finally, the cantilever beam was studied under both cyclic bending and cyclic axial force, Fig. 2.21, with the following maximum load and displacement excursions,

$$P_x = -75 \pm 30 \; kips, \quad p_y = \pm 0.96 \; in, \quad p_z = \pm 0.96 \; in$$

Load and deformation increments were applied such that all three quantities reached their maximum and minimum values at the same time. This example is particularly important in showing that the proposed element is capable of representing softening and stiffness degradation without numerical convergence problems.

2.5 Analysis of Buildings with Poor Detailing

2.5.1 Introduction

As already mentioned earlier, the most difficult step in the dynamic analysis of concrete structures is the modeling of the deterioration of the structural elements. In well-designed buildings we expect little distress caused by bond and shear and, therefore, the hysteresis loops are expected to remain stable during the design-level earthquake. However, in the evaluation of existing buildings, which is an important problem in earthquake engineering today, we often find that the details are not adequate, and deterioration occurs after only a few load cycles. Such buildings are rather flexible to begin with, and the loss of stiffness causes additional displacements, though perhaps accompanied by lower forces. Most of these buildings have strong beams and weak columns. Thus, the *evaluation* of existing buildings is usually more difficult and less reliable than the analysis of buildings in conjunction with their *design*.

2.5.2 Experimental Program

A seven-year coordinated project on the evaluation of buildings, that were designed only for gravity loads, has been completed. It was sponsored by the National Center for Earthquake Engineering Research and has involved four institutions, with most of the work performed at SUNY/Buffalo and Cornell University [2.42-2.44]. The project started with a survey of building design practices in the 1940-1970 time period and with the identification of non-seismic detailing practices. Many full-size beam-column joint regions have been tested to determine the force-displacement characteristics of these elements. Analytical models have been developed and incorporated into program IDARC, a nonlinear dynamic analysis program for concrete structures. Shake-table tests were performed to test the analytical tools. Since the performance of buildings depends greatly on their initial stiffness, geometry, and the ground motion, it is essential to have an acceptable computer program for the prediction of dynamic responses.

Typical details in gravity-load designed buildings, Fig. 2.22, included little or no column ties in the joints; large tie spacings in the columns and beams; bottom beam reinforcement that is not continuous through the joint but with an embedment length of about 150 mm; construction joints above and below the beams; column splices right above the floor level (where the maximum seismic moment occurs); and no extra column ties.

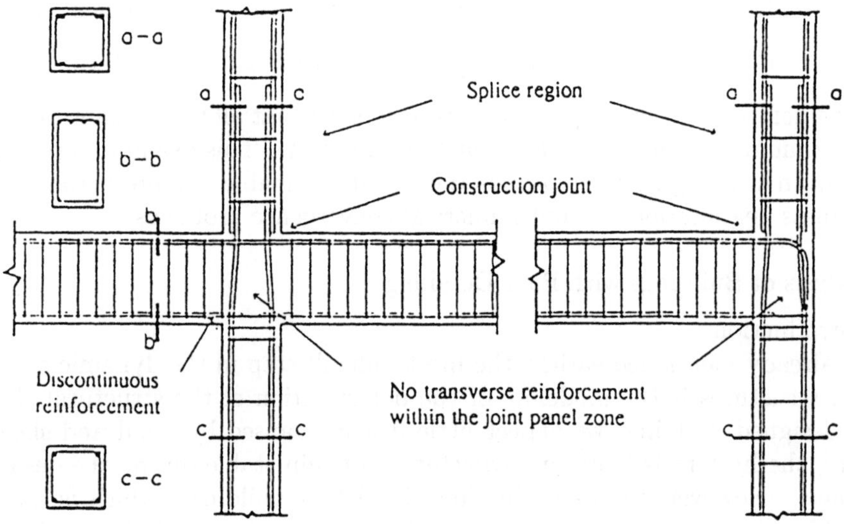

Fig. 2.22 Frame with typical non-seismic details

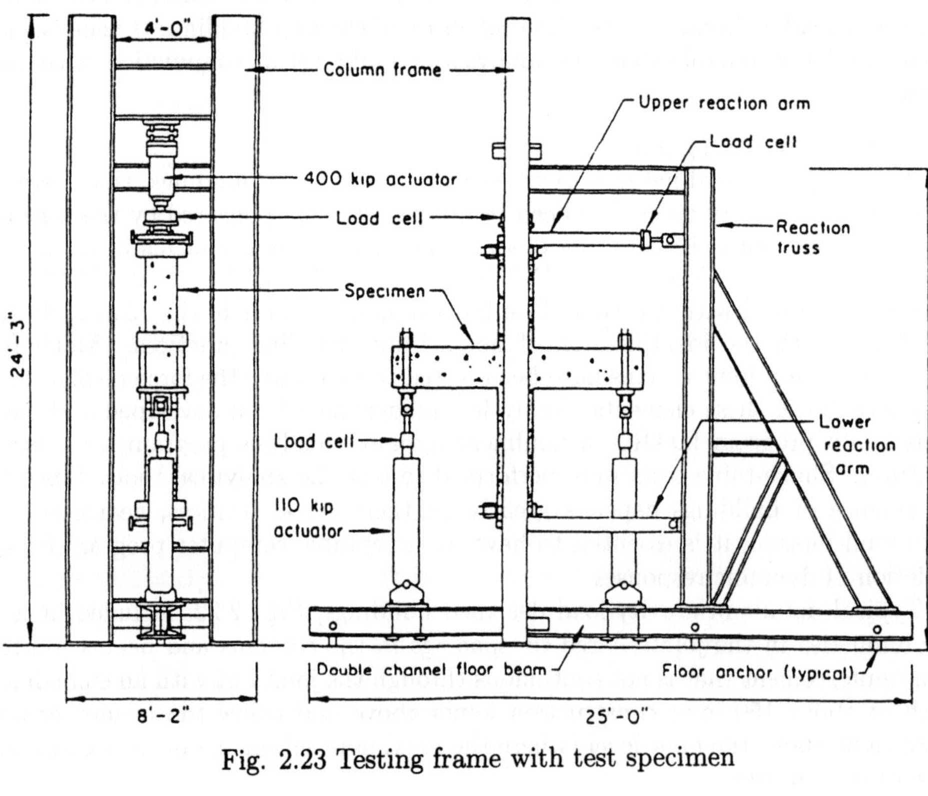

Fig. 2.23 Testing frame with test specimen

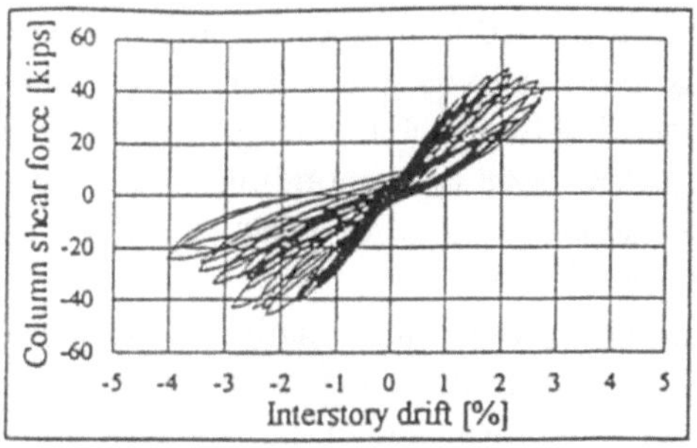

a. Interior joint, continuous positive bars

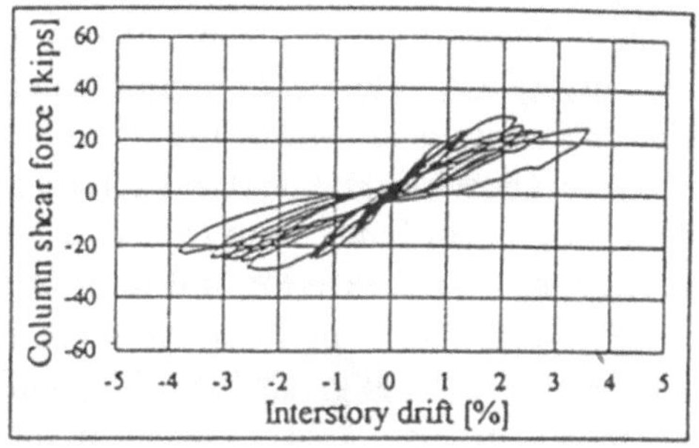

b. Interior joint, discont. positive bars

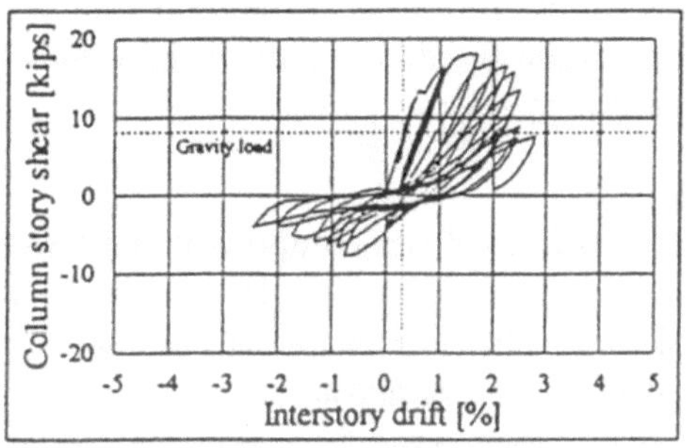

c. Exterior joint

Fig. 2.24 Typical column shear force vs story drift curves

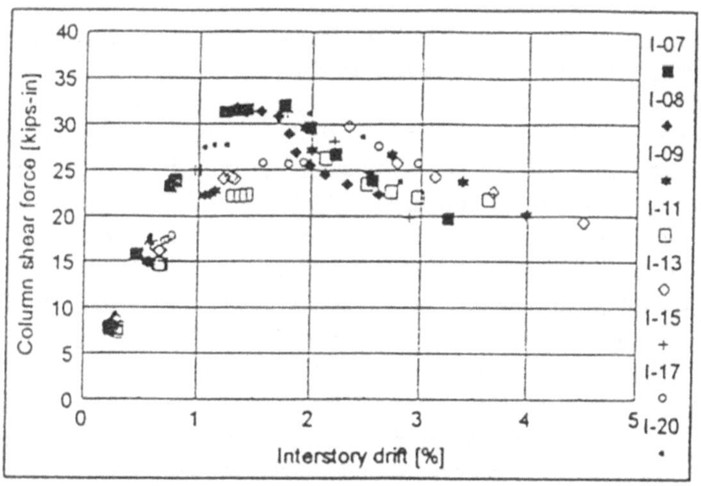

Fig. 2.25 Strength degradation

Full-scale beam-column tests evaluated the effects of these variables on their cyclic behavior. Typical beam sizes were 350 mm by 610 mm, and the columns were 400 mm square. The beam reinforcement consisted of two 19 mm or two 25 mm bars. The test setup is shown in Fig. 2.23. The initial loading included a fixed axial force of up to 1500 kN and vertical forces at the beam ends corresponding to the dead load. The beam loads were cycled to produce moments equal to those caused by an earthquake. Both exterior and interior joints were tested.

Extensive force and displacement measurements were taken; only a representative set of results is given here. Typical plots of column shear versus interstory drift are shown in Fig. 2.24. The most interesting differences indicated by these plots are the lower stiffness and higher drift for joints with discontinuous bottom bars and the highly unsymmetrical hysteresis loops for exterior joints. Fig. 2.25 shows the strength degradation as a function of drift. It is seen that a reasonably high level of shear of about 110 kN was maintained. In terms of joint shear, it is about 2/3 of the shear capacity of a well-detailed seismic joint.

2.5.3 Dynamic Analysis

The hysteresis curves obtained from component tests were used to establish the parameters in both the trilinear and the smooth models. Current work attempts to define these curves based on mechanical properties and to extend the applicability of the results to cases not studied experimentally.

Nine earthquake records were used to study the performance of two typical buildings, a three-story and a ten-story building, both designed primarily for gravity loads. These included the Nahanni ground motion (judged to be typical for the Eastern United States) with peak accelerations of 0.15g and 0.20g, the Taft earthquake with 0.20g and 0.30g, and the El Centro earthquake with 0.20g, 0.35g and 0.52g. Details are given in [2.45,2.46].

These buildings were quite flexible because of their small columns, and they lost stiffness during the motion mainly due to the slip of the discontinuous bottom bars. Therefore, they shifted to the right on the response spectrum curve, with the result that the earthquake imparted relatively little energy into the structure. Thus the forces were not large, though the displacements may be considered excessive. The maximum story drifts are summarized in Table 7.1. Although these are preliminary results, we see that the drifts do not increase much for higher level input, mainly because of the large hysteretic energy absorption and flexibility.

The question of acceptable performance levels for existing buildings is a perplexing one, which is being discussed extensively within the earthquake engineering community. What are the acceptable drift and force levels? What risk should we tolerate against collapse? How much damage to the structure and to the contents is acceptable? These questions are more difficult for existing buildings because any rehabilitation or retrofit is relatively expensive. In some geographical regions, such as the Eastern part of the United States, major earthquakes occur at very long intervals (about 1000 years), which complicates the question of acceptable risk level even further.

Table 7.1 Story Drifts, (%)

	Story 1	Story 3	Story 10
Three-story building			
Nahanni, hard soil, 0.15g	.09	.10	
Nahanni, soft soil, 0.20g	.17	.15	
Taft 0.02g	.09	.06	
Taft 0.20g	.84	.45	
Taft 0.35g	1.48	.66	
El Centro 0.20g	.85	.39	
Ten-story building			
Taft 0.20g	.35	.61	.65
Taft 0.35g	.89	.92	1.11

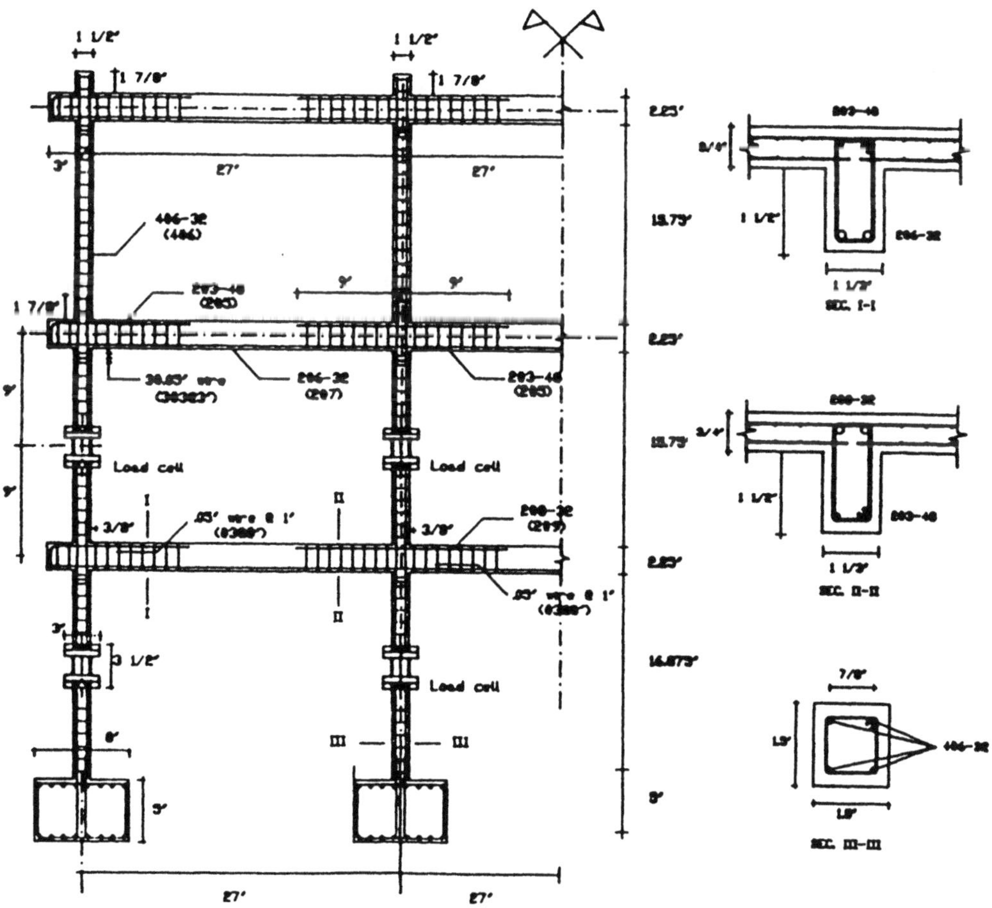

Fig. 2.26 Details of three-story building

2.5.4 Comparison with Shake Table Tests

Several shake table tests were performed at Cornell University and SUNY/Buffalo to evaluate the analytical predictions made with the program IDARC. The results for the 1/8 scale test and a 1/3 scale test are summarized below. A sketch of the smaller frame is shown in Fig. 2.26. The other frame was similar.

The 1/8 scale frame tested at Cornell University [2.45] was subjected to 1952 Taft earthquakes with increasing intensity. The natural periods of the building were measured before and after each test. Also, extensive acceleration and displacement measurements were taken. The columns in the lower two stories had load cells, which allowed the calculation of the moments and shear forces in the columns.

As part of the program input, one has to specify the modulus of elasticity of the concrete, estimate the effect of initial cracking, and select an effective width of the slab. These assumptions affect the initial stiffness and periods of the structure. Based on a survey of the literature, a modulus of elasticity of 0.7E was used to account for initial cracking and an effective slab width of 1/4 of the span. The measured table accelerations and initial damping ratios were used as input for the nonlinear dynamic analysis.

The analytical and experimental story shears and story drifts are compared in Fig. 2.27 for the Taft 0.18g motion. The agreement is seen to be satisfactory. The maximum amplitudes are close and the motions are in phase. This means that the hysteretic properties, including degradation, have been properly modelled in the analysis. However, we should realize that the analytical results depend greatly on the initial structure stiffness because the initial period strongly influences the phase of the motion. This means that the dynamic analysis of structures can only be as accurate as the estimation of the initial period. In the present case, the 0.70 factor for the gross member stiffnesses worked well. If that were not the case, the predicted motion would have a different periodicity (phase) from the measured one, and the two curves would not be as close. Of course, in terms of design use, the peak quantities are not likely to be affected much by the phase error (except, perhaps, for nearly harmonic input). The stiffnesses of actual buildings are influenced by walls, partitions, staircases, and fassade elements.

The prediction for a 0.35g Taft run was not as good, as can be seen in Fig. 2.28. However, the agreement is still reasonable. The analytical results did not include the P-delta effect, which was considerable in this case. (The P-delta effect was subsequently added to the program). Hinges formed at the ends of the first and second story columns in the experiments, and the building was very close to collapse at this stage.

The analysis also predicted a few column hinges in the third story and a few hinges in the first two stories; the effective width of the slab assumed for the analysis was probably too small.

Typical results for the 1/3 scale model, tested at SUNY/Buffalo, are shown in Fig. 2.29 for a 0.30g level Taft input. The agreement and the conclusions are similar to those discussed above [2.44].

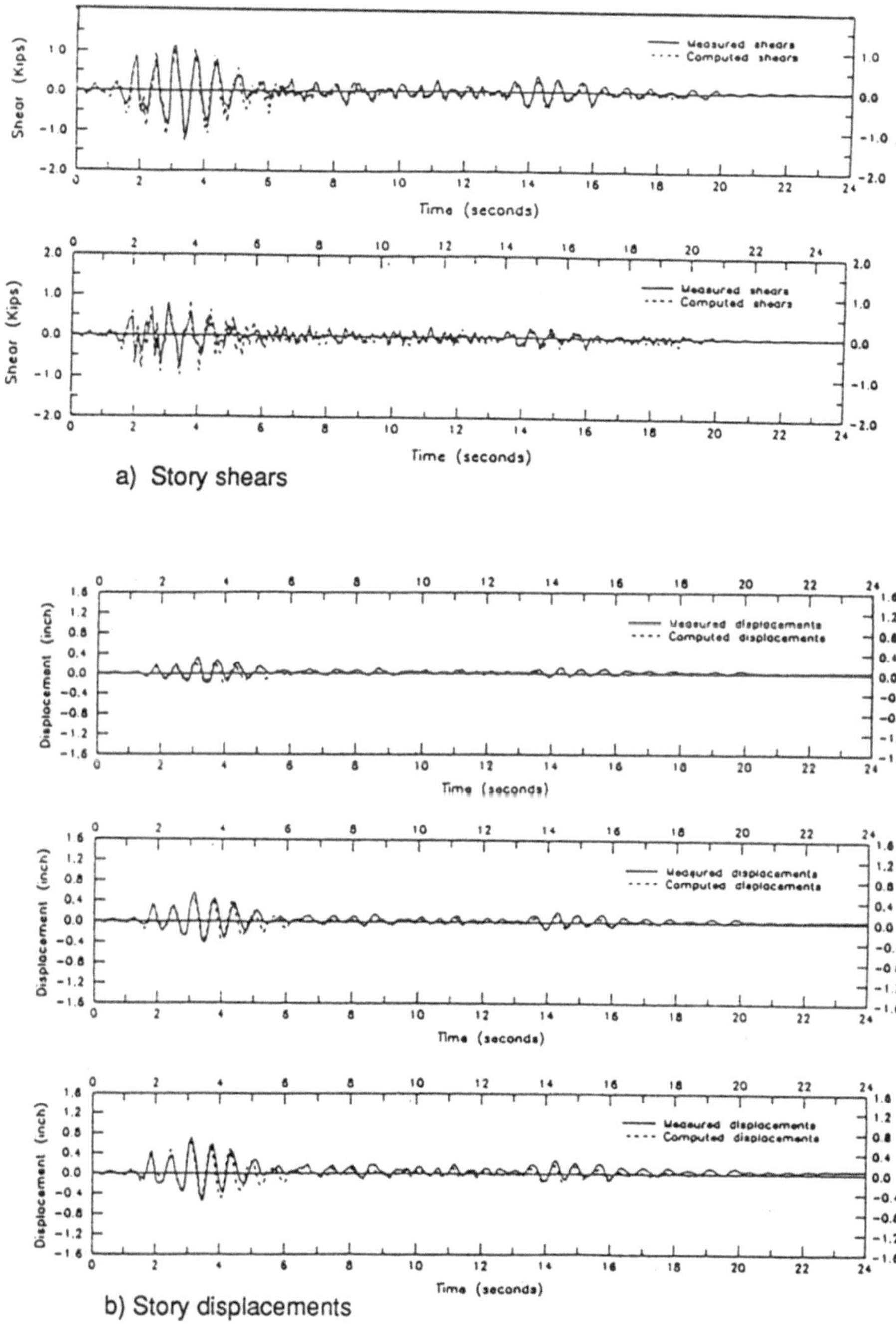

Fig. 2.27 Computed and measured story shears and displacements (Taft 0.18g)

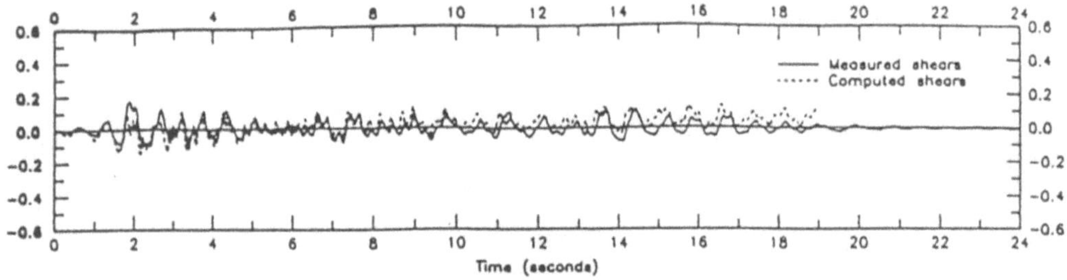

(a) Column 1-A.

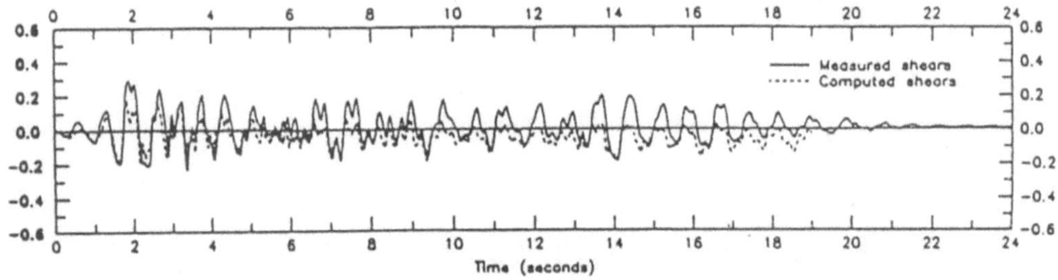

(b) Second Story.

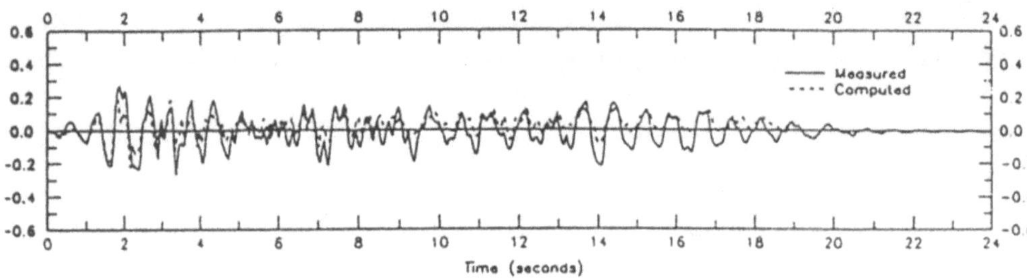

(c) Exterior Column 1-C.

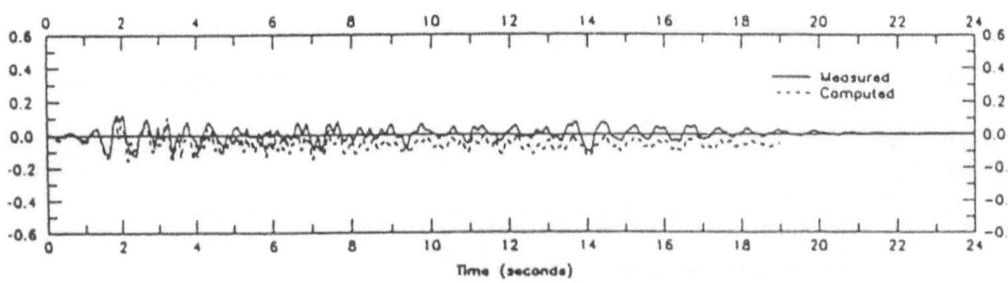

(d) Interior Column 1-D.

Fig. 2.28 Computed and measured story shears (Taft 0.35g)

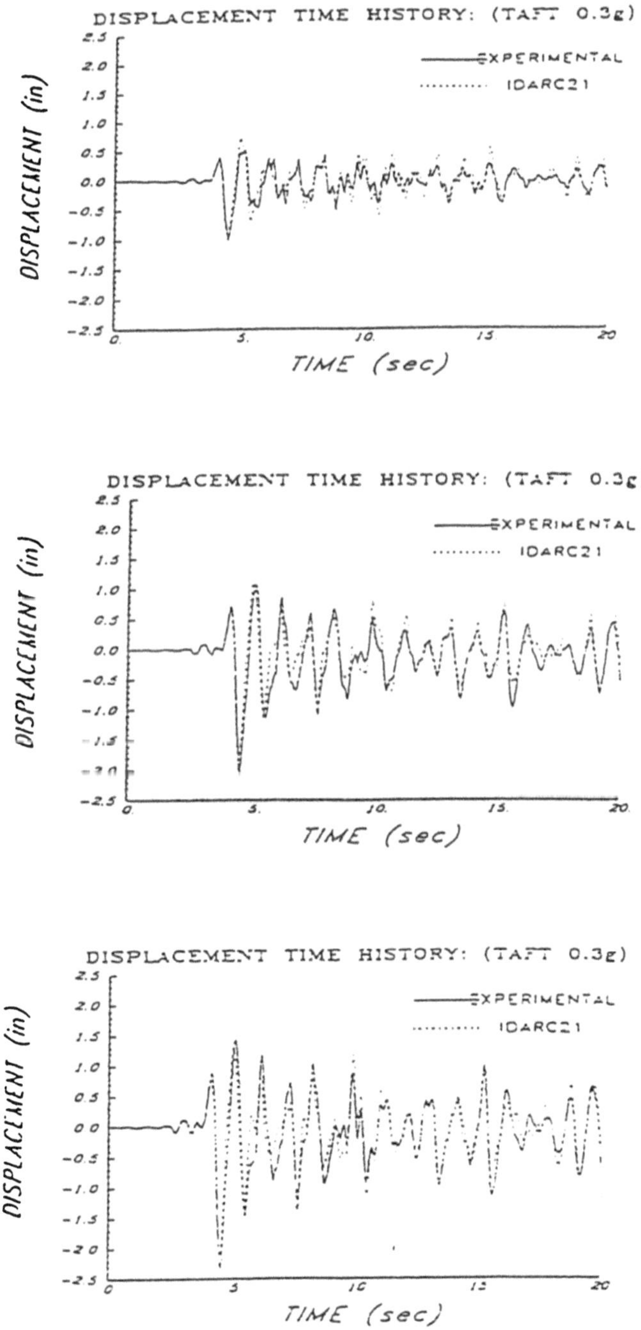

(a) Story 3

(b) Story 2

(c) Story 1

Fig. 2.29 Computed and measured displacements, 1/3 scale model (Taft 0.3g)

These and other comparisons indicate that a macro-level program such as IDARC can give a reasonably good estimate of the global response of frames, even with weak and deteriorating elements. The reliability of the analysis depends mainly on the estimation of the rate of strength and stiffness deterioration of the elements, and the time history of the response also depends on the accuracy of the initial period.

2.5.5 Modelling of Frames with Infills

The prediction of the dynamic response of building frames with concrete or masonry infill walls is of great interest because most buildings have such infill walls. Micromodels are often used, especially in the linear range and for walls with openings. Macromodels may be better suited for the nonlinear analysis of tall buildings. A relatively simple macromodel has been developed [2.47]. This consists of three diagonal struts in each direction, which act only in compression, Fig. 2.30. When a wall crushes in the corners, as may happen for weak infills made of hollow tiles or blocks, the center strut loses stiffness, and the other two parallel struts pick up the force. This model allows the formation of hinges away from the ends of beams and columns. Analysis of multistory buildings has shown that this type of behavior is possible. This model is efficient and permits the global analysis of large structures. The nonlinear properties of the struts are obtained from experiments. The use of this approach in walls with openings will require input from analyses using micromodels or experiments. Work is continuing on the development of this model.

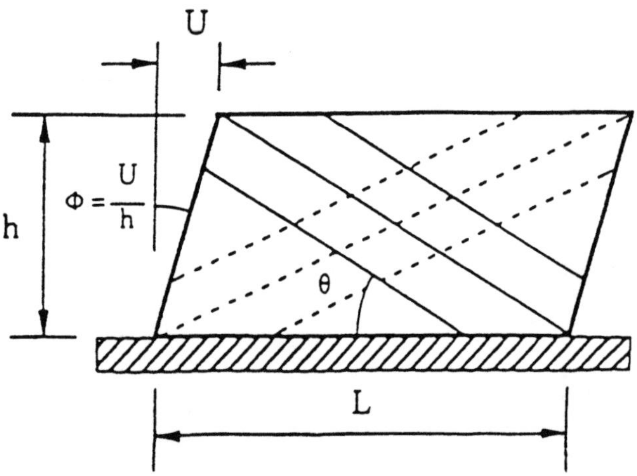

Fig. 2.30 Three-strut model for frames with infills

2.6 References

[2.1] Chung, Y.S., C. Meyer, and M. Shinozuka: Seismic Damage Assessment of Reinforced Concrete Members, Report No. NCEER-87-0022, National Center for Earthquake Engineering Research, Buffalo, New York, October 1987.

[2.2] Paskova, T.: Low-Cycle Fatigue Behavior of Concrete With and Without Fiber Reinforcement, Ph. D. Dissertation, Columbia University, New York 1994.

[2.3] Park, Y.J. and A.H.S. Ang: Mechanistic seismic damage model for reinforced concrete, Journal of Structural Engineering, ASCE, Vol. 111, No. 4, April 1985.

[2.4] Grzybowski, M. and C. Meyer: Damage accumulation in concrete with and without fiber reinforcement, ACI Materials Journal, Nov-Dec 1993.

[2.5] Fang, L.: Low-Cycle Fatigue and Damage Mechanics of Fiber Reinforced Concrete Under Biaxial Stress States, Ph.D. Dissertation, Columbia University, New York 1996.

[2.6] Clough, R.W. and S. Johnston: Effect of stiffness degradation on earthquake ductility requirements, Trans. Japan Earthq. Eng. Symp., Tokyo, Oct. 1966, 195-198.

[2.7] Takizawa, H. and H. Aoyama: Biaxial effects in modelling earthquake response of RC structures, Earthq. Engin. and Struct. Dyn. (1976), Vol 4, 523-552.

[2.8] Giberson, M.: The Response of Nonlinear Multi-Story Structures Subjected to Earthquake Excitations, Earthq. Engin. Res. Lab., Calif. Inst. of Tech., Pasadena 1967.

[2.9] Clough, R.W. and L. Benuska: Nonlinear earthquake behavior of tall buildings, J. of the Engin. Mech. Div, ASCE, Vol 93, No. EM3, June 1967.

[2.10] Takeda, T., M.A. Sozen and N. Nielsen: Reinforced concrete response to simulated earthquakes, J. of the Struct. Div., ASCE, Vol 96, No ST12, Dec. 1970.

[2.11] Banon, H., J. Biggs and M. Irvine: Seismic damage in reinforced concrete frames, J. of the Struct. Div., ASCE, Vol 107, No ST9, Sept 1981.

[2.12] Otani, S.: Inelastic analysis of R/C frame structures, J. of the Struct. Div., ASCE, Vol 100, No ST7, July 1974.

[2.13] Filippou, C.F., E.P. Popov and V.V. Bertero: Effects of Bond Deterioration on Hysteretic Behavior of Reinforced Concrete Joints, Earthq. Engin. Res. Center, Univ. of California Berkeley, Report EERC 83-19, Aug. 1983.

[2.14] Ozdemir, H.: Nonlinear Transient Dynamic Analysis of Yielding Structures", Ph.D. Dissertation, Dept. of Civil Engin., Univ. of California Berkeley, 1981.

[2.15] Iwan, W.: Application of Nonlinear Analysis Techniques, in: Applied Mechanics in Earthquake Engineering (W. Iwan, ed.), ASME, AMD, Vol 8, Nov.

[2.16] Ciampi, V. and M. Nicoletti: Parameter identification for cyclic constitutive models with stiffness and strength degradation, 8th European Conf. on Earthq. Engin., Lisbon, Sept. 1986.

[2.17] Prager, W. and P. Hodge: Theory of Perfectly Plastic Solids, John Wiley and Sons, New York 1951.

[2.18] Takayanagi, T. and W. Schnobrich: Nonlinear analysis of coupled wall systems, Earthq. Engin. and Struct. Dyn. (1979), Vol 7, 1-22.

[2.19] Charney, F. and V.V. Bertero: An Evaluation of the Design and Analytical Seismic Response of a Seven Story Reinforced Concrete Frame-Wall Structure, Earthq. Engin. Res. Center, Univ. of California Berkeley, Report EERC 82-08, Aug. 1982.

[2.20] Bertero, V.V., A. Aktan, F. Charney and R. Sause: Earthquake simulator tests and associated experimental, analytical and correlation studies on one-fifth scale model, in: Earthquake Effects on Reinforced Concrete Structures, ACI Special Publication SP-84, 1984.

[2.21] Anagnostopoulos, S.: Inelastic beams for seismic analysis of structures, J. of the Struct. Div., ASCE, Vol 107, No ST7, July 1981.

[2.22] Lai, S., G. Will and S. Otani: Model for inelastic biaxial bending of concrete members, J. of the Struct. Div., ASCE, Vol 107, No ST11, Nov. 1984.

[2.23] Hellesland, J. and A.C. Scordelis: Analysis of RC Bridge Columns under Imposed Deformations", IABSE, Delphi, 1981, 545-559.

[2.24] Mari, A. and A.C. Scordelis: Nonlinear Geometric Material and Time Dependent Analysis of Three Dimensional Reinforced and Prestressed Concrete Frames, Dept. of Civil Engin., Univ. of California Berkeley, Report SESM 82-12, June 1982.

[2.25] Bazant, Z.P. and P. Bhat: Prediction of hysteresis in reinforced concrete members, J. of the Struct. Div., ASCE, Vol 103, No ST1, Jan. 1977.

[2.26] Menegotto, M. and P.E. Pinto: Slender RC compressed members in biaxial bending, J. of the Struct. Div., ASCE, Vol 103, No ST3, Mar. 1977.

[2.27] Mahasuverachai, M.: Inelastic Analysis of Piping and Tubular Structures, Earthq. Engin. Res. Center, Univ. of California Berkeley, Report EERC 82-27, 1982.

[2.28] Ciampi, V. and L. Carlesimo: A nonlinear beam element for seismic analysis of structures, 8th European Conf. on Earthq. Engin., Lisbon, Sept. 1986.

[2.29] Spacone, E.: A Beam Model for Damage Analysis of Reinforced Concrete Structures Under Seismic Loads, Dept. of Civil Engin., Univ. of California Berkeley, Report CE 299, 1992.

[2.30] Brancaleoni, F., V. Ciampi and R. Di Antonio: Rate-yype models in nonlinear hysteresis structural behavior", EUROMECH Colloq., Palermo, Italy, Oct. 1983.

[2.31] Roufaiel, M.S.L. and C. Meyer: Analytical modeling of hysteretic behaviour of R/C frames, Journal of Structural Engineering, ASCE, Vol. 113, No. 3, March 1987.

[2.32] Chung, Y.S., C. Meyer and M. Shinozuka: Automatic seismic design of RC building frames, ACI Structural Journal, May-June 1990.

[2.33] Ma, S-Y. M., V.V. Bertero and E.P. Popov: Experimental and Analytical Studies on the Hysteretic Behaviour of Reinforced Concrete Rectangular and T-Beams, Earthq. Engin. Res. Center, Univ. of California Berkeley, Report EERC 76-2, 1976.

[2.34] Scribner, C.F. and J.K. Wight: Delaying Shear Strength Decay in Reinforced Concrete Flexural Members under Large Load Reversals, Report No. UMEE 78R2, Dept. of Civil Engin., University of Michigan, Ann Arbor, May 1978.

[2.35] Kaba, S. and S.A. Mahin: Refined Modeling of Reinforced Concrete Columns for Seismic Analysis, Earthq. Engin. Res. Center, Univ. of California Berkeley, Report EERC 84-03, 1984.
[2.36] Zeris, C. and S.A. Mahin: Analysis of reinforced concrete beam-columns under uniaxial excitation, Journal of Structural Engineering, ASCE, Vol 114, No ST4, April 1988.
[2.37] Zeris, C. and S.A. Mahin: Analysis of reinforced concrete beam-columns subjected to biaxial excitation, Journal of Structural Engineering, ASCE, Vol 117, No ST11, Sept. 1991.
[2.38] Clough, R.W. and J. Gidwani: Reinforced Concrete Frame 2: Seismic Testing and Analytical Correlation, Earthq. Engin. Res. Center, Univ. of California Berkeley, Report EERC 76-15, 1976.
[2.39] Taucer, F.F., E. Spacone and F.C. Filippou: A Fiber Beam-Column Element for Seismic Response Analysis of Reinforced Concrete Structures, Earthq. Engin. Res. Center, Univ. of California Berkeley, Report EERC 91-17, 1991.
[2.40] Zienkiewicz, O.C. and R.L. Taylor: The Finite Element Method, Vol 1, Basic Formulation and Linear Problems, 4th Ed., McGraw Hill, London 1989.
[2.41] Low, S.S. and J.P. Moehle: Experimental Study of Reinforced Concrete Columns Subjected to Multiaxial Cyclic Loading, Earthq. Engin. Res. Center, Univ. of California Berkeley, Report EERC 87-14, 1987.
[2.42] Pessiki, S.P., C.H. Conley, P. Gergely and R.N. White: Seismic Behavior of Lightly-Reinforced Concrete Column and Beam-Column Joint Details, Report No. NCEER-90-0014, National Center for Earthquake Engineering Research, Buffalo, New York, 1990.
[2.43] Beres, A., R.N. White and P. Gergely: Seismic Behavior of Reinforced Concrete Frames with Nonductile Details, Report No. NCEER-92-0024, National Center for Earthquake Engineering Research, Buffalo, New York, 1992.
[2.44] Hoffman, G.W., S.K. Kunnath, J.B. Mander and A.M. Reinhorn: Gravity Load Designed R/C Buildings, Report No. NCEER-92-0016, National Center for Earthquake Engineering Research, Buffalo, New York, 1992.
[2.45] El-Attar, A.G., R.N. White and P. Gergely: Shake Table Test of a 1/8 Scale Three-Story Lightly Reinforced Concrete Building, Report No. NCEER-91-0018, National Center for Earthquake Engineering Research, Buffalo, New York, 1991.
[2.46] El-Borgi, S., R.N. White, P. Gergely and A. Beres: Evaluation and analysis of existing lightly reinforced concrete buildings, National Earthquake Engineering Conference, Memphis, TN, Vol. II, p 111, May 1993.
[2.47] Gergely, P., R.N. White, D. Zawilinski and K. Mosalam: The interaction of masonry infill and steel or concrete frames, National Earthquake Engineering Conference, Memphis, TN, Vol. II, p 183, May 1993.

MODELLING OF PANEL STRUCTURES

C. Meyer
Columbia University, New York, NY, USA

J. Blaauwendraad, P.H. Feenstra and R. de Borst
Delft University of Technology, Delft, The Netherlands

3.1 Modelling Aspects

3.1.1 Introduction

The terms "panel structures" and "planar structures" are used interchangeably to designate two-dimensional structural elements that are loaded primarily in their own plane. They include walls, deep beams, and to some extent also shell elements that are subjected to only membrane stresses.

Structural walls constitute maybe the most important type of panel structures. Without loss of generality, we may therefore restrict our discussion here to such wall components. They constitute important integral parts of concrete buildings, providing stiffness against lateral loads and controlling interstory drifts, thereby limiting damage suffered by nonstructural building components. They also provide strength for resisting horizontal and gravity loads. Finally, if designed properly, they offer the ductility and energy dissipation capacity needed to mitigate the effects of severe earthquake ground motions.

The difficulty of the analysis task derives from the walls' nonlinear behavior, the complex geometries often found in buildings, and the interaction between the various structural components. To accurately model the constitutive relationships of wall elements, their hysteresis behavior under load reversals and the various potential failure modes need to be considered.

This section will review important aspects of dynamic wall behavior and summarize some of the mathematical models that have been proposed to simulate their response to both monotonic and cyclic loads.

3.1.2 Behavior of Structural Walls

Walls are acting either individually (isolated walls), or they interact with each other (coupled walls), or with other systems such as frames (integrated walls), Fig. 3.1. In steel buildings, isolated walls can be used as shear walls. Although less common in reinforced concrete construction, their behavior characteristics are fundamental and are found in coupled and integrated walls as well.

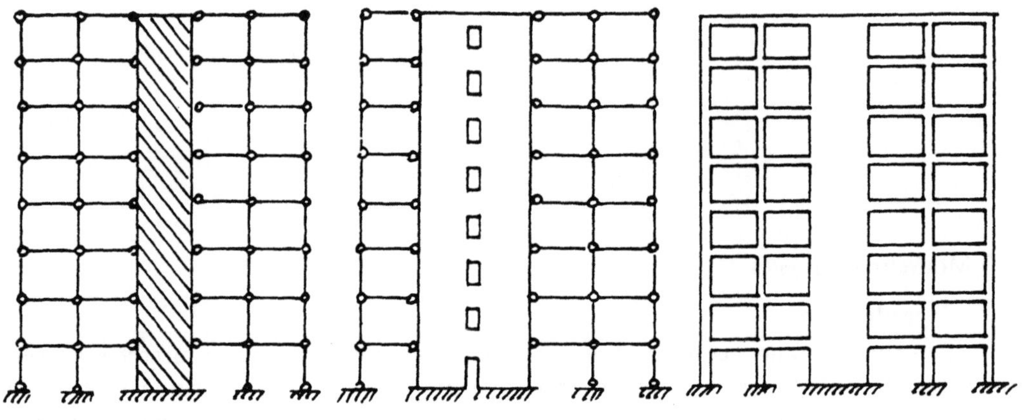

a) isolated b) coupled c) integrated

Fig. 3.1 Wall types

Isolated walls are either slender, medium-short, or short. Slender walls have height-to-width ratios of at least 2 or 3 and behave like vertical cantilever beams, except for the axial load effect. For optimum effect, the reinforcement is concentrated near the wall edges, with nominal steel distributed evenly throughout the wall. Because of the large cross-sectional areas of walls, the axial forces due to gravity loads are usually well below the balanced point, so that gravity loads increase the moment capacity, but the ductility is reduced. Walls are often nominally over-reinforced, i.e. the concrete on the compression side is likely to crush before the tension steel yields. Utilizing the large moment capacity of a wall with heavy flexural reinforcement leads to high shear stresses and potential shear failures. By controlling the flexural reinforcement, the designer can control the failure mode to be either of shear or flexural type. The addition of flanges increases the flexural strength and the potential of a shear failure in the web.

Walls of thin rectangular cross section can buckle on the compression side. Floor slabs offer sufficient lateral restraint to limit the unsupported length to the interstory height. Stiffening the wall with flanges or barbells effectively eliminates such lateral buckling. But if flexural reinforcement on the tension side yields, large cracks will open, and upon stress reversal only the reinforcing bars will resist the compression force until the cracks close again. While the cracks are open, the effective moment of inertia to resist lateral buckling is greatly reduced. The possible consequence is instability or large out-of-plane displacements. This potential failure mode can be eliminated with proper confinement of the edge reinforcement that creates ductile edge members, which also serve as vertical dowels if the plastic zone contains a horizontal construction joint [3.1].

The load-deformation behavior of medium-high walls (height-to-width ratios of 1 to 3) depends strongly on the relative amount of shear, which is controlled by the amount of flexural reinforcement. For walls with low nominal shear stress ($v_c < 3\sqrt{f_c'}$), a crack pattern such as shown in Fig. 3.2a might be expected [3.2]. Because of the horizontal cracks near the base, little truss action can develop, rendering the horizontal reinforcement virtually ineffective in resisting shear forces. The shear is resisted primarily by interface shear and dowel action of the vertical steel, especially in the edge members. The shear stiffness clearly degrades with the number of load cycles. But since the shear stresses are relatively low, the shear strength is generally adequate.

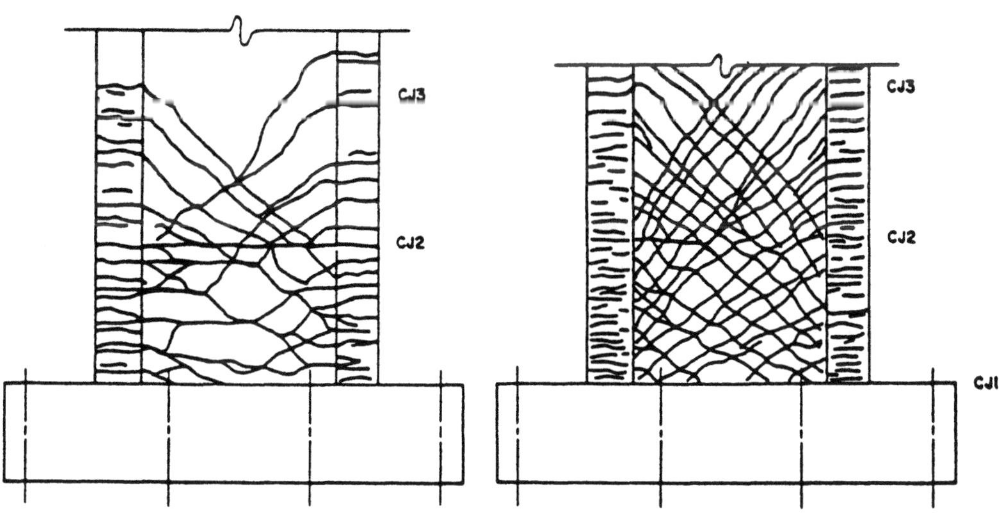

a) low shear b) high shear

Fig. 3.2 Crack patterns in shear walls [3.2]

For walls with high nominal shear ($v_c > 7\sqrt{f_c'}$), a typical crack pattern is shown in Fig. 3.2b [3.2], which permits truss action to create a stiff shear-resistant system. But repeated load cycling causes abrasion in the diagonal struts, reducing their compressive strength and increasing shear deformations. Eventually, failure occurs by web crushing. (The ACI Code prevents this failure mode by limiting the amount of shear reinforcement allowed.) The contribution of the concrete to resist shear is greatly diminished in the plastic range, where large cracks open up when the vertical steel yields.

Short walls with height-to-width ratios less than 1 behave more like corbels or brackets than beams. Here, flexural strength is seldom a problem and ductility demand small. Such walls have typically sufficient strength to resist seismic loads elastically. If a ductile, i.e. flexural failure mode is desired, the nominal shear stresses must be moderate, say less than $6\sqrt{f_c'}$. This can be readily achieved by controlling the wall thickness. The shear strength depends heavily on truss action, which requires sufficient horizontal as well as vertical reinforcement. Because a flexural failure mechanism is associated with large cracks, not much reliance should be placed on the contribution of concrete toward the shear strength. For short walls, there is an increased likelihood of sliding across construction joints, partially because of the negligible axial loads, which in taller walls provide the clamping forces necessary for aggregate interlock.

A number of reinforced concrete walls have been tested under cyclic load at the Portland Cement Association [3.2,3.3]. These experiments contributed greatly to the understanding of wall panels under such loading. Of equal if not greater importance were the experiments conducted by Collins and his co-workers at the University of Toronto [3.4-3.7] with a test rig that permits the application of arbitrary 2-D states of stress to panels of about 1.5 m by 1.5 m size, Fig. 3.3. Four of these panel tests served as experimental benchmarks in an international competition to predict the response of reinforced concrete panels, Fig. 3.4. Twenty-seven teams of researchers from 13 different countries submitted their analysis results. Of these, 20 also predicted the deformational response in addition to making strength predictions [3.5].

Panel A was isotropically reinforced ($\rho_x = \rho_y = 0.018$) and loaded in pure shear up to failure. Panel B was nominally identical to panel A except that it was subjected to combined biaxial compression and shear in order to determine the effect of the biaxial compression on the panel's shear strength. Panel C was again tested in pure shear, but was nonisotropically reinforced ($\rho_x = 2.5\rho_y = 0.018$) to assess the effect of the lower steel capacity in the y-direction on the shear strength. Panel D was also nonisotropically reinforced ($\rho_x = 2\rho_y = 0.018$) and subjected to nonproportional loading. This panel was initially loaded in pure shear up to a stress of 3.9 MPa (565 psi). Further shear stress increments were accompanied by biaxial compressive stresses of equal magnitude [3.5].

The comparison between analytical and experimental results is summarized in Fig. 3.5. Panel A failed at a shear stress of 6.3 MPa (913 psi) in a sliding failure of the concrete. The biaxial compression applied to panel B increased the shear strength to 9.1 MPa (1320 psi); i.e. the increase was 45%. The large scatter exhibited by Fig. 3.5 indicates the wide range of accuracy of the analytical models. However, it should be added that experimentalists are familiar with a similar problem, which they refer

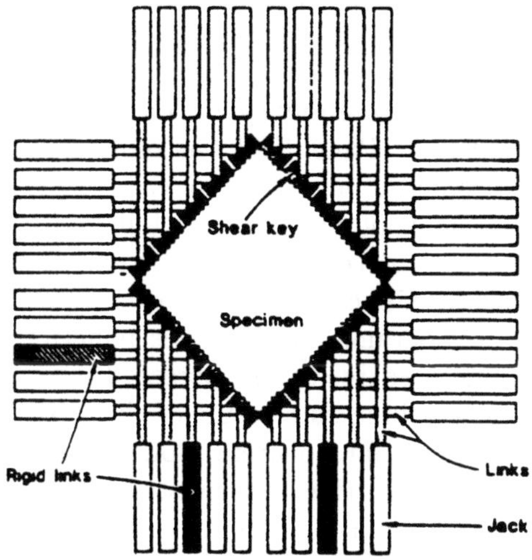

Fig. 3.3 University of Toronto panel test frame [3.4]

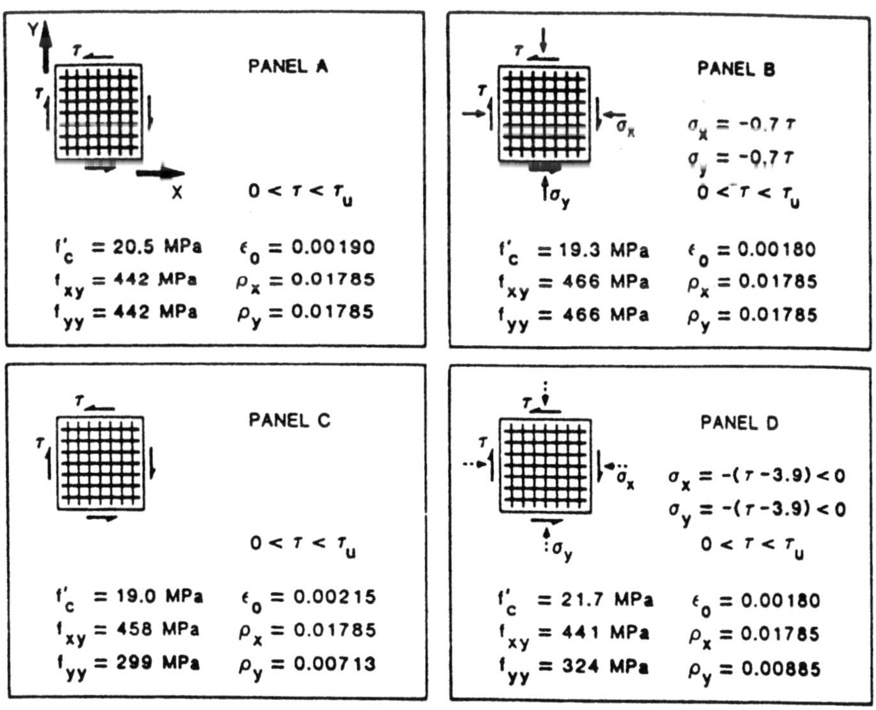

Fig. 3.4 Test panels for international competition [3.5]

to as "repeatability". This involves statistical scatter of possibly comparable order of magnitude and is the result of material properties, different laboratory equipment, testing methods, and quality control.

The same testing rig was used to apply cyclic loads to three reinforced concrete panels of comparable size [3.6,3.7]. Panel SE8 was reinforced anisotropically ($\rho_x = 3\rho_y = 0.01$) and loaded three times in pure shear for each of the increasing load levels. Panel SE9 was subjected to the same load history, but had isotropic reinforcement ($\rho_x = \rho_y = 0.03$). Panel SE10 had the same reinforcement as panel SE8 but was loaded in shear together with biaxial compression ($\sigma_x = \sigma_y = -|\tau_{xy}|/3$).

The shear stress-strain histograms for the three panels are shown in Fig. 3.6. They illustrate the strength increase due to the increased steel area in the y-direction (panel SE9) and the delay in cracking as a result of the biaxial compression (panel SE10). The stress-strain diagram for concrete alone, Fig. 3.7, was obtained by subtracting the steel contribution to the load resistance of the reinforced concrete composite. Also included for comparison is the uniaxial stress-strain curve for concrete to illustrate the large drop in compressive strength that can be attributed to the transverse (principal) tensile strains. These transverse strains increase under cyclic loading and are a manifestation of the damage accumulation as we have seen earlier, which progressively degrades the concrete's compressive strength.

3.1.3 Modelling of Reinforced Concrete for Monotonic Loading

Several theories have been proposed to model the constitutive behavior of the reinforced concrete composite under monotonic load [3.8-3.10]. By addressing the steel-concrete composite, we are now dealing with what was referred to earlier as mezo- or macro-models. We shall limit ourselves to only one of these theories, namely the modified compression field theory of Collins and Vecchio [3.11].

The problem is to determine the relationship between the three in-plane stress components, $\sigma_x, \sigma_y, \tau_{xy}$, and the corresponding strains, $\epsilon_x, \epsilon_y, \gamma_{xy}$. More specifically, since this is a nonlinear problem, we are interested in the stress increments $\Delta\sigma_x, \Delta\sigma_y, \Delta\tau_{xy}$, resulting from strain increments $\Delta\epsilon_x, \Delta\epsilon_y, \Delta\gamma_{xy}$. The following assumptions are made:

1. Stresses and strains are averaged over areas large enough to include several cracks.
2. Perfect bond exists between steel and concrete.
3. Steel reinforcement is uniformly distributed in both the x- and y-direction.
4. Stress-strain relationships for the constituent materials are unique and unaffected by load history, i.e. restricted to monotonic loading.
5. Stress-strain relationships for steel and concrete are independent of each other.
6. The dowel shear strength of steel bars is ignored.
7. Principal stress axes in concrete coincide with principal strain axes.

Strain *compatibility* between concrete and steel requires,

$$\epsilon_{sx} = \epsilon_{cx} = \epsilon_x, \qquad \epsilon_{sy} = \epsilon_{cy} = \epsilon_y \tag{3.1}$$

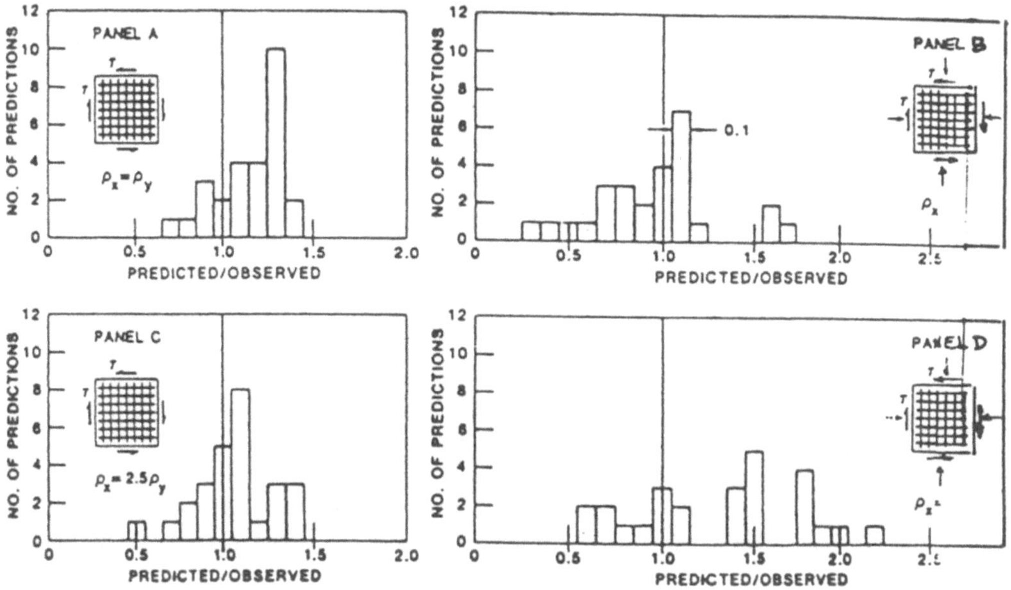

Failure Shear Stress (MPa)	A	B	C	D
Experimental	6.3	9.1	4.0	5.9
Analytical				
Low	4.4	2.5	2.1	3.4
High	8.8	15.2	5.7	16.5
Mean	7.2	8.7	4.1	8.1
Standard Deviation	1.2	3.0	0.9	3.2

Fig. 3.5 Statistical scatter of predicted results [3.5]

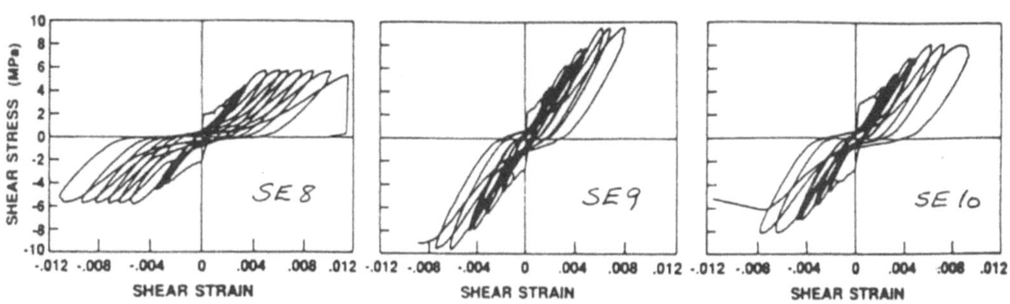

Fig. 3.6 Shear stress-strain histograms [3.7]

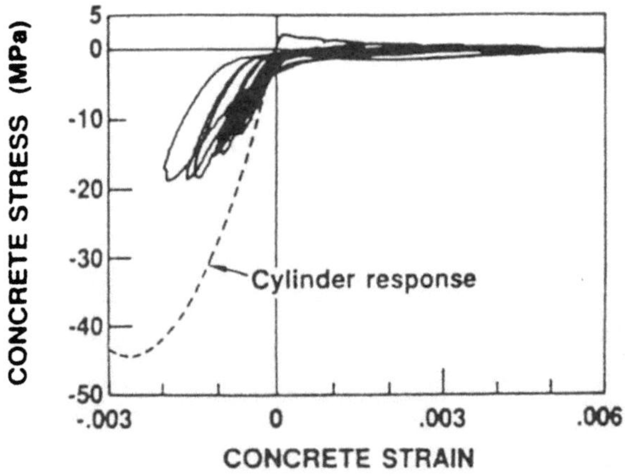

Fig. 3.7 Concrete stress-strain curve for panel SE9 [3.7]

where subscripts s and c refer to steel and concrete, respectively. If strains are known in one coordinate system, they can be determined for any other direction (including principal strains ϵ_1 and ϵ_2) by standard transformation.

Equilibrium requires that the stresses in a reinforced concrete element are equal to the sum of those in the steel and in the concrete, e.g. in the x-direction,

$$\int_A \sigma_x dA = \int_{A_c} \sigma_{cx} dA_c + \int_{A_s} \sigma_{sx} dA_s$$

or

$$\sigma_x = \sigma_{cx} + \rho_{sx}\sigma_{sx} \tag{3.2}$$

and similarly,

$$\sigma_y = \sigma_{cy} + \rho_{sy}\sigma_{sy} \tag{3.3}$$

$$\tau_{xy} = \tau_{cxy} = \tau_{cyx} \tag{3.4}$$

i.e., $\tau_{sxy} = \tau_{syx} = 0$.

Regarding *stress-strain relationships*, a simple elasto-plastic formulation suffices for steel. For concrete, the principal compressive stress σ_{c2} was found to be not only a function of ϵ_2, but also of the coexisting principal tensile strain ϵ_1, and the following relationship was determined experimentally,

$$\sigma_{c2} = \sigma_{c2,max}\left[2\left(\frac{\epsilon_2}{\epsilon_c'}\right) - \left(\frac{\epsilon_2}{\epsilon_c'}\right)^2\right] \tag{3.5}$$

where

$$\frac{\sigma_{c2,max}}{f_c'} = \frac{1}{0.8 - 0.34\frac{\epsilon_1}{\epsilon_c'}} \leq 1.0 \tag{3.6}$$

and $\epsilon_c' \approx -.002$ is the strain at maximum stress f_c'.

In tension, the following relationships were proposed for loading before and after cracking, respectively,

$$\sigma_{c1} = E_c \epsilon_1, \qquad \sigma_{c1} = \frac{f_{cr}}{1 + \sqrt{200 \epsilon_1}} \tag{3.7}$$

where f_{cr} is the concrete's tensile or cracking strength.

Recalling that the purpose of this analysis is to determine the state of stress for a given state of strain, the following steps constitute the solution process.

1. From given strains ϵ_x and ϵ_y compute the steel stresses, σ_{sx} and σ_{sy}.
2. Transform the given strains to principal axes, i.e. find ϵ_1 and ϵ_2 and the principal direction angle θ.
3. From ϵ_1 and ϵ_2 determine the principal concrete stresses σ_{c1} and σ_{c2} (Eqs 3.5-3.7).
4. Check whether the average concrete tensile stresses can be transmitted across cracks and reduce them if necessary.
5. Transform concrete stresses back into the x,y coordinate system.
6. The *reinforced concrete* stresses follow from superposition (Eqs. 3.2-3.4.).

For details, refer to Refs. [3.4,3.11].

3.1.4 Modelling for Cyclic Loading

For the case of general loading histories, including load reversals, the following solution procedure is suggested [3.6,3.7] to determine the stress increments $\Delta\sigma_x, \Delta\sigma_y, \Delta\tau_{xy}$ for a given set of strain increments, $\Delta\epsilon_x, \Delta\epsilon_y, \Delta\gamma_{xy}$.

1. For some load increment i, compute principal strain increments, $\Delta\epsilon_1$ and $\Delta\epsilon_2$, by transforming the strain increments from x,y coordinates to principal directions determined at the end of the previous load increment $i-1$, i.e. the angle $\Delta\theta^i$ between the x-axis and the principal direction associated with $\Delta\epsilon_1$.
2. Transfer the states of stress and strain at the end of the previous load step $i-1$ to the new principal axes, so that,

$$\epsilon_1^i = \epsilon_1^{i-1} + \Delta\epsilon_1, \qquad \epsilon_2^i = \epsilon_2^{i-1} + \Delta\epsilon_2$$

   Since, by definition, $\Delta\gamma_{12} = 0$, we have $\gamma_{12}^i = \gamma_{12}^{i-1}$. The concrete stresses are likewise transformed to the new principal axes to give, $\sigma_{c1}^{i-1}, \sigma_{c2}^{i-1}, \tau_{12}^{i-1}$.
3. Determine the new concrete stress state, using an appropriate constitutive relationship, the state of stress at the end of the previous load step, $\sigma_{c1}^{i-1}, \sigma_{c2}^{i-1}, \tau_{12}^{i-1}$, peak strain responses of the past load history, and the principal strain increments, $\Delta\epsilon_1, \Delta\epsilon_2$. Note that, by definition, $\Delta\tau_{12} = 0$, so that $\tau_{12}^i = \tau_{12}^{i-1}$.
4. Compute the new principal direction θ^i and the corresponding principal concrete stresses, $\sigma_{c1}^i, \sigma_{c2}^i$, associated with the current load step.

5. Check whether the new principal stresses are within their strength envelopes.
6. Transform the new principal stresses back to the x,y coordinate system and combine them with the steel stresses.

The most difficult task is step 3, that is, establishing the concrete principal stress increments associated with given strain increments. This relationship is shown qualitatively in Fig. 3.8 for the idealized case where the principal strain directions remain constant (a simplification). An initial loading in compression (OA) follows a parabola likely to be softer than that for a cylinder loaded in uniaxial compression, because of the coexisting transverse tensile strain. First-time unloading (AB) follows a steep curve until the concrete cracks in tension (point B). This cracking stress is likely to be smaller than the concrete's virgin tensile strength, because of the damage caused by the prior compressive loading. After cracking, the response follows the tensile envelope (BC). Upon stress reversal, there is an initial stiff response, which levels off rapidly, as the tension in the concrete between cracks is released (CD). After leveling off, the response remains very soft, as cracks begin to close, and becomes stiffer, as the cracks close completely. Thereafter, the response follows again a compression parabola, offset slightly from the initial parabola as a result of damage (and strain) accumulation. Unloading of previously cracked concrete (DE) is initially similar to unloading of uncracked concrete, but at a certain stress level, the response softens and is almost linear until it reaches the tensile envelope at about the previous maximum tensile strain. Upon further tensile load increments, the response follows the tensile envelope.

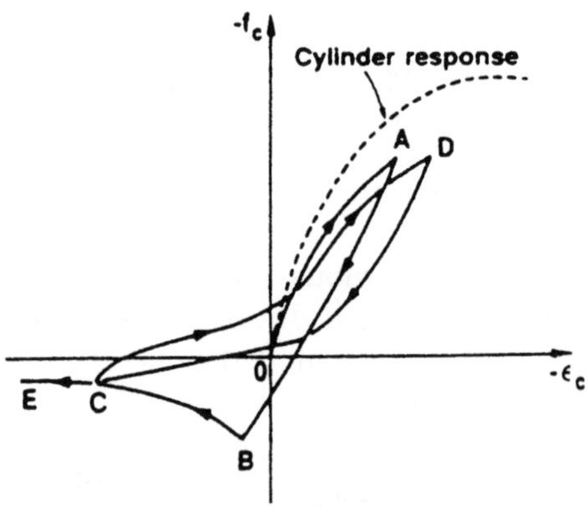

Fig. 3.8 Model for concrete stress-strain behavior [3.7]

Quantitative expressions for each of these loading and unloading branches are given in Ref. [3.6]. They were derived empirically from the data obtained from the tests referred to earlier. Unfortunately, the state of the art of damage mechanics has not yet advanced enough to realistically simulate such behavior solely on the basis of continuum mechanics principles. But the response of concrete to cyclic loading can already be reproduced numerically within limits [3.12], provided the state of stress remains compressive, i.e. no stress reversals from compression to tension and vice versa are involved. Efforts are currently under way to simulate also this case [3.13,3.14].

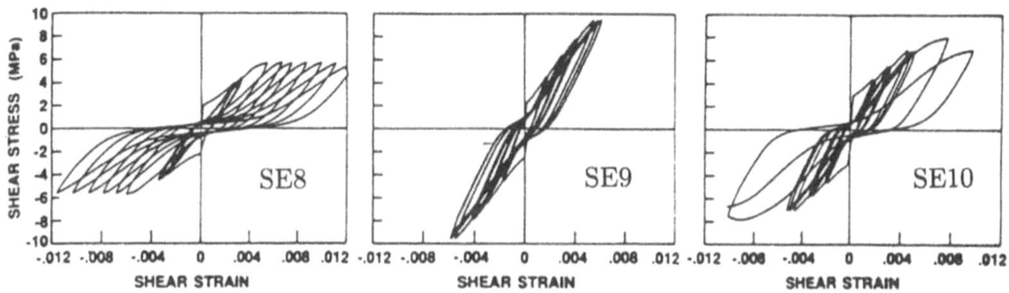

Fig. 3.9 Analytical shear stress-strain histograms [3.7]

Figure 3.9 shows the response obtained numerically by Stevens et al [3.7] for the three test panels. The agreement with the experimentally observed behavior, Fig. 3.6, is very good. However, nothing less should be expected, since the data obtained from these three panels were used to calibrate the numerical models.

3.2 Modelling of Panels with DIANA

3.2.1 Approach Outline

For the modelling of reinforced concrete shear walls one needs appropriate models for concrete, reinforcement and interface action due to bond. The concrete model must describe proper behavior both in compression and tension, taking into account post-peak softening in either state of stress. Many computer programs handle the failure of concrete in compression as a plasticity problem and failure of concrete in tension as a fracturing problem. Earlier implementations of program DIANA did likewise. A major problem then occurs at integration points where a compressive plastic strain in one direction occurs simultaneously with a tensile fracturing strain in the direction perpendicular to it. The solution algorithm easily becomes unstable in that case, and a spurious solution may occur. Therefore, the decision was made to substantially improve the robustness of DIANA by handling the constitutive behavior of concrete in tension in the softening regime as a plastic problem as well. This unified approach indeed leads to the desired robustness.

The reinforcement steel is considered as an elastic-plastic material, and compression and tension are dealt with in the same way.

Perfect bond between steel and concrete is assumed, with no relative slip. The ultimate tensile failure load of a reinforced concrete element is determined by the yield tensile force of the steel bar and is not influenced by the bond. However, the strength does increase due to bond in the strain interval between the initiation of cracks in concrete and the start of steel yielding, i.e. in the tension-stiffening domain. The interaction of steel and concrete increases the stress at a given strain in that domain, compared with the sum of the stress σ_c in plain concrete and the stress σ_s in the steel at that strain. The additional stress which can be carried due to interaction is called σ_{ia}. So DIANA sums the stress contributions as follows,

$$\sigma = \sigma_c + \sigma_s + \sigma_{ia} \tag{3.8}$$

The following two sections will explain in detail how the contributions σ_c and σ_{ia} of this composite stress model can be quantified. The approach guarantees a smooth transition from tension-stiffening in reinforced concrete (all quantities σ_c, σ_s and σ_{ia} present) to tension-softening in plain concrete (only σ_{ia} present).

3.2.2 Constitutive Model for Plain Concrete

The accumulated damage in the concrete material under compression and/or tension is represented by an internal parameter κ which is assumed to be determined by the inelastic work using a work-hardening hypothesis. The inelastic work rate $\dot{W}_c$ is defined by

$$\dot{W}_c = \sigma^T \dot{\epsilon}_c = \overline{\sigma}(\kappa) \cdot \dot{\kappa} \tag{3.9}$$

in which $\dot{\epsilon}_c$ is the inelastic strain rate vector and $\overline{\sigma}(\kappa)$ an equivalent stress as a function of the internal parameter κ. In the case of tension the inelastic work g_f is then defined by the following integral,

$$g_f = \int_0^\infty \dot{W}_c dt = \int_0^\infty \overline{\sigma}(\kappa) d\kappa \tag{3.10}$$

Assuming that the inelastic work g_f is distributed uniformly over an equivalent length h, the relation between the fracture energy G_f and the work g_f (see Fig. 3.10) is given by

$$g_f = \frac{G_f}{h} \tag{3.11}$$

The equivalent length h is representative for the element length and is

$$h = \alpha_h \sqrt{A_e} \tag{3.12}$$

where the modification factor α_h is equal to one for quadratic elements and equal to $\sqrt{2}$ for linear elements [3.15]. The resulting material model in tension is related to the energy which has to be dissipated due to the irreversible damage in the material.

The same concept is also used for concrete in compression. The modelling of compressive softening is based on an energy approach and again, the equivalent length will be used, although it is recognized that the mechanism in compression is perhaps more related to the volume of the elements. The concept further presupposes that the compressive softening of concrete can be represented by a compressive fracture energy, denoted as G_c, see Fig. 3.11. This energy is assumed to be a material parameter. From experiments [3.16] values for G_c have been found ranging from 10 to 25 N/mm, which is about 50 to 100 times the tensile fracture energy. With this energy-based approach the compressive and tensile softening can be described within the same context, which is plausible because the underlying failure mechanisms are identical, viz. continuous crack growth on the micro-level.

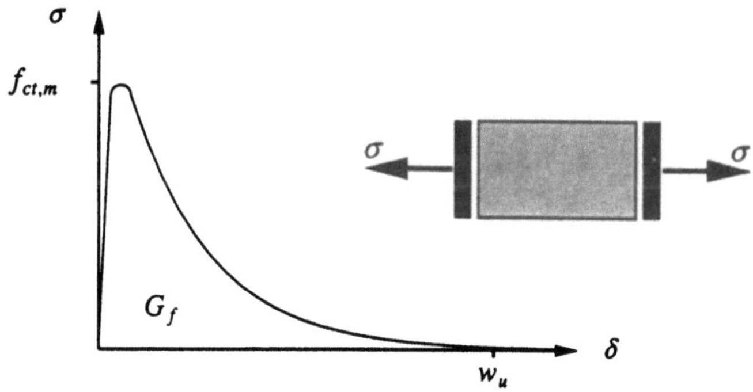

Fig. 3.10 Concrete specimen under uniaxial tensile loading

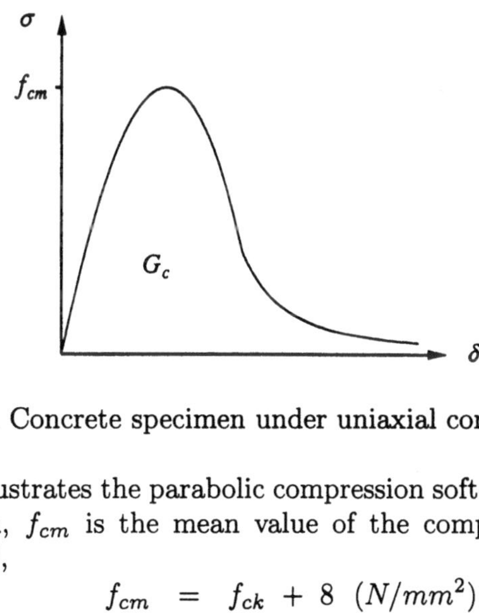

Fig. 3.11 Concrete specimen under uniaxial compressive loading

Figure 3.12 illustrates the parabolic compression softening model at the disposal in DIANA. Herein, f_{cm} is the mean value of the compressive strength, which is estimated by [3.17],

$$f_{cm} \;=\; f_{ck} + 8 \;\; (N/mm^2) \tag{3.13}$$

For Young's modulus, the following expression is used,

$$E_c \;=\; 10^4\, f_{cm}^{1/3} \;\; (N/mm^2) \tag{3.14}$$

The compressive stress reaches its maximum at an equivalent strain κ_e, which is material dependent only,

$$\kappa_e \;=\; \frac{4 f_{cm}}{3 E_c} \tag{3.15}$$

and the maximum equivalent strain κ_u is related to material and element size,

$$\kappa_u = \frac{3G_c}{2hf_{cm}} \tag{3.16}$$

To avoid possible snap-back on the constitutive level, the ultimate equivalent strain κ_u is limited by,

$$\kappa_u \geq 1.75\,\kappa_e \tag{3.17}$$

Figure 3.13 shows the two tension softening models at disposal in DIANA. The characteristic value of the tensile strength can be estimated by the relationship given in the CEB-FIP model code [3.17],

$$f_{ct,m} = 0.3\,f_{ck}^{2/3}\ (N/mm^2) \tag{3.18}$$

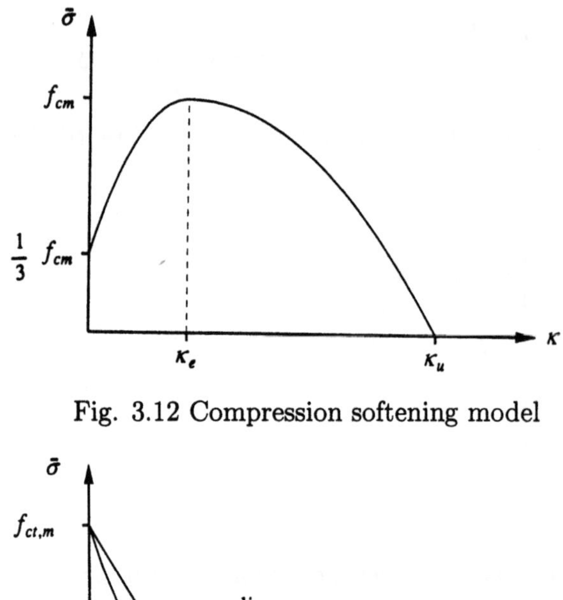

Fig. 3.12 Compression softening model

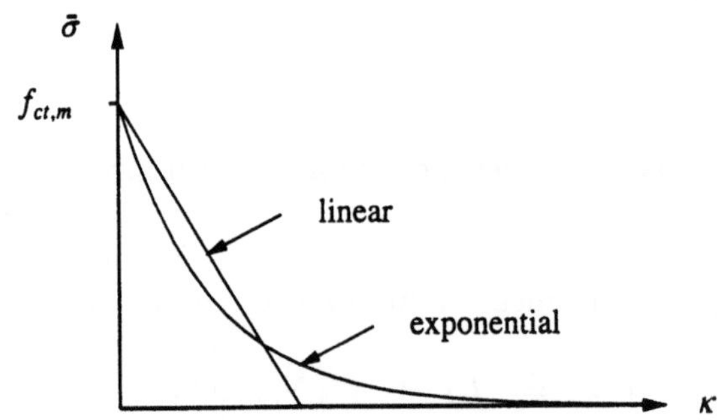

Fig. 3.13 Tension softening models

The value of G_f is given in the CEB-FIP model code in relation to the compressive strength of the material, f_{cm}, and the maximum aggregate size, d_{max}. The shape of the assumed softening curve is linear,

$$\overline{\sigma}(\kappa) = f_{ct,m}\,(1 - \frac{\kappa}{\kappa_u}) \tag{3.19}$$

or exponential,

$$\overline{\sigma}(\kappa) = f_{ct,m} \exp(-\frac{\kappa}{\kappa_u}) \tag{3.20}$$

Herein the ultimate damage parameter κ_u is calculated by,

$$\kappa_u = k \frac{G_f}{h f_{ct,m}} \tag{3.21}$$

with k=2 for linear-softening and k=1 for exponential softening. In case of reinforced concrete one has to replace the equivalent length h by the average crack spacing ℓ_s, assuming that ℓ_s is smaller than h. This spacing is two-third of the maximum crack spacing and according to the CEB-FIP model code [3.17] reads,

$$\ell_s = \frac{2}{3}\ell_{s,max} = \frac{2}{3}(2S_o + \frac{\phi_s}{\alpha\rho_s}) \tag{3.22}$$

where S_o = the minimum bond length, ϕ_s = the average diameter of the reinforcement, α = a factor equal to 4 for deformed bars and 2 for plain bars, and ρ_s = the reinforcing ratio. The minimum bond length S_o is usually taken equal to 25 mm, in the absence of more precise data.

In plane two-dimensional structures like panels, reinforcing grids of bars may be present in two orthogonal directions near the two faces. The CEB-FIP model code recommendations should be used in that case to find the effective reinforcing ratio $\rho_{s,eff}$ and the equivalent bar diameter ϕ_{eq}. When the cracks form at inclined angles with the reinforcement directions, Eq. 3.22 for the crack spacing cannot be used directly. The CEB-FIP model code recommends in this case to compute ℓ_s from,

$$\frac{1}{\ell_s} = \frac{|cos\alpha|}{\ell_{s,p}} + \frac{|sin\alpha|}{\ell_{s,q}} \tag{3.23}$$

where α denotes the angle between the reinforcement in the p-direction and the direction of the principal tensile stress at incipient cracking.

The failure surface in stress space can be written in the form of a yield function,

$$f(\sigma, \overline{\sigma}(\kappa)) = 0 \tag{3.24}$$

The inelastic strain vector is assumed to be given by an associated flow rule,

$$\epsilon_c = \dot{\lambda} \frac{\partial f}{\partial \sigma} \tag{3.25}$$

where $\dot{\lambda}$ is the inelastic multiplier which is determined by the additional constraint that $\dot{f} = 0$.

The yield function consists of two different parts, one bounding the tensile stresses, the other applicable in the compressive region:

$$\begin{aligned} f_1 &= 0 \\ f_2 &= 0 \end{aligned} \tag{3.26}$$

In this composite yield criterion the Rankine yield function is chosen for f_1 (tension), and the Von Mises yield function is applied to f_2 (compression). Figure 3.14 compares such a Rankine-Von Mises yield surface with experimental data [3.18].

For a complete description of f_1 and f_2 see Ref. [3.19]. The uni-axial compressive strength is chosen as 80 percent of the cube compressive strength, irrespective of the value of the transverse strain.

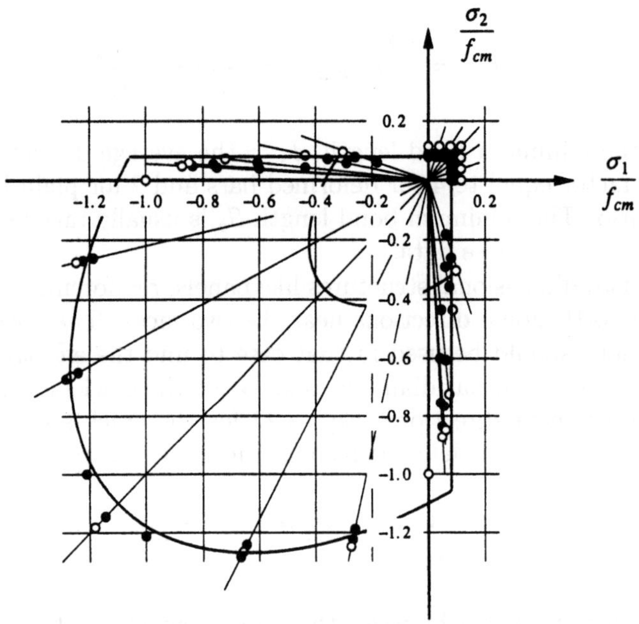

Fig. 3.14 Comparison of Rankine-Von Mises yield surface with experimental data [3.18], $\overline{\sigma}_1 = f_{ct,m}$, $\overline{\sigma}_2 = 1.1 f_{cm}$

3.2.3 Tension-Stiffening Model

After a stabilized crack pattern has developed, stresses are still transferred from reinforcement to concrete between cracks due to bond action, which increases the total stiffness of the structure. As indicated in Sect. 3.2.1, it is assumed that the stiffness of cracked reinforced concrete loaded in tension can be obtained by superposing the stiffnesses of plain concrete and the reinforcement, and an additional stiffness due to bond between concrete and steel. These three separate contributions are shown in Fig. 3.15. The tension-stiffening part is the interaction stress contribution σ_{ia}, which is assumed to be a function of strain in the direction of the reinforcement and will be active on an effective tension area as indicated in CEB-FIP model code [3.17]. In panels with two reinforcement layers the effective tension area is determined by the effective thickness h_{eff}, which reads,

$$h_{eff} = \min\left\{\frac{t}{2},\ 2.5(c+\frac{\phi_{eq}}{2})\right\} \tag{3.27}$$

where t = panel thickness, c = concrete cover on the reinforcement, and ϕ_{eq} = equivalent bar diameter.

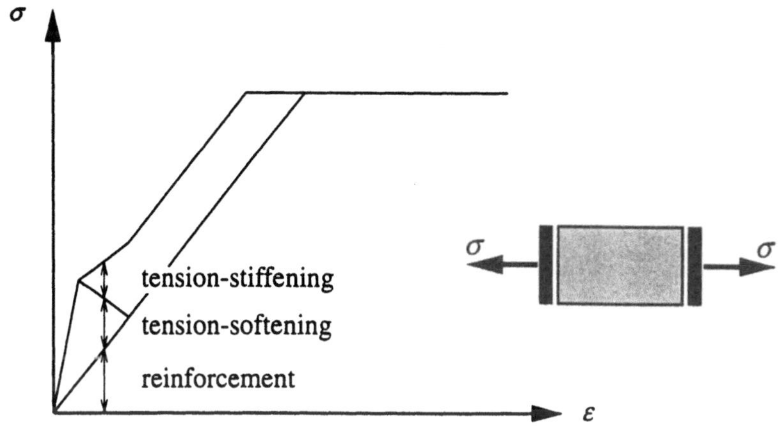

Fig. 3.15 Constitutive model of reinforced concrete
Schematical representation

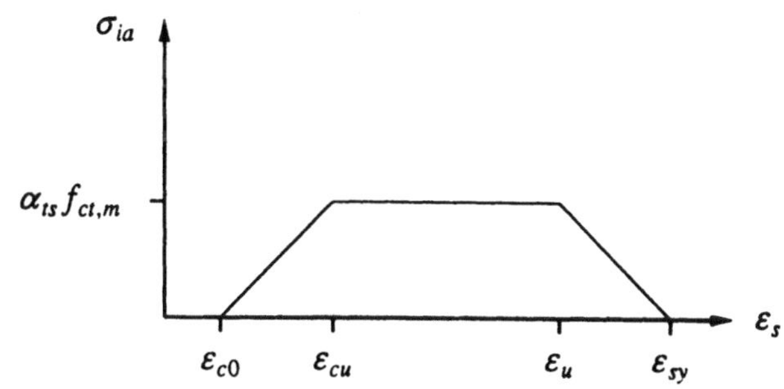

Fig. 3.16 Tension stiffening model

The interaction stress is assumed to be given by a tri-linear function according to Cervenka et al [3.20], see Fig. 3.16. The strain ϵ_{c0} in this figure is determined by,

$$\epsilon_{c0} = \frac{f_{ct,m}}{E_c} cos^2\alpha \tag{3.28}$$

where α = angle between the direction of the reinforcement and the direction of the principal stress at incipient cracking. The strain ϵ_{cu} in Fig. 3.16 is computed from

κ_u of Eq. 3.21,

$$\epsilon_{cu} \quad = \quad \kappa_u \; cos^2\alpha \tag{3.29}$$

The constant part of the diagram is approximating the tensile strength, i.e. $\alpha = 1.0$.

The strain at which the tension-stiffening starts to decrease is given by,

$$\epsilon_u \quad = \quad \epsilon_y - \frac{\alpha_{ts} f_{ct,m}}{\rho_{s,eff} E_s} \tag{3.30}$$

where ϵ_{sy} = yield strain of the reinforcing steel.

This tension-stiffening model has been validated by simulating a large number of panel tests [3.4,3.21,3.22], with different degrees of anisotropy of the reinforcement, defined by the ratio ω_r of the yield capacities of the two reinforcements. The model is satisfactory up to $\omega_r = 7$. A typical result is shown in Fig. 3.17 for panel pv 19 out of Vecchio/Collin's test series, for which ω_r is about 4.

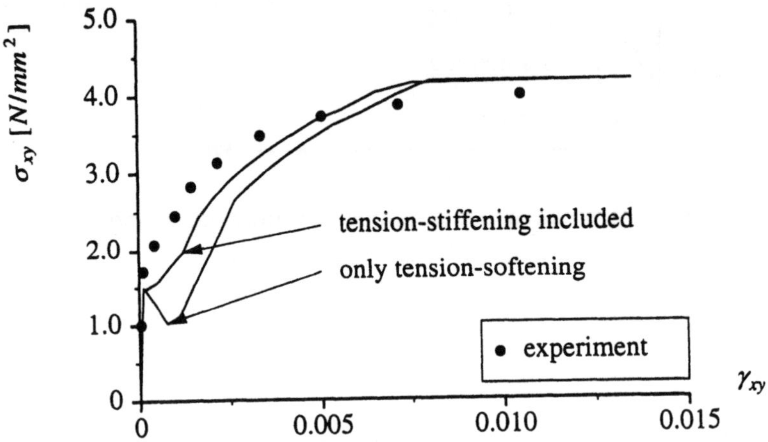

Fig. 3.17 Influence of tension-stiffening on panel pv19 of Vecchio and Collins [3.4]

3.2.4 Application to Panel

The potential of the panel model will be illustrated with the analysis of panel tests conducted by Maier and Thürlimann at the ETH Zürich [3.23]. The experimental set-up is shown in Fig. 3.18. The panels are square and have rigid top and bottom beams, of which the bottom beam is fixed. A horizontal force F_h is applied at the horizontal top beam, and a vertical load F_v can be applied simultaneously. Some panels in the series have been provided with flanges along the vertical edges. The finite element discretizations are shown in Fig. 3.19a for two panels with flanges and in Fig. 3.19b for two panels without flanges.

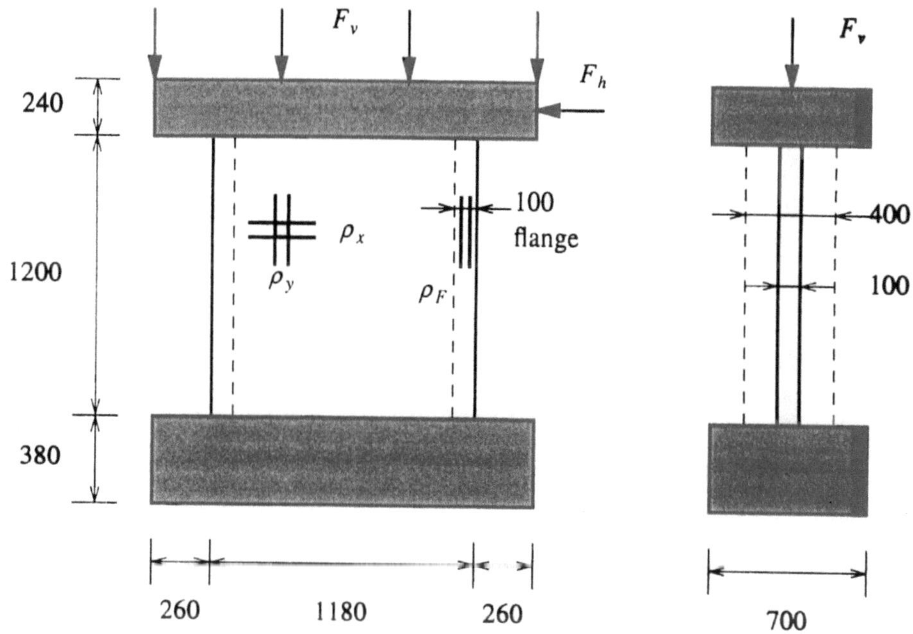

Fig. 3.18 Experimental set-up (dimensions in mm)

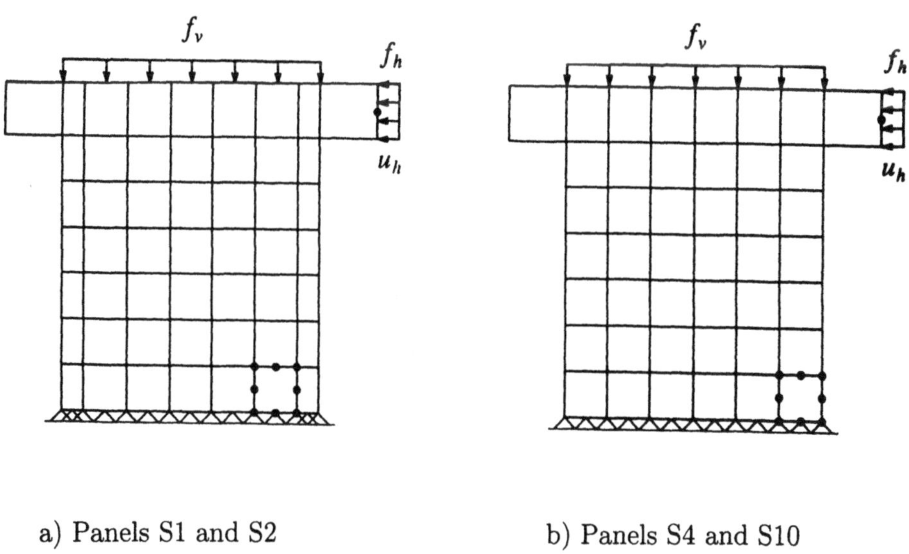

Fig. 3.19 Finite element discretization of Maier and Thürliman panels [3.23]

The material properties have been averaged from the experimental data provided by Maier and Thürlimann for the four panels, and the compressive strength was reduced by 20%. The material properties used in the analyses are as follows:

for concrete -
$f_{cm} = 27.5\ N/mm^2$
$E_c = 3000\ N/mm^2$
$\nu = 0.15$
$f_{ct,m} = 2.2\ N/mm^2$
$G_f = 0.07\ Nmm/mm^2$ (linear softening)
$G_c = 50.0\ Nmm/mm^2$ (parabolic softening)
for reinforcing steel
$f_{sy} = 574 - 764\ N/mm^2$ (hardening of steel with $\epsilon_{sv} = 24.6 \times 10^3$)
$E_s = 200,000\ N/mm^2$.

The material properties were averaged to simulate the behavior of the panels in a qualitative manner. In this way, it is easier to study the influence of the different reinforcing ratios, initial vertical stress and geometry. The reinforcing ratios and the initial vertical forces are as follows:

Panel	ρ_x	ρ_y	$\rho_F[10^{-3}]$	$F_v[kN]$
S1	10.3	11.6	11.6	-433
S2	10.3	11.6	11.6	-1653
S4	10.3	10.5	10.5	-262
S10	9.8	10.0	57.1[1)]	-262

1) additional reinforcement in tension area over 197 mm.

The 8 mm reinforcing bars were arranged in two orthogonal grid layers. The top slab has been modelled with linear-elastic elements without reinforcement, whereas the supporting block has been simulated by fixed supports in the x- and y-direction.

The load-displacement diagrams of the four panels are shown in Figs. 3.20 to 3.23. Panels S1 and S2 are identical flanged panels, but differ in the applied vertical loads that result in a vertical stress of 2.5 N/mm^2 in S1 and about 10.0 N/mm^2 in S2. Because of the higher vertical load panel S2 will be stronger but dramatically more brittle (with a rather explosive experimental failure mode).

The panels S4 and S10 without flanges have the same vertical load, but S10 has additional "hidden flange" reinforcement. As a consequence, panel S10 is stronger and slightly less ductile. In general the agreement between the analyses and tests is satisfactory to good.

The analysis results for the flanged panel S1 will be examined in more detail. The results are presented by plotting the active cracks, the integration points which are in a compressive plastic state, and the principal stresses which are in general compressive. The active cracks are defined as those cracks for which the internal parameter κ is equal to or greater than 0.5 κ_u. The results for a top slab displacement

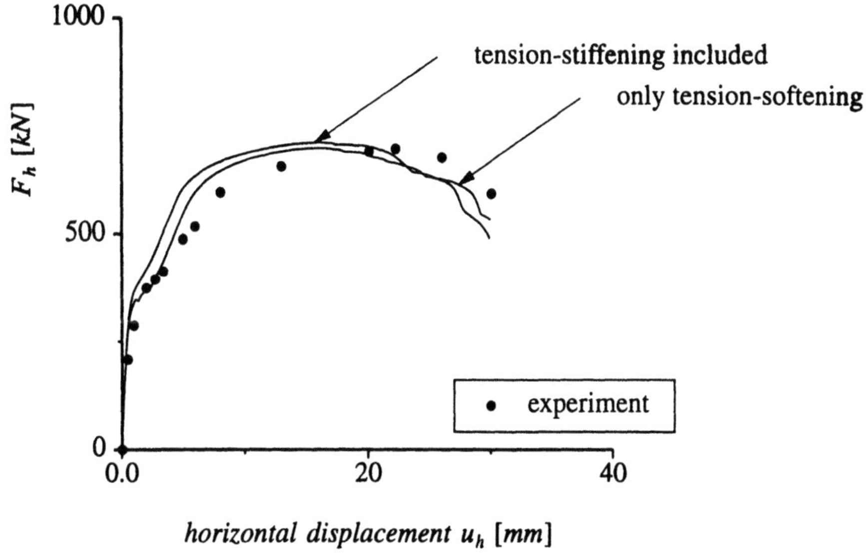

Fig. 3.20 Panel S1. Load-displacement diagram

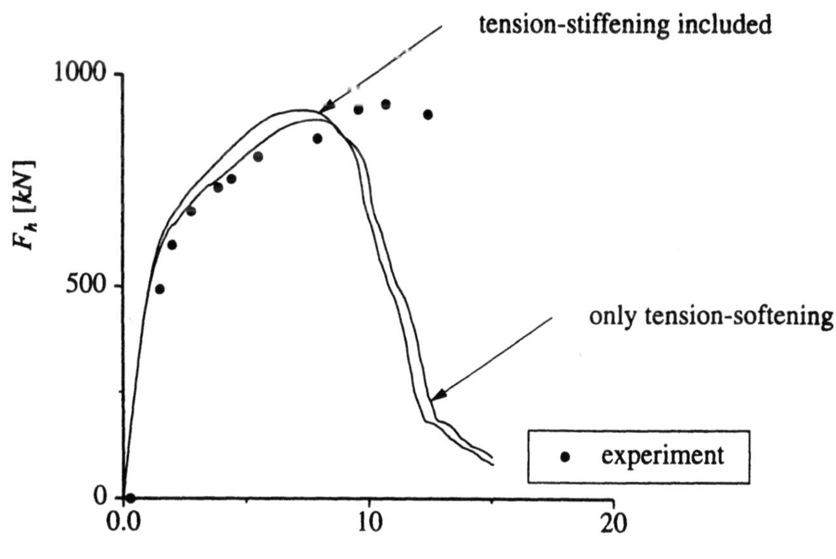

Fig. 3.21 Panel S2. Load-displacement diagram

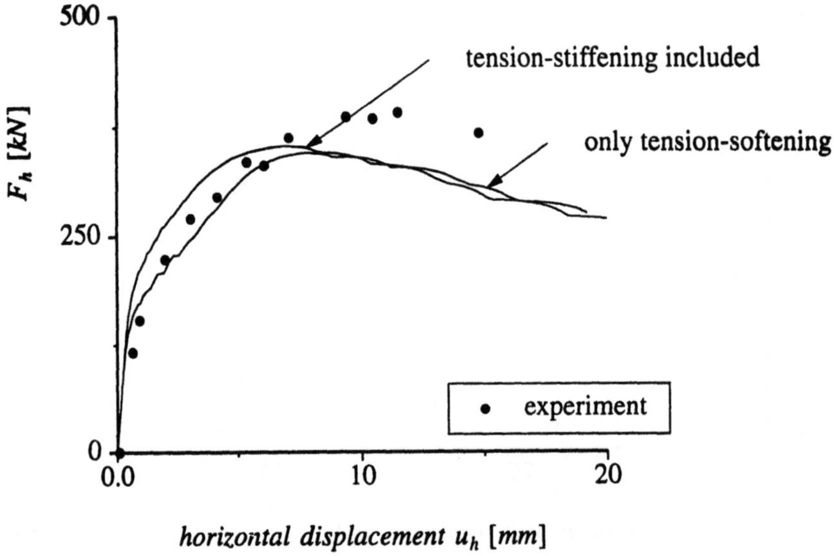

Fig. 3.22 Panel S4. Load-displacement diagram

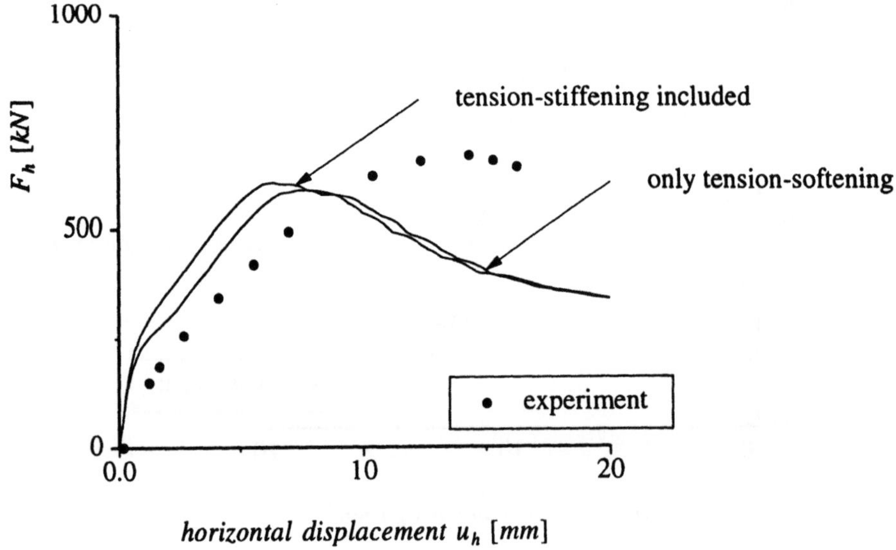

Fig. 3.23 Panel S10. Load-displacement diagram

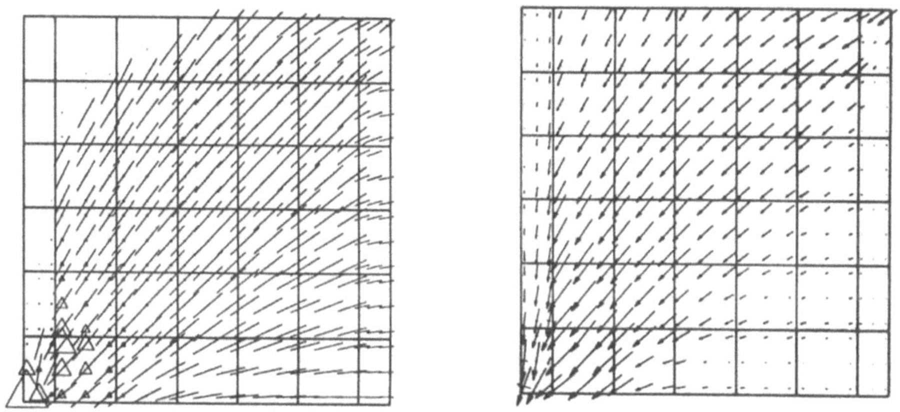

a) active cracks and plastic points b) principal stresses

Fig. 3.24 Panel S1. Analysis results at 10 mm displacement

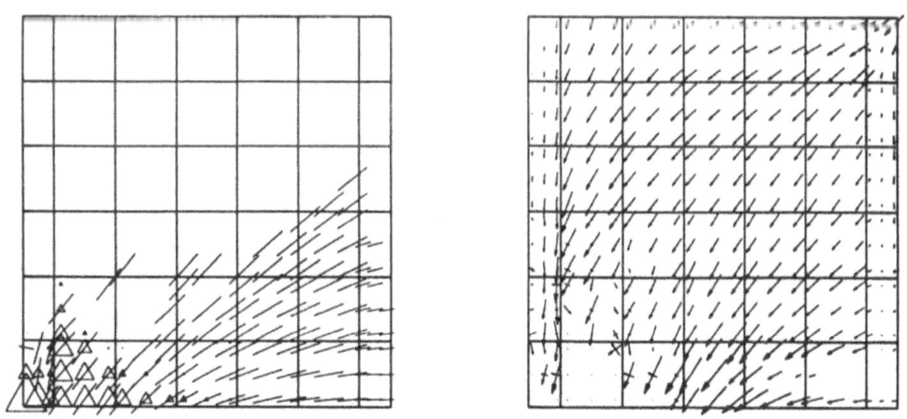

a) active cracks and plastic points b) principal stresses

Fig. 3.25 Panel S1. Analysis results at 30 mm displacement

of 10 mm are shown in Fig. 3.24. The panel is densely cracked with plastic points in the bottom-left corner of the panel. The concrete in the bottom-right corner does not transfer stress anymore, as can be seen from Fig. 3.24b. The principal stresses are clearly forming a compressive strut as the load-carrying mechanism. At the final displacement of 30 mm the panel exhibits a failure mechanism which is governed by compressive softening of the concrete and yielding of the reinforcing steel both in tension and compression, see Fig. 3.25. In this ultimate state, the concrete in the bottom-left corner transfers no stress anymore. This is in agreement with the experimentally observed failure mechanism in which the concrete was crushed in the bottom-left corner of the panel and in the flange on the compression side.

3.2.5 Summary and Conclusions

The structural engineer needs more and more reliable numerical tools to analyze the post-failure behavior of structures in order to assess their structural safety. In general, a reliable numerical tool consists of an accurate material description in combination with a robust solution strategy. In this study an attempt has been made to provide such a tool for the analysis of reinforced concrete structures.

A material model for plain and reinforced concrete in a plane-stress situation has been discussed. It is assumed that the failure mechanism of concrete loaded in tension and compression is governed by crack growth at the micro-level. Furthermore it is assumed that the internal damage caused by these micro-cracks can be modelled using internal parameters which are related to a fracture energy in tension and to a fracture energy in compression. The material properties have been estimated using CEB-FIP recommendations which are based on the compressive strength of the concrete. The comparisons between numerical analyses using material properties based on these recommendations and experiments show that estimated material properties are adequate. The biaxial behavior of concrete is assumed to be governed by the failure surface of Kupfer and Gerstle with a tension-softening and a compression-softening constitutive behavior.

The mechanical response of reinforced concrete is assumed to be given by a superposition of the elasto-plastic behavior of reinforcement, the tension-softening behavior of the concrete and an additional stiffness due the interaction between concrete and reinforcement. The total fracture energy due to distributed cracking is assessed using the average crack spacing based on CEB-FIP recommendations.

Application of the composite plasticity model to reinforced concrete structures with predominantly biaxial stress states shows that this model results in a stable and accurate procedure. The entire pre- and post-failure regime can be traced in a stable manner, which gives a good insight into the failure characteristics of these types of structures.

3.3 References

3.1 Park, R. and T. Paulay: Reinforced Concrete Structures, John Wiley and Sons, New York 1975.

3.2 Oesterle, R.G. et al: Earthquake Resistant Structural Walls - Tests of Isolated

Walls, Portland Cement Association, Skokie, IL, Nov. 1976.

3.3 Oesterle, R.G. et al: Earthquake Resistant Structural Walls - Tests of Isolated Walls, Phase II, Portland Cement Association, Skokie, IL, Oct. 1979.

3.4 Vecchio, F. and M.P. Collins: The Response of Reinforced Concrete to In-Plane Shear and Normal Stresses, Dept. of Civil Engin., University of Toronto, Publication No. 82-03, March 1982.

3.5 Collins, M.P., F.J. Vecchio, and G. Mehlhorn: An international competition to predict the response of reinforced concrete panels, Canadian Journal of Civil Engineering, 12 (1985), no 3, 624-644.

3.6 Stevens,N.J., S.M. Uzumeri,and M.P. Collins: Analytical Modelling of Reinforced Concrete Subjected to Monotonic and Reversed Loadings", Dept. of Civil Engin., University of Toronto, Publication No. 87-1, January 1987.

3.7 Stevens,N.J., S.M. Uzumeri,and M.P. Collins: Reinforced concrete subjected to reversed cyclic shear - experiments and constitutive model, ACI Structural Journal, March-April 1991.

3.8 Nilson, A. H., ed.: Finite Element Analysis of Reinforced Concrete, Special Publication, ASCE, New York, 1982.

3.9 Meyer, C. and H. Okamura, eds: Finite Element Analysis of Reinforced Concrete Structures, Special Publication, ASCE, New York, 1986.

3.10 Gupta, A.K. and R. Maestrini: Post-cracking behavior of membrane reinforced concrete elements including tension-stiffening, ASCE, Journal of Structural Engineering, 115 (1989), no 4.

3.11 Vecchio, F.J. and M.P. Collins: The modified compression field theory for reinforced concrete elements subjected to shear, ACI Journal, March-April 1986.

3.12 Simo, J.C. and J.W. Ju: Strain- and stress-based continuum damage models - II Computational aspects, Int. J. Solids and Structures 23 (1987), no 7, 841-869.

3.13 Xiong, S.: Anisotropic Damage Model for Inelastic Solids, Ph.D. Thesis, Columbia University, 1993.

3.14 Paskova, T.I.: Low-Cycle Fatigue Behavior of Concrete With and Without Fiber Reinforcement, Ph.D. Thesis, Columbia University, 1994.

3.15 Rots, J.G.: Computational Modelling of Concrete Fracture, Dissertation, Delft University of Technology, The Netherlands, 1988.

3.16 Vonk, R.A.: Softening of Concrete Loaded in Compression, Dissertation, Eindhoven University of Technology, The Netherlands, 1992.

3.17 CEB-FIP Model Code 1990, Bulletin d'Information, Comité Euro-International du Béton, 1990.

3.18 Kupfer, H.B. and K.H. Gerstle: Behavior of concrete under biaxial stresses, ASCE, J. Eng. Mech. 99 (1973), No. 4, 853-866.

3.19 Feenstra, P.H. and R. De Borst: Aspects of robust computational modelling for plain and reinforced concrete, Heron 38 (1993), no. 4.

3.20 Cervenka, V., R. Pukl, and R. Eligehausen: Computer simulation of anchoring technique in reinforced concrete beams, in: Computer Aided Analysis and Design of Concrete Structures (eds. N. Bicanic et al.), Pineridge Press, Swansea, 1990, 1-21.

3.21 Bhide, S.B. and M.P. Collins: Reinforced Concrete Elements in Shear and Tension, Publication 87-02, University of Toronto, Canada, 1987.
3.22 Kollegger, J. and G. Mehlhorn: Experimentelle Untersuchungen zur Bestimmung der Druckfestigkeit des gerissenen Stahlbetons bei einer Querzugbeanspruchung, Heft 413, Deutscher Ausschuss für Stahlbeton, Germany, 1990.
3.23 Maier, J. and B. Thürlimann: Bruchversuche an Stahlbetonscheiben, Report 8003-1, Eidgenössische Technische Hochschule Zürich, Switzerland, 1985.

MODELLING FOR DESIGN

T.P. Tassios
National Technical University of Athens, Athens, Greece

4.1 Modelling of RC Columns for a Given Ductility Factor

4.1.1 Introduction

In several current aseismic codes, the design of structures takes into account the so-called *behavior factor*, that is an estimator of the structure's ability to undergo large post-yield cyclic deformations with an acceptable, relatively small reduction of its bearing capacity. Under given conditions of redundancy and regularity the behavior factor assumed in design can be translated into required curvature ductility factors at the section level [4.1], which are normally ensured by adequate layout and sufficient quantity of transverse confining reinforcement (closed hoops). Herein, a model is presented for the calculation of the required transverse reinforcement of columns to achieve a given curvature ductility, $\mu_{1/r}$. In this model, the required transverse reinforcement is related to the main parameters that influence the curvature ductility, namely the normalized axial force, ν_d, the mechanical characteristics of the materials, the transverse reinforcement arrangement, as well as the ratio between the column's gross cross-sectional area A_c and its confined part, A_o.

In what follows, the curvature ductility of a square column section is defined as the ratio between the curvature at failure, $(d/r)_u$, and the curvature at yield, $(d/r)_y$, where r is the radius of curvature and d the conventional effective section depth. Design values for actions and resistances are used throughout this presentation.

4.1.2 Determination of Curvature at Yield

Two cases are distinguished, Fig. 4.1:

1. For relatively small axial load (below the balanced point), the tensioned steel

reaches first its yield strain, ϵ_{yd}, while the strain in the extreme concrete compression fiber is less than 0.0035, Fig. 4.1a. Assuming equal compression and tension reinforcement and expressing the stress of the compressed reinforcement as $\lambda' f_{yd}$ ($\lambda' \leq 1.0$), the yield curvature may be calculated as,

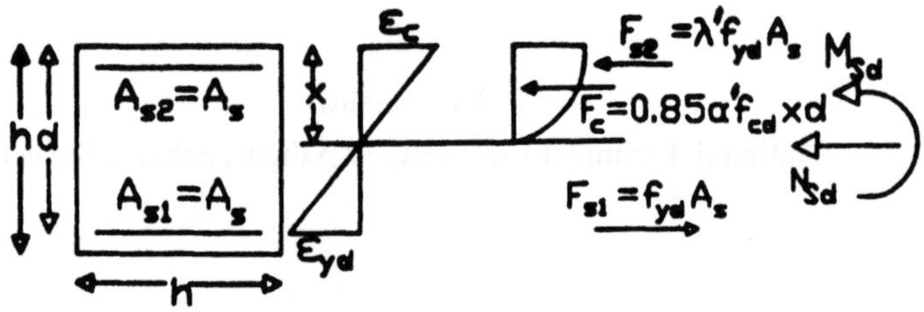

a) steel critical case

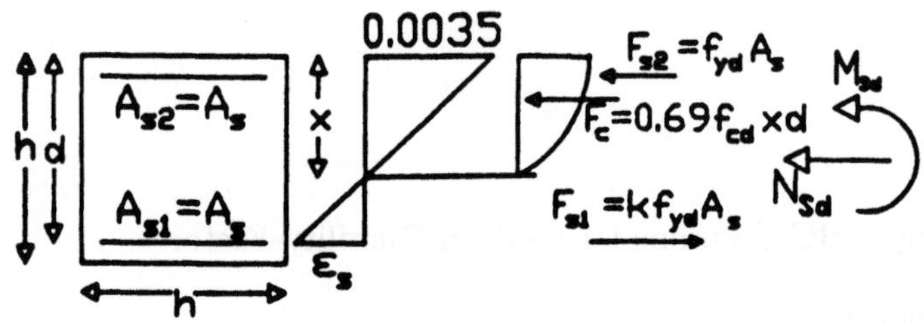

b) concrete critical case

Fig. 4.1 Internal forces and strains of square, symmetrically reinforced column at yield

$$(d/r)_y = \epsilon_c + \epsilon_{yd} \tag{4.1}$$

The axial design force follows as,

$$N_{sd} = \nu_d A_c f_{cd} = 0.85\alpha' f_{cd} x d - A_s f_{yd}(1 - \lambda') \tag{4.2}$$

where f_{cd} is the design value of the unconfined concrete's compressive strength, and x locates the neutral axis. Defining $\gamma = f_{yd}/f_{cd}$, $\rho = A_s/A_c$, and setting $d \approx 0.9h$, Eq. (4.1) can be rewritten as

$$(d/r)_y = \epsilon_{yd}\left(1 - \frac{\nu_d + \gamma\rho(1-\lambda')}{0.7\alpha'}\right)^{-1} \tag{4.3}$$

which indicates that, for given material and section properties, the yield curvature increases when the axial load increases.

2. For moderate to high axial loads, the compressed concrete reaches the strain 0.0035 before the tensioned steel is strained to ϵ_{yd}, Fig. 4.1b. Denoting the stress of the tensioned reinforcement as kf_{yd} ($k \leq 1.0$), it follows that,

$$(d/r)_y = \epsilon_s + 0.0035 \tag{4.4}$$

and with $\alpha' = 0.81$,

$$N_{sd} = \nu_d f_{cd} A_c = 0.69 f_{cd} x d + A_s f_{yd}(1-k) \tag{4.5}$$

Similarly to the preceding case, Eqs. (4.4) and (4.5) lead to the following expression for the curvature at yield,

$$(d/r)_y = \frac{0.002}{\nu_d - \rho\gamma(1-k)} \tag{4.6}$$

Contrary to Eq. (4.3), the yield curvature decreases now when the axial load increases.

Equations such as (4.3) and (4.6) have the advantage of describing the effect of the main parameters on $(d/r)_y$. The yield curvature depends also on the unknown strains, ϵ_c and ϵ_s, since α' and λ' (for Eq. (4.3)) and k (for Eq. (4.6)) depend on them. Furthermore, intermediate reinforcing bars, which are usually present even in columns with rather small cross-sectional dimensions, are not taken into account in the above expressions. Appropriate approximate improvements of those expressions can be made by means of practical simplifications. However, it is preferable to follow a less direct but essentially more precise method. An extensive parametric investigation was carried out for the derivation of a simplified yet safe expression for $(d/r)_y$. More than 50 combinations of parameters such as A_c/A_o, cross-sectional dimensions, arrangement of longitudinal reinforcement, reinforcement percentage, and properties of constituent materials were considered. The results of this parametric investigation are illustrated in Fig. 4.2. On the basis of these results, $(d/r)_y$ can be approximated as follows:

$$(d/r)_y = \epsilon_{yd}(1.4 + 3.2\nu_d) \quad \text{for } 0.10 \le \nu_d \le 0.50 \tag{4.7}$$

$$(d/r)_y = \epsilon_{yd} + 0.0035 \quad \text{for } 0.50 \le \nu_d \le 0.75 \tag{4.8}$$

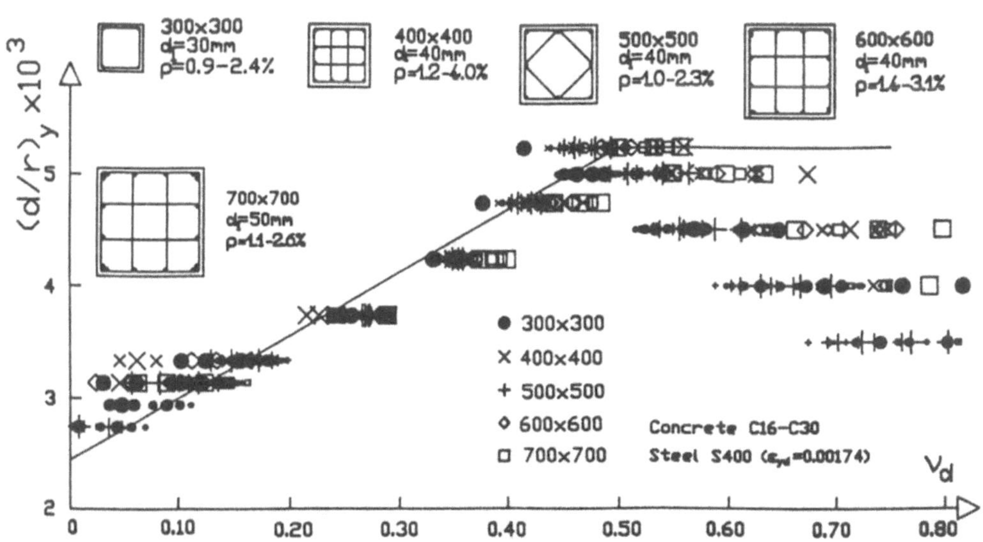

Fig. 4.2 Yield curvatures for square sections, derived from parametric investigation

4.1.3 Determination of Curvature at Failure

To determine the failure curvature of a section, the following assumptions are made:

1. The concrete has spalled, since the unconfined concrete has a strain larger than 0.0035;
2. The confined core concrete has the following characteristics (for the respective model, see [4.2]),
   - the maximum compressive strength of concrete is given by

$$f^*_{cd} = (1 + 2.5\alpha\omega_{wd})f_{cd} \quad \text{for } \alpha\omega_{wd} < 0.10 \qquad (4.9)$$
$$f^*_{cd} = (1.125 + 1.25\alpha\omega_{wd})f_{cd} \quad \text{for } \alpha\omega_{wd} > 0.10 \qquad (4.10)$$

   where ω_{wd} is the design value of the mechanical volumetric confinement reinforcement ratio, which is defined as $\omega_w = (V_s/V_c)(f_{sy}/f_{cd})$, with V_s = volume of confinement steel and V_c = volume of confined concrete; α is the confinement effectiveness factor and thus depends on the tie spacing and in-section configuration;
   - the compressive strain corresponding to f^*_{cd} is

$$\epsilon^*_{co} = \epsilon_{co}\left(\frac{f^*_{cd}}{f_{cd}}\right)^2 \qquad (4.11)$$

   where $\epsilon_{co} = 0.002$ for unconfined concrete;
   - the compressive strain at failure is

$$\epsilon^*_{cu} = \epsilon_{cu} + 0.1\alpha\omega_{wd} \qquad (4.12)$$

   where $\epsilon_{cu} = 0.0035$ for unconfined concrete;
3. The longitudinal reinforcement is placed symmetrically in the section, $\rho = \rho' = \rho_{tot}/2$. Furthermore, the conservative assumption is made that $\rho_{tot} = 0.04$ (the maximum allowed percentage of longitudinal steel);
4. Since the strains in both the compressed and tensioned steel may be large, hardening of both groups of bars is considered: for compressed steel, $f_{sc} = \delta f_{yd}$, and for tensioned steel, $f_{st} = \theta f_{yd}$;
5. Although the concrete cover has spalled at failure, the section continues to carry load, thanks to the contribution of the confinement steel, which increases the strength of the concrete core, as well as because of steel hardening; thus, the equation

$$N_{sd} = F_{cc} + F_{sc} - F_{st} \qquad (4.13)$$

   is valid for spalled sections as well.

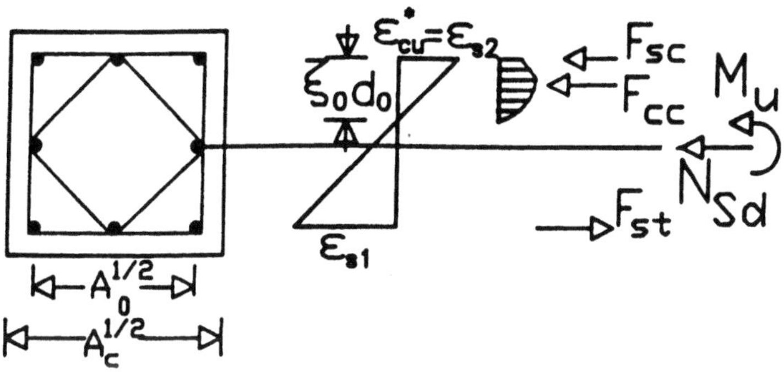

Fig. 4.3 Internal forces and strains after spalling, at failure

The failure curvature of the section after spalling (see Fig. 4.3) can be calculated as follows: $(1/r)_u = (\epsilon^*_{cu}/\xi_o d_o)$, i.e., $(d/r)_u = (\epsilon^*_{cu}/\xi_o)(d/d_o)$. With $d_o = \sqrt{A_o}$ and $d = (\sqrt{A_c}+\sqrt{A_o})/2$, we obtain $(d/d_o) = 0.5(\sqrt{A_c}+\sqrt{A_o})/\sqrt{A_o} = 0.5(\sqrt{\lambda}+1)$, where $\lambda = A_c/A_o$. Hence,

$$(d/r)_u = 0.5(\frac{\epsilon^*_{cu}}{\xi_o})(\sqrt{\lambda}+1) \tag{4.14}$$

An expression for ξ_o, which locates the neutral axis, may be found from Eq. (4.13). To this end, the compressive force carried by the confined concrete, F_{cc}, should be computed first (Fig. 4.4),

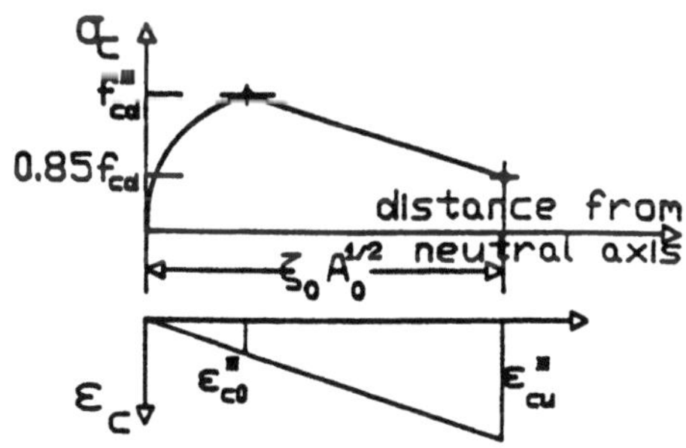

Fig. 4.4 Stress and strain distribution across compressed zone after spalling

$$F_{cc} = \frac{2}{3}\frac{\epsilon^*_{co}}{\epsilon^*_{cu}}\xi_o\sqrt{A_o}f^*_{cd}\sqrt{A_o} + 0.5(f^*_{cd}+0.85f_{cd})\sqrt{A_o}\xi_o\sqrt{A_o}(1-\frac{\epsilon^*_{co}}{\epsilon^*_{cu}}) \tag{4.15}$$

or

$$F_{cc} = \beta\xi_o A_o f_{cd} \tag{4.16}$$

where the coefficient

$$\beta = 0.5\frac{\epsilon^*_{co}}{\epsilon^*_{cu}}\left(\frac{1}{3}\frac{f^*_{cd}}{f_{cd}}+0.85\right)+0.5\left(\frac{f^*_{cd}}{f_{cd}}+0.85\right)$$

depends on the confining conditions of the section.

On the basis of Eq. (4.16) and with $\rho = \rho' = 0.02$, Eq. (4.13) becomes,

$$\lambda \nu_d = \beta \xi_o + 0.02\lambda(\delta - \theta)\frac{f_{yd}}{f_{cd}} \tag{4.17}$$

Thus, ξ_o can be calculated as follows,

$$\xi_o = \frac{\lambda}{\beta}\left[\nu_d - 0.02(\delta - \theta)\frac{f_{yd}}{f_{cd}}\right] \tag{4.18}$$

By substituting Eq. (4.18) into (4.14), the desired expression for the failure curvature is derived,

$$(d/r)_u = \frac{1}{2\lambda}\frac{(\sqrt{\lambda}+1)\epsilon^*_{cu}\beta}{\nu_d - 0.02(\delta - \theta)\frac{f_{yd}}{f_{cd}}} \tag{4.19}$$

4.1.4 Required Volumetric Mechanical Ratio of Transverse Reinforcement

The curvature ductility $\mu_{1/r}$ is defined as the ratio $(\frac{d}{r})_u/(\frac{d}{r})_y$, or

$$\mu_{1/r} = \frac{1}{2\lambda}\frac{(\sqrt{\lambda}+1)\beta(0.0035 + 0.1\alpha\omega_{wd})}{[\nu_d - 0.02(\delta - \theta)\frac{f_{yd}}{f_{cd}}](d/r)_y} \tag{4.20}$$

where the ductility curvature ratio $(d/r)_y$ was given by Eqs. (4.7) and (4.8).

The required confining reinforcement ratio, $\alpha\omega_{wd}$, follows then as,

$$\alpha\omega_{wd} = \frac{20\mu_{1/r}\lambda}{(\sqrt{\lambda}+1)\beta}[\nu_d - 0.02(\delta - \theta)\frac{f_{yd}}{f_{cd}}](d/r)_y - 0.035 \tag{4.21}$$

4.1.5 Application of the Model and Simplifications

The general equations of the preceding sections were used to calculate the required amounts of confinement reinforcement for the three levels of ductility specified in Eurocode 8, Part 1.3.2: high ("H"), medium ("M"), and low ("L"), with the respective curvature ductility ratios 13, 9, and 5. To evaluate Eq. (4.21) for these three ductility levels, the following simplifications are made:

1. As shown in Eqs. (4.15) and (4.16), the coefficient β can be calculated if the confinement ratio is known, but $\alpha\omega_{wd}$ is the unknown in the model. To overcome this problem, a parametric investigation was carried out. Recognizing that higher ductility levels require higher $\alpha\omega_{wd}$-values, which in turn are associated with higher β-values, the following values are suggested:

$$\begin{aligned} \beta &= 1.15 \quad \text{for ductility category H} \\ & 1.0 \quad \text{for ductility category M} \\ & 0.90 \quad \text{for ductility category L} \end{aligned} \tag{4.22}$$

2. $\alpha\omega_{wd}$ depends on the material properties, Eq. (4.21). At this stage, C20 and S400 were considered, as they represent a widely used combination of concrete and steel, while giving rather conservative $\alpha\omega_{wd}$-values.
3. The required confinement reinforcement depends also on the stress-strain relationship of the longitudinal steel. Adopting the relationship of Fig. 4.5, it can be shown that the term $(\delta - \theta)$ which indicates the level of hardening of the tensioned and compressed steel, varies between -0.03 and +0.04, increasing for increasing ν_d-values.
4. The above assumptions 2 and 3 lead to a simplification in that the term $[\nu_d - 0.02(\delta-\theta)\frac{f_{yd}}{f_{cd}}]$ in Eq. (4.21) can be approximated satisfactorily by the normalized axial force n_d.
5. The term $\lambda/(\sqrt{\lambda}+1)$ is very close to the straight line $(0.35\lambda + 0.15)$.

   Thus, the following simplified equation is derived,

$$\alpha\omega_{wd} = \frac{20\mu_{1/r}}{\beta}(0.35\lambda + 0.15)\nu_d(d/r)_y - 0.035 \tag{4.23}$$

This result is plotted in Fig. 4.6a for S400 steel and $\lambda = 1.40$ for the three ductility categories. If in addition, the term $0.35\lambda + 0.15$ is approximated by $0.5\ \lambda$ $(1.1 \leq \lambda \leq 1.5)$, the relationships plotted in Fig. 4.6b are obtained, which are described by the equation,

$$\alpha\omega_{wd} = k_o\mu_{1/r}\epsilon_{yd}\lambda n_d - 10\epsilon_{cu} \tag{4.24}$$

where

$\nu_d \leq 0.55,\quad k_o = 24$ for ductility category H

$\nu_d \leq 0.65,\quad k_o = 27$ for ductility category M

$\nu_d \leq 0.75,\quad k_o = 30$ for ductility category L

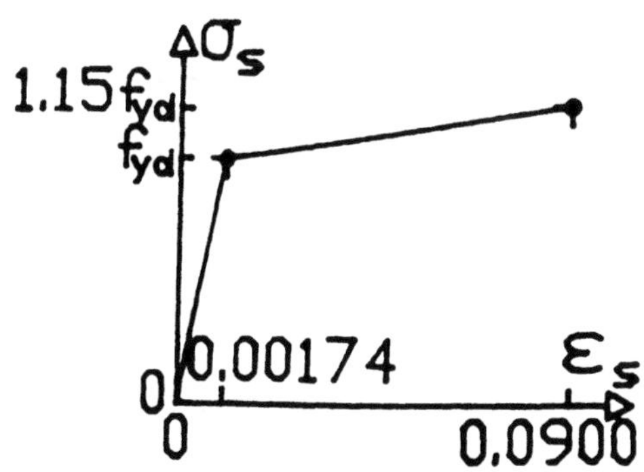

Fig. 4.5 Stress-strain curve for longitudinal steel

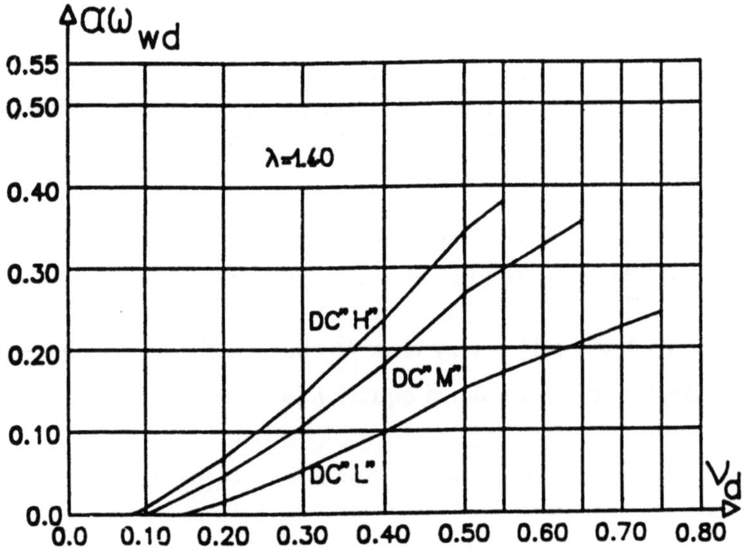

a) derived from model

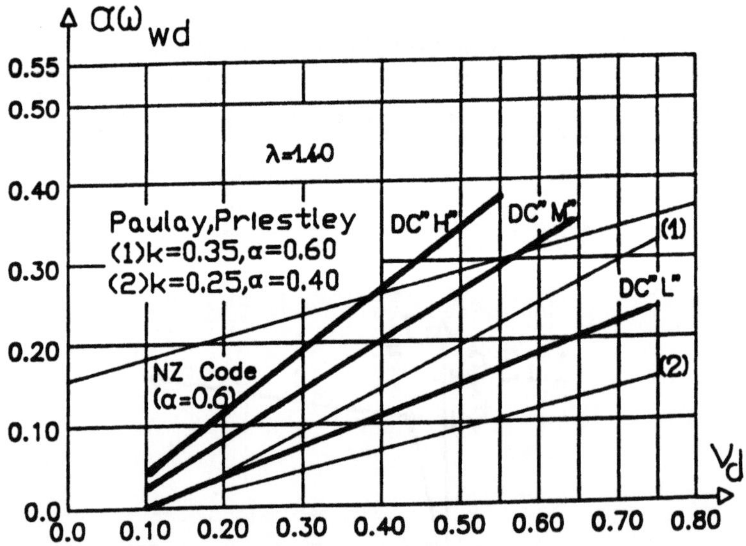

b) comparison between EC8, NZ Code, and formula by Paulay and Priestley, 1993

Fig. 4.6 Required confinement reinforcement vs normalized axial force

Also shown in Fig. 4.6b are the corresponding requirements of the New Zealand Code [4.3] and the values resulting from a formula proposed by Paulay and Priestley [4.4]. One may observe that the results of the model presented here seem to be considerably more conservative than those of Paulay and Priestley, and that the New Zealand Code formula, which was derived on the basis of results for columns with $\mu_{1/r}$ values of at least 20, seems to require high confining reinforcement ratios at low axial force values.

4.2 Modelling of Beam-Column Joints

4.2.1 The Vulnerability of Beam-Column Joints

One of the most vulnerable regions of RC frame structures under seismic loading are the beam-column joints. Their importance was underestimated for many decades. Even now, seismic code provisions for joint design are not coordinated internationally. The complexity of the performance of such joints cannot be overemphasized. In fact, the overall behavior of a joint depends on,

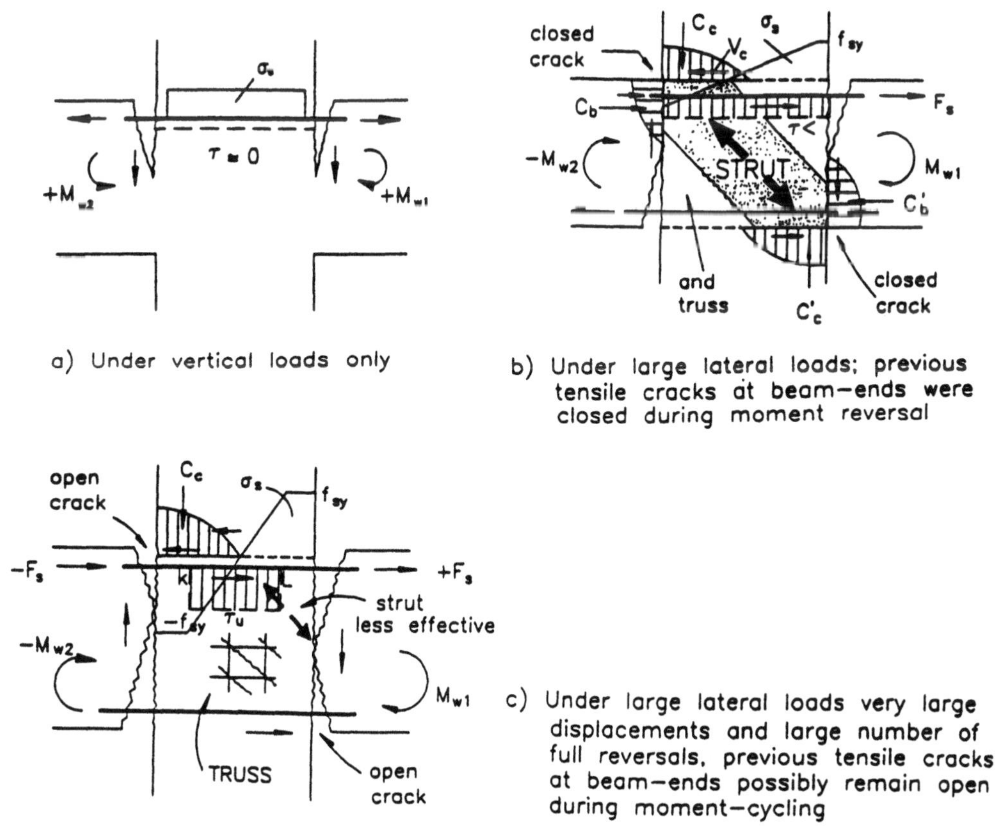

a) Under vertical loads only

b) Under large lateral loads; previous tensile cracks at beam−ends were closed during moment reversal

c) Under large lateral loads very large displacements and large number of full reversals, previous tensile cracks at beam−ends possibly remain open during moment−cycling

Fig. 4.7 Beam-column joint (schematic)

a) the anchorage of longitudinal bars of the beams, under cyclic push/pull action;
b) the integrity of the joint core itself, under cyclic shear.

Figure 4.7 is an attempt to summarize this issue in a schematic way. Under moderate seismic load (or in the case that structural walls do not allow large floor displacements and corresponding moment reversals at the beam-ends) compressive forces may be transferred through the concrete after full closure of tensile cracks created by a previous load cycle (Fig. 4.7b). In such a case, equilibrium of the longitudinal bars of the beams may be ensured thanks to sufficient bond developed within the joint; transverse compression due to the column axial load is beneficial for satisfying the relatively high demand on bond strength, even if cyclic action causes this strength to degrade.

On the other hand, shear transfer through the joint core is ensured by a diagonal concrete strut, since the two force components C_b and C_c enable this strut to form, Fig. 4.7b. In this situation, two favorable consequences can be utilized for design: adequate bond may be available without extraordinary measures, and truss mechanisms for shear transfer through the joint are not pronounced. Therefore, low shear reinforcement percentages are required in the joint core.

For more severe loadings, i.e., if large displacements are imposed together with a large number of full load reversals, tensile cracks at the beam-joint interface may not close during subsequent load cycles, Fig. 4.7c. In this case compressive forces at the beam-end will be transferred to the joint only by means of the reinforcement, and cyclic pull-out/push-in action of the longitudinal bars may have the following doubly unfavorable results:

a) due to large slip reversals, bond strength may degrade rapidly;
b) the total force acting on the longitudinal bars is now almost twice as high as in the previous situation.

As a consequence, yielding will rapidly penetrate the joint, and the local bond demand over the remaining length from K to L, Fig. 4.7c, may exceed the available strength. In addition, the horizontal compressive force C_b does not exist anymore, so that the diagonal shear transfer strut cannot form. Exacerbated by the local concentration of high bond stresses, this lack of truss action requires drastic measures of joint design for such unfavorable conditions. Very small diameters should be used for the longitudinal bars and considerable amounts of shear reinforcement provided in the joint core.

For actual designs, both situations should be considered and verified appropriately. To this end, the following analytical model may be helpful.

4.2.2 Modelling of the Diagonal Strut

The diagonal strut of Fig. 4.7b is subject to axial compression σ_{cc} and transverse tension σ_{ct}, acting simultaneously, Fig. 4.8a. The failure criterion may be expressed by the following simplified equation, Fig. 4.8b,

$$\frac{\sigma_{cc}}{f_{cc}} + \frac{\sigma_{ct}}{f_{ct}} = 1 \tag{4.25}$$

Assuming that in this case, $\sigma_{ct} = \sigma_{cc}/3$ and $f_{ct} = 0.10 f_{cc}$, one obtains $\sigma_{cc} \approx \frac{1}{4.4} f_{cc}$, and with $f_{ct} = 1.5 f_{ct,0.05}$,

$$(\tau_{jh})_R = 20\tau_{Rd}$$

Thus, it should be verified that

$$\overline{\tau}_{jh} \leq 20\tau_{Rd} \tag{4.26}$$

where according to the Eurocode 2 [4.5],

$$\tau_{Rd} = \frac{1}{4}\frac{f_{ct,0.05}}{\gamma_c}$$

and

$$\overline{\tau}_{jh} = \frac{\gamma_{Rd}(\frac{M_{w1,R}}{z_1} + \frac{M_{w2,R}}{z_2}) - V_c}{b_j h_{jc}} \tag{4.27}$$

with $M_{w1,R}, M_{w2,R}$ = flexural moment capacities of the beam ends; z_1, z_2 = internal lever arms; γ_{Rd} = a factor accounting for possible steel hardening; V_c = shear force at the column ends. $\overline{\tau}_{jh}$ is the total shear stress acting in the joint; this has to be distinguished from τ_h, the "nominal" shear stress used in the next section.

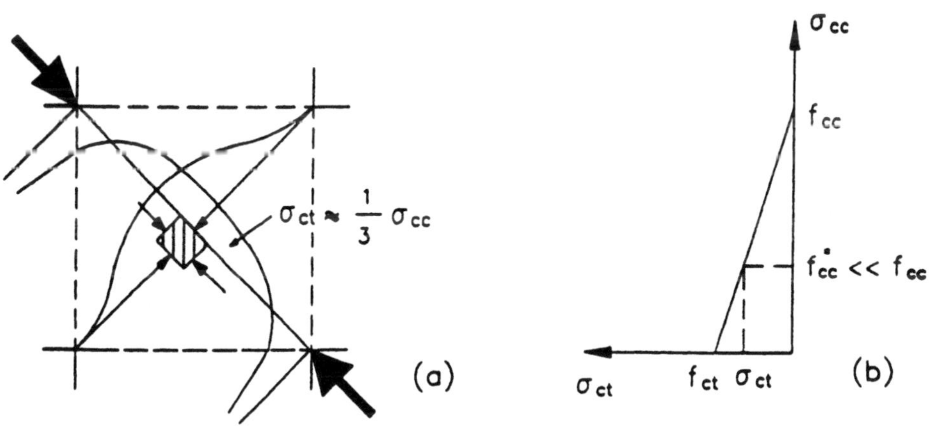

Fig. 4.8 Diagonal strut mechanism of force transfer through a beam-column joint

4.2.3 Forces Acting On Joint Core When Flexural Cracks Remain Open

Let us consider the case of a joint when the flexural cracks remain open during a load cycle. As mentioned earlier, equilibrium in this case cannot be achieved with a one-strut mechanism alone; spread shear forces are needed within the joint, which cause diagonal tension and, eventually, undesirable diagonal cracks. We shall formulate the conditions under which such cracks can form.

Referring to Fig. 4.9, the stresses acting on the joint can be categorized as horizontal shear stresses, vertical normal stresses, and horizontal normal stresses.

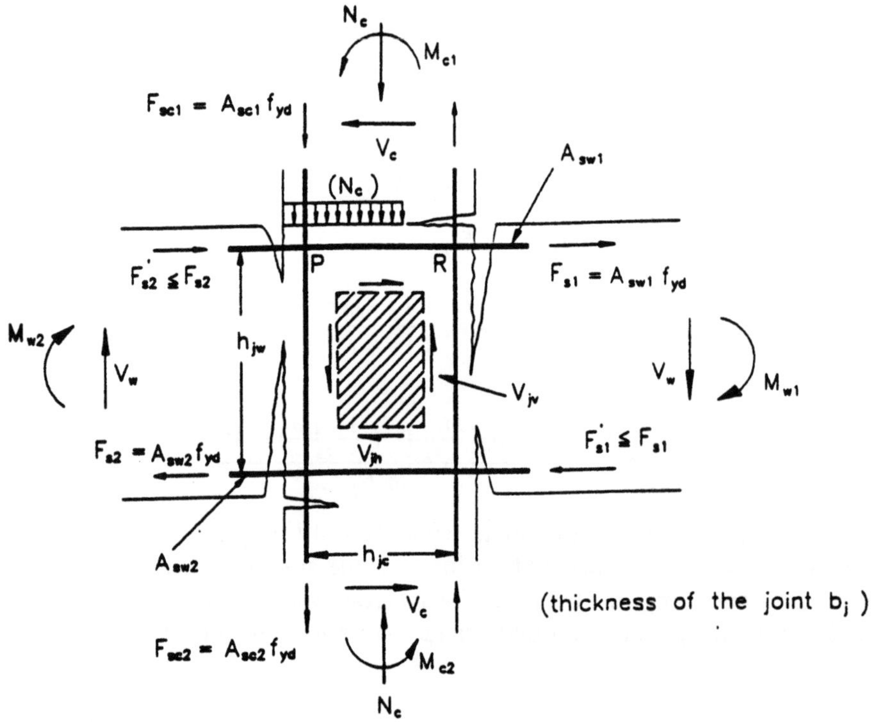

Fig. 4.9 Joint with open flexural cracks

Horizontal shear stresses. Bond stresses along the horizontal bar PR (Fig. 4.9) must equal the sum of the forces acting on the bar at P and R, $F_{s1} + F'_{s2}$. Part of these bond stresses spread into the upper column, where they are equilibrated by the column stirrups and reenter the joint via the inclined strut. They are eventually directed closer to the corner point R, where they are more easily resisted by dowel action of the horizontal and vertical bars. For this reason, we shall reduce the demand on the bond stresses by a factor 2/3. Thus the joint shear force becomes

$$V_{jh} = \frac{2}{3}(F_{s1} + F'_{s2}) - V_c \tag{4.28}$$

with $F_{s1} = A_{sw1} f_{yd}$ and $F'_{s2} = \lambda A_{sw2} f_{yd}$. λ is a coefficient that accounts for the limited compression force resisted by an open rough crack. It tends to be unity for large plastic excursions and is approximated by

$$\lambda = \frac{q}{q_{max}} = \frac{q}{5} \tag{4.29}$$

where q is the behavior factor used in the analysis. With this, the "nominal" horizontal shear stress becomes

$$\tau_h = \frac{V_{jh}}{b_j h_{jc}} = \gamma_{Rd} \frac{\frac{2}{3}(A_{sw1} + \frac{q}{5} A_{sw2}) f_{yd} - V_c}{b_j h_{jc}} \tag{4.30}$$

where γ_{Rd} is the steel hardening factor, and b_j, h_{jc} are the dimensions of the joint core.

Vertical normal stresses. In the case that the column reinforcement is stressed to yield on both sides (equilibrating each other), it will be assumed that half of the axial force N_c causes shear stresses along the vertical side of the core, while the other half causes vertical normal stresses across the joint core.

The Poisson effect of the horizontal normal stresses acting on the joint, as well as its dilatancy due to some limited shear slip along the cracks that will eventually appear, tend to produce vertical expansion of the core concrete. Such expansion is resisted by the vertical column reinforcement, without taking into account corner bars. Thus, intermediate longitudinal bars of area A_{sci} provided on the two opposite faces of the column, offer a vertical "confining" force of approximately $A_{sci} f_{yd}$. With this, the vertical normal stresses become,

$$\sigma_v \approx \frac{0.5N_c + A_{sci} f_{yd}}{b_j h_{jc}} = \frac{1}{2}\nu_d f_{cd} + \frac{A_{sci} f_{yd}}{b_j h_{jc}} \tag{4.31}$$

where the normalized axial force, $\nu_d = N_c/A_c f_{cd}$, had been introduced in Sect. 4.1.

Horizontal normal stresses. The Poisson effect of vertical normal stresses, together with the dilatancy of the core due to shear slip along the cracks, tends to cause horizontal expansion of the core concrete, which is resisted by the horizontal reinforcement A_{sh} in the joint. Thus, a horizontal confining force of $A_{sh} f_{yd}$ is generated, which corresponds to a nominal horizontal stress,

$$\sigma_h \approx \frac{A_{sh} f_{yd}}{b_j h_{jw}} \tag{4.32}$$

4.2.4 Conditions for Limitation of Diagonal Cracking

Because of the importance of the joint as a force transfer zone, its integrity must be maintained. This calls for quasi-elastic behavior, which is deemed to be satisfied if the principal tension σ_{ct} in the core concrete does not exceed the mean value of the concrete's tensile strength, f_{ctm}, i.e.,

$$\sigma_{ct} \leq \frac{f_{ctm}}{\gamma_c} \tag{4.33}$$

It is expected that under this condition cracking of only limited extent will appear.

The requirement of Eq. (4.33) may be rewritten in terms of τ_{Rd} [4.5] as,

$$\sigma_{ct} \leq \frac{f_{ctm}}{\gamma_c} \approx \frac{1.5 f_{ct,0.05}}{\gamma_c} \approx 6\tau_{Rd} \tag{4.34}$$

The principal tension, σ_{ct}, has now to be expressed in terms of stress components σ_h, σ_v, and τ_h. According to Fig. 4.10, Eq. (4.34) becomes

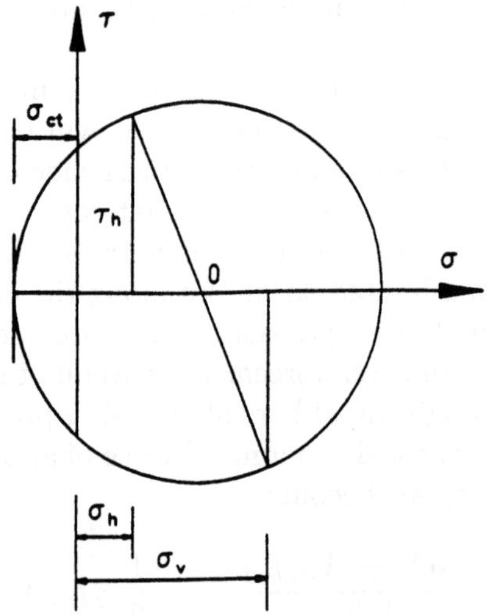

Fig. 4.10 Mohr's circle for concrete state of stress in joint core

$$\sigma_{ct} \;=\; \sqrt{(\frac{\sigma_v - \sigma_h}{2})^2 + \tau_h^2} \;-\; \frac{\sigma_v + \sigma_h}{2} \;\leq\; 6\tau_{Rd}$$

or

$$\tau_h^2 \;\leq\; 36\tau_{Rd}^2 + 6\tau_{Rd}(\sigma_h + \sigma_v) + \sigma_v\sigma_h \tag{4.35}$$

We shall now assume that an adequate number of intermediate bars is provided on the column sides, such that

$$\frac{A_{sci}}{h_{jc}} \;\geq\; \frac{A_{sh}}{h_{jw}}$$

In this case we can set conservatively,

$$\frac{A_{sci}}{h_{jc}} \;=\; \frac{A_{sh}}{h_{jw}} \tag{4.36}$$

Equation (4.31) can now be written, with Eq. (4.32), as

$$\sigma_v \;=\; \frac{1}{2}\nu_d f_{cd} + \sigma_h \tag{4.37}$$

Thus, Eq. (4.35) becomes

$$\tau_h^2 \;\leq\; 36\tau_{Rd}^2 + 6\tau_{Rd}(\frac{1}{2}\nu_d f_{cd} + 2\sigma_h) \;+\; (\frac{1}{2}\nu_d f_{cd} + \sigma_h)\sigma_h \tag{4.38}$$

or, setting $\tau_{Rd} \approx 0.015 f_{cd}$,

$$\tau_h^2 \leq \sigma_h^2 + \sigma_h (0.18 + \frac{1}{2}\nu_d) f_{cd} + 0.045 f_{cd}^2 (0.18 + \nu_d) \tag{4.39}$$

A parametric investigation of Eq. (4.39) for various values of f_{cd} and n_d has shown that the relationship between σ_h and τ_h is practically linear. On the basis of this finding, the following simplified formula is proposed for a safe relationship between τ_h and σ_h,

$$\sigma_h \geq \tau_h - 1.5\sqrt{\tau_{Rd}(12\tau_{Rd} + \nu_d f_{cd})} \tag{4.40}$$

The second term on the right-hand side of Eq. (4.40) expresses the contribution of concrete to the shear resistance of the joint. For design purposes, this contribution should be reduced (say, by one third) to allow for the degradation of shear transfer mechanisms during load reversals. With that, the following equation is proposed for the design of joints,

$$\sigma_h = \frac{A_{sh} f_{yd}}{b_j h_{jw}} \geq \tau_h - \frac{2}{3}\, 1.5\sqrt{\tau_{Rd}(12\tau_{Rd} + \nu_d f_{cd})}$$

that is, the horizontal shear reinforcement requirement becomes,

$$\frac{A_{sh} f_{yd}}{b_j h_{jw}} \geq \tau_h - \sqrt{\tau_{Rd}(12\tau_{Rd} + \nu_d f_{cd})} \tag{4.41}$$

It should be recalled that, because of the assumption of Eq. (4.36), the following requirement for vertical reinforcement is to be verified as well,

$$A_{sci} \geq \frac{h_{jc}}{h_{jw}} A_{sh} \tag{4.42}$$

Note that instead of the confining action of the joint reinforcement, the following truss mechanism could be considered: spread diagonal compression and tension forces along the joint core are resisted by the horizontal and vertical reinforcement. However, rather pronounced diagonal cracking is needed to activate such a mechanism, and such cracking is not desirable in beam-column joints. For this reason it was considered preferable to model the relevant structural behavior through the confinement mechanism.

4.2.5 Beam Reinforcement Anchorage in the Joint

As mentioned in Sect. 4.2.1, considerable yield penetration and cyclic bond degradation along the bars anchored within a beam-column joint may lead to considerable bar slip. In this case, the building will undergo excessively large displacements without significant energy dissipation and with unforseeable second-order effects. For this reason, it is of paramount importance to verify the beam reinforcement anchorage within the width of the column. A relatively simple model is proposed below.

Referring to Fig. 4.11, the most adverse condition exists when the beam-end cracks are open. For a completely open crack, the compressive force acting on the upper reinforcement and to be transferred is,

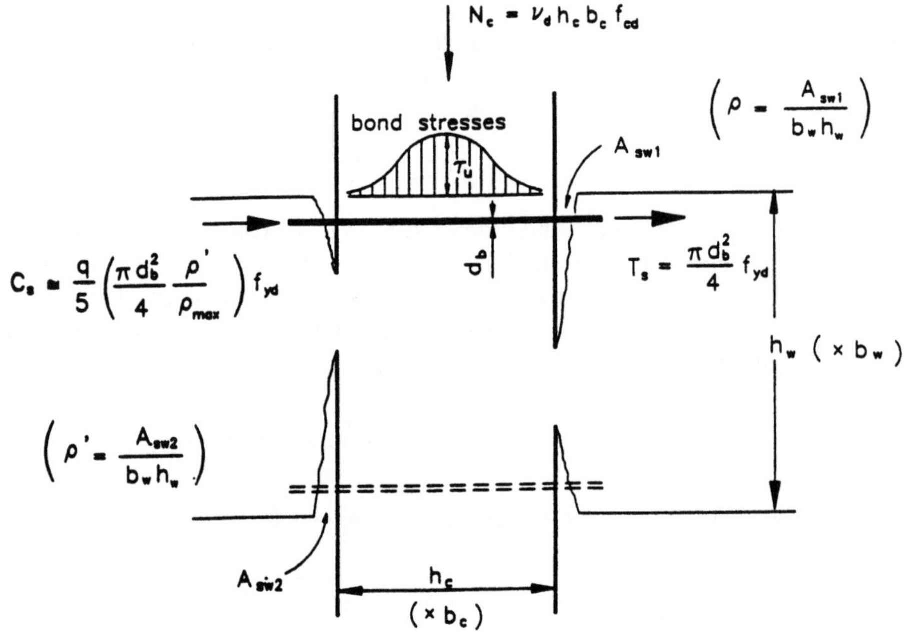

Fig. 4.11 Loading and anchorage conditions of longitudinal beam reinforcement

$$C = \frac{\pi d_b^2}{4}\frac{\rho'}{\rho} f_{yd} \tag{4.43}$$

If we consider the crack re-closure to be accompanied by small plastic excursions (i.e. small q/q_{max}-values, with $q_{max} = 5$) and small steel ratios (i.e. small ρ/ρ_{max}-values, where ρ_{max} denotes the maximum allowable steel ratio for the beam), then we can account for the small compressive stresses in the concrete due to the roughness of the crack. Thus, the compression force of Eq. (4.43) to be transferred may be modified as follows,

$$C_s = \frac{\pi d_b^2}{4}\left(\frac{q}{5}\frac{\rho}{\rho_{max}}\right)\frac{\rho'}{\rho} f_{yd} \tag{4.44}$$

and the total force to be transferred becomes,

$$F = \gamma_{Rd}(C_s + T_s) = \gamma_{Rd}\frac{\pi d_b^2}{4} f_{yd}\left(1 + \frac{q}{5}\frac{\rho'}{\rho_{max}}\right) \tag{4.45}$$

where again, γ_{Rd} reflects the possible hardening of steel.

The resisting force has two components. With the regular bond strength $\tau_u = 3.5 f_{ctm}$, and column depth h_c, the resisting force due to bond is $\frac{2}{3}\tau_u 0.8 h_c \pi d_b$. The axial force of the column, N_c, causes the frictional resistance $\mu_f(\frac{N_c}{h_c b_c})0.8 h_c d_b$, where μ_f is the friction coefficient, and it was accounted for the fact that limited yield penetration reduces the effective resisting bar length to $0.8h_c$. The total resisting force becomes then,

$$\begin{aligned} R &= \frac{2}{3}\tau_u 0.8 h_c \pi d_b \; + \; \mu_f \left(\frac{N_c}{h_c b_c} \cdot \frac{d_b}{b_w} \right) 0.8 h_c d_b \\ &= 0.8 h_c d_b f_{ctm} \left(\frac{2}{3}\, 3.5\pi + \mu_f \nu_d \frac{f_{cd}}{f_{ctm}} \right) \end{aligned} \tag{4.46}$$

For a safe joint design, the resistance R needs to be at least equal to the force demand F. After some manipulation of the resulting expression and setting $\mu_f \approx 1$, we obtain the following approximate design requirement,

$$\frac{d_b}{h_c} \;\le\; \frac{7.5 f_{ctm}}{\gamma_{Rd} f_{yd}} \frac{1 + 0.8\nu_d}{1 + \frac{q}{5}\frac{\rho'}{\rho_{max}}} \tag{4.47}$$

Substituing practical values, $\gamma_{Rd} = 1.35$, $f_{ctm} = 2.5$MPa, $f_{yd} = 350$MPa, $n_d = 0.4$, $q = 3$, $\rho'/\rho_{max} = 0.5$, one finds,

$$d_b \;\le\; \frac{h_c}{25}$$

4.3 Estimation of Behavior Factors of RC Structures

4.3.1 The Concept of Behavior Factor

Definition: The so-called "behavior factor", q, is an oversimplified means to produce approximate elastoplastic spectra, by simply dividing the ordinates of corresponding elastic spectra by such a constant factor.

Despite its practicality and wide use, it has to be expected that such an approach cannot be as general and as simple as it used to be thought of a decade ago. In other words, for the actual state of the art, it has to be admitted that such an oversimplification entails considerable uncertainties as well as some inevitable inaccuracies. The following discussion will address some of these points.

Qualitatively speaking, the q-factor depends on the following conditions:

a) Structural properties
   - ductility of critical regions;
   - possibilities for redistribution of action effects (redundancy);
   - regularity of mass and stiffness distribution;
   - availability of second lines of defense (additional energy dissipation mechanisms, not accounted for in the analysis);

b) Characteristics of the input accelerograms
   - frequency content (vs the natural frequency of the structure considered);
   - number of largest amplitude cycles leading to post-yield excursions.

This is a large number of input parameters that influence the q-value to be selected in each case. This fact tends to increase the uncertainties and inaccuracies mentioned earlier. By way of consequence, it is maintained that q-values are bound to have a nominalistic conventional significance; related liabilities should be taken explicitly by the national Codes.

In areas of low seismicity, elastic seismic spectra might be used directly, without any reduction by means of a q-factor, or only with a minor reduction. This might be of economical interest, because the favorable conditions listed under a) above may be much more costly than a mere strengthening of cross-sections of the structure, so that it will remain elastic even under the strongest probable earthquake of the region. The oversimplified pseudo-numerical application which follows is indicative of this fact.

Let us consider a one-degree-of-freedom structure (a RC column) with the following characteristics: height, h; dead load, G; live load, $Q = G$; eccentricity of all loads, $e = 0.05\ h$; average combination factor, $\psi = 0.3$; seismic coefficient, $\epsilon = \frac{a/q}{g}$, where a is the maximum elastic spectral acceleration.

Case 1: Static combination of actions,

$$M_{s1} = (1.35G + 1.5Q)e = 2.85Qe$$

Case 2: Seismic combination of actions,

$$M_{s2} = (G + 0.3Q)(e + \epsilon h) = 1.3Qe(1 + 20\epsilon)$$

As very rough estimators of construction costs we will assume here the values of applied bending moments, neglecting the role of axial and shear forces. Only a correction factor of $(1 + \frac{q-1}{20})$ will be applied to account for the additional costs associated with implementing a given q-factor, for morphological constraint detailing, capacity design, etc. Then the ratio of the costs corresponding to Case 2 versus Case 1 is written as follows:

$$\frac{1.3\ Qe(1 + 20\epsilon)(1 + \frac{q-1}{20})}{2.85Qe}$$

Equating this ratio with unity gives $(1 + 20\frac{a/q}{q})(1 + \frac{q-1}{20}) = 2.18$. We now define an optimum q-value, q_{opt}, Fig. 4.12. A q-value higher than that is useless; the economic savings it offers make no sense since the static combination of loads has already imposed a construction cost increase. The optimum q-values determined by this simplified method ($q'_{opt} = 1.4$ for $a/g = 0.08$ and $q''_{opt} = 8.0$ for $a/g = 0.25$) should be considered only for their relative significance. The higher the seismicity, the higher the economically optimum q-value should be. (However, disproportionately high q-values in highly seismic regions may not be feasible, so that a cost increase should be expected if a q-value is selected that is lower than the one determined by this simplified method). It was precisely this idea which led the CEB Seismic Model Code to introduce the notion of *Ductility Level*, *DL*, i.e. the average q-value to be aimed at in a specific country.

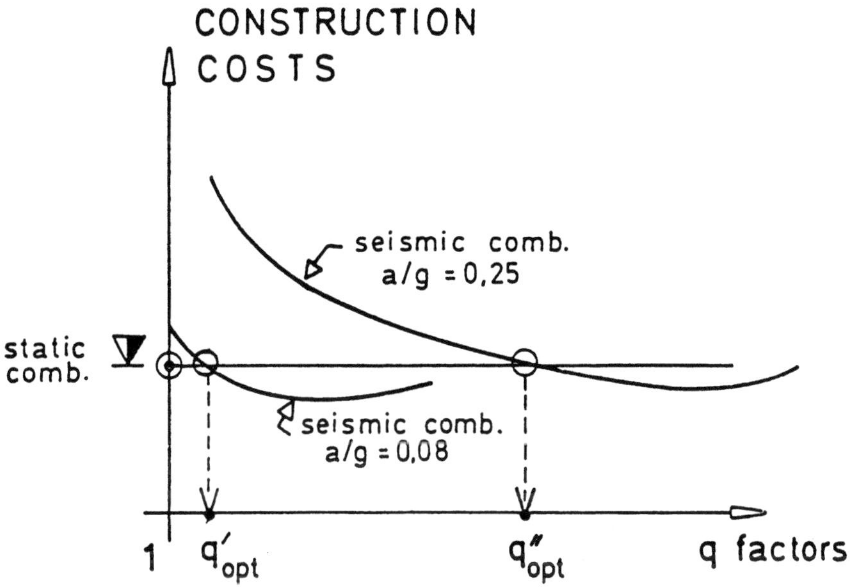

Fig. 4.12 Construction costs

4.3.2 Methods to Estimate q-Factors

Behavior factors can be estimated by one of the following direct or indirect methods, or a combination thereof.

a) Experience from past earthquakes

In some regions of the world with high seismicity, it is possible that several strong earthquakes of similar magnitude are generated at the same focus during the lifetime of a specific structure. In such situations, a systematic study of damage and failures will form a valuable basis for the classification of at least the different behavior levels. Gradual improvements in design and construction methods induced by successive Code modifications are then tested during each subsequent seismic event and translated into better q-values. Of course, in real situations such a procedure will not be as apparent as one might wish. Nevertheless, some of the early behavior factors suggested in national Codes seem to be of empirical origins.

b) Experimental methods

Although not yet utilized systematically, the possibilities offered by modern experimental facilities should not be underestimated. Testing full structure models on shaking tables with appropriate sets of acceleration time histories has a great potential for estimating q-values directly. Another potential tool for such investigations is the hybrid pseudo-dynamic simulation of earthquake loads applied to large-scale models, by means of computerized actuators.

c)Numerical parametric studies

These are the techniques actually used in the investigations for the Draft Eurocode 8.

Elastoplastic Spectra. In the usual simulations, one-degree-of-freedom models (with given elastoplastic behavior and damping) are subjected to an appropriate set of accelerograms. Time domain integrations are carried out for different frequencies of the model, and the peak accelerations determined as functions of these frequencies. An elastoplastic spectrum envelope is drawn, and the ratios between these ordinates and the corresponding ordinates of the elastic spectrum envelope are considered as approximations of the q-factors. Inspite of its fundamental and quite general character, this technique is limited by the given or assumed elastoplastic features of the model, which cannot be translated directly into specific ductility, redundancy and other properties of a real structure.

Indirect Method. Here, a complete structure is considered, designed conventionally to satisfy given Code provisions, such as maximum ground acceleration a_g, normalized spectral acceleration values $\beta(T)$, and behavior factor q. Reasonable elastoplastic models are utilized to describe the behavior of its critical regions. Time-domain integrations are performed for a given accelerogram with gradually increasing intensity. Collapse is assumed to occur when the rate of increase of the roof displacement exceeds a prescribed limit.

Direct Method. A given structure is dimensioned and analyzed by means of conventional Code methods, taking into account a specific q-factor. Subsequently, the same structure is numerically subjected to several accelerograms of ground motions expected in the geographic area in which the Code provisions are meant to be applicable. Time domain integrations are performed, considering an elastoplastic hysteretic behavior of critical regions. Whenever the structure endures this "test" without collapse (or without unacceptable plastic hinge patterns), the initial q-factor may be increased, and vice versa.

The rather limited significance of all these methods cannot be overemphasized. Some of their handicaps are recalled below:

- Formalistic modelling of hysteretic behavior of critical regions; sometimes even the role of actual axial loads is not considered appropriately.
- It is difficult to systematically cover all possible structural irregularities.
- Correct modelling of infill walls in buildings is not feasible.
- While the frequency contents of input excitations are somehow known for each earthquake generating focus and mechanism, this is not the case regarding the number of expected large amplitude post-yield load reversals.

However, it is very important to apply these rational methods and try to gradually improve the validity of their outputs. Meanwhile, these outputs should be considered with caution and be calibrated appropriately.

4.3.3 Numerical Application of the Direct Method

In what follows, the Direct Method described above is applied in order to assess the q-value of a specific building. The following steps are taken to estimate the behavior factor q:

a) The member dimensions are chosen.
b) The static action effects from dead and live load are calculated.
c) The elastic seismic action effects are calculated with a response spectrum analysis, using the spectrum of a suitable earthquake record.
d) The elastic seismic action effects are divided by the estimated behavior factor q to give the first estimate of design seismic action effects.
e) The design action effects are obtained by combining static action effects (a) and design seismic action effects (d). All members are designed according to the pertinent national Code, observing all requirements of seismic design. In case the cross sections of certain building elements need to be changed, steps (b) through (e) need to be repeated.
f) An elastoplastic analysis of the structure is performed by direct integration for the acceleration record of the chosen earthquake. Plastic deformations are compared with available deformation capacities. If the deformation limits are exceeded, the estimated behavior factor is considered too optimistic, and steps (d) through (f) are repeated with a lower value of q. If the deformations are much smaller than the allowable, the design is not economic, and steps (d) through (f) may be repeated with a somewhat higher value of q.

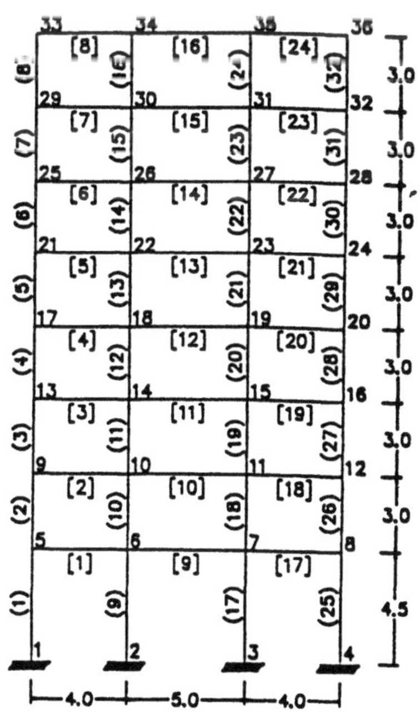

Fig. 4.13 8-story example frame with joint and member numbers

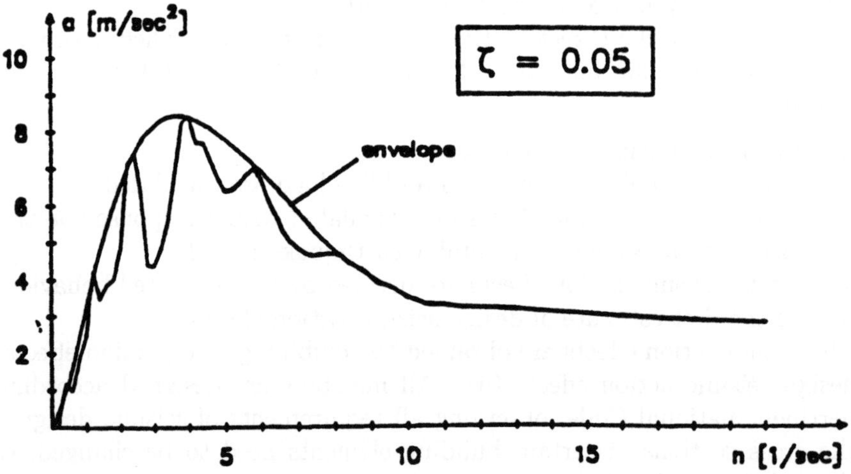

Fig. 4.14 Response spectrum accelerogram recorded in Corinth, 1981

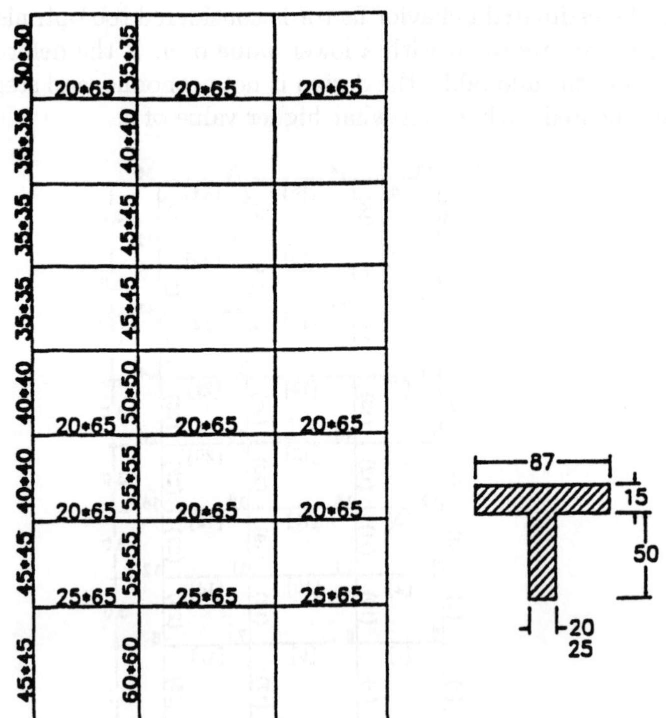

Fig. 4.15 Member dimensions for q= 4

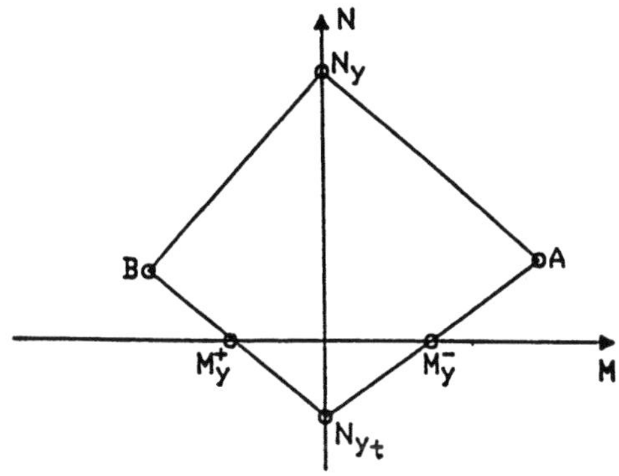

Fig. 4.16 Simplified M-N interaction diagram used by DRAIN-2D

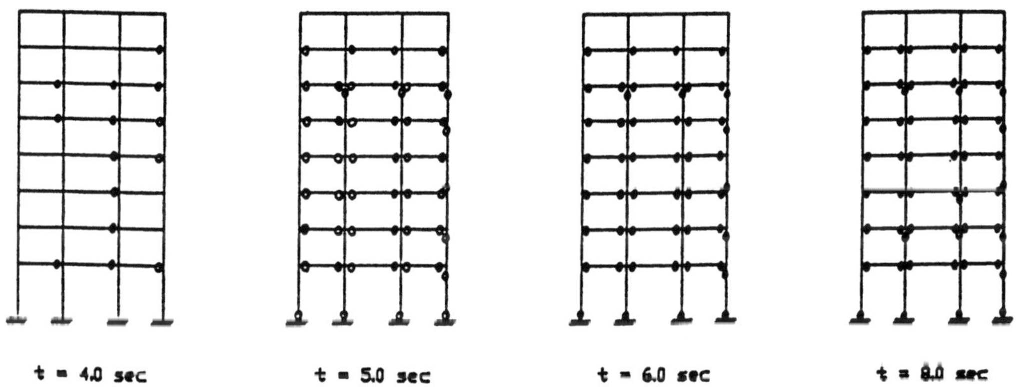

Fig. 4.17 Formation of plastic hinges at different instances for $q= 4$
(∘ - new plastic hinges, • - plastic hinges of previous time steps)

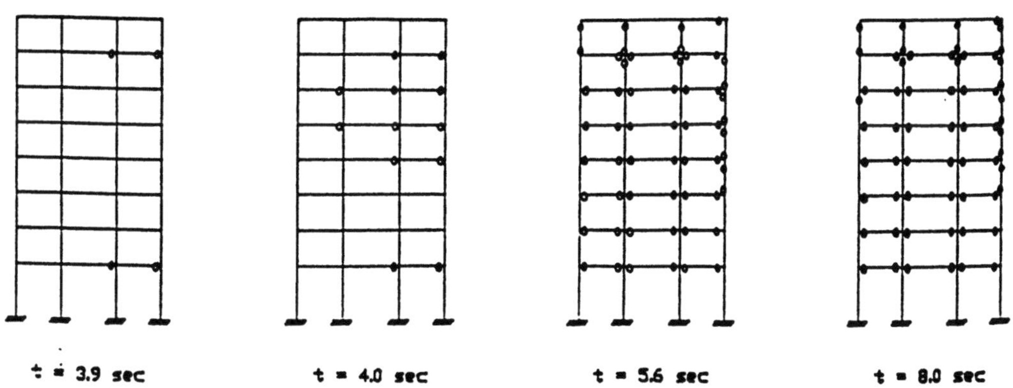

Fig. 4.18 Formation of plastic hinges at different instances for $q= 3$
(∘ - new plastic hinges, • - plastic hinges of previous time steps)

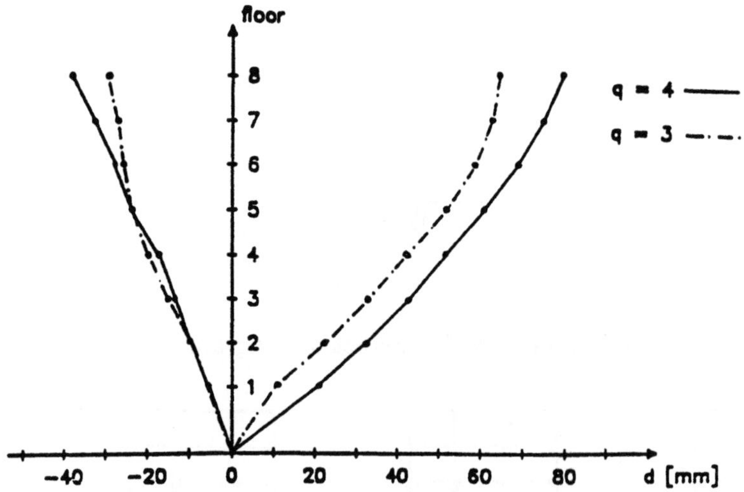

Fig. 4.19 Maximum floor displacements

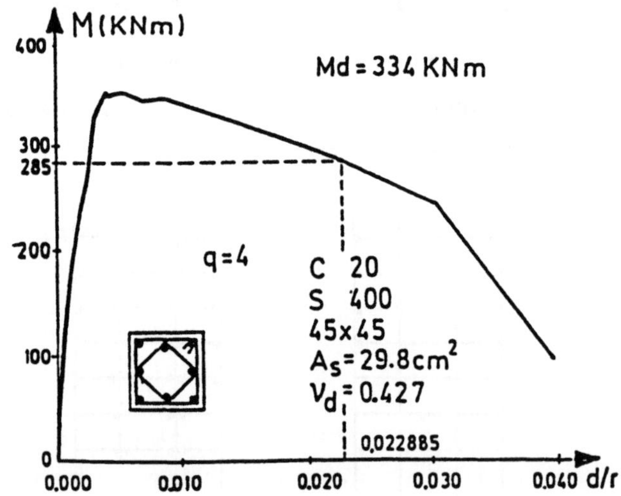

Fig. 4.20 Typical column moment-curvature diagram

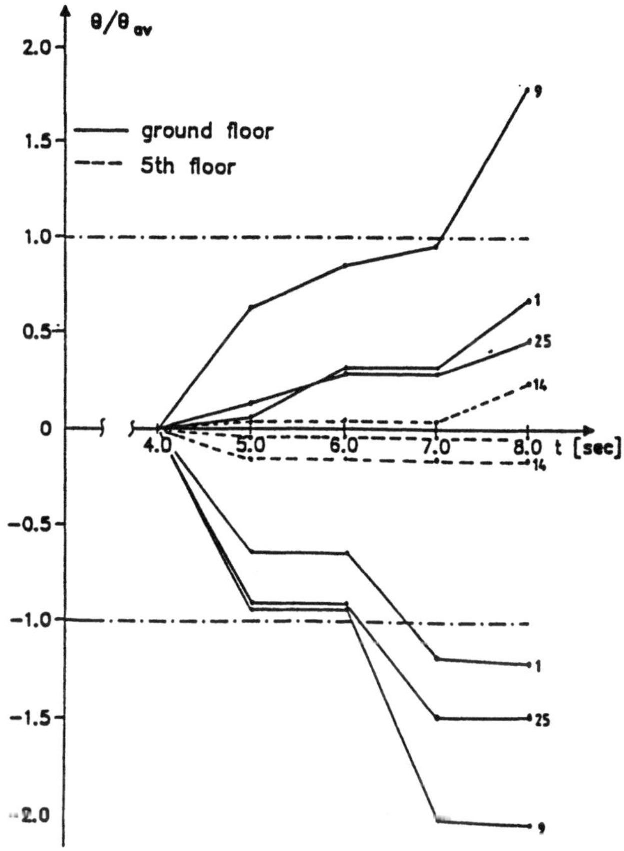

Fig. 4.21 Normalized plastic hinge rotations for $q= 4$

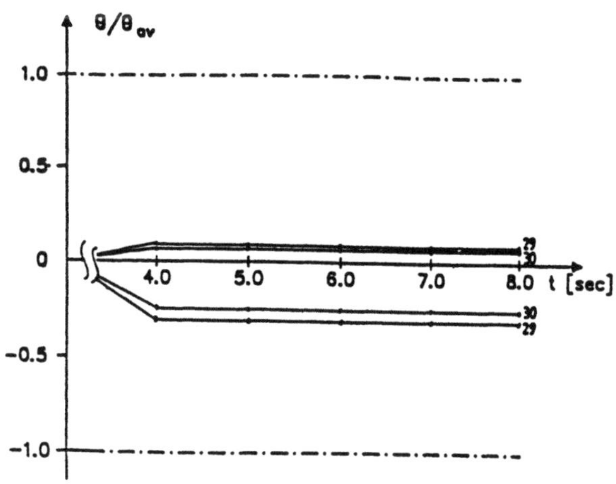

Fig. 4.22 Normalized plastic hinge rotations for $q= 3$

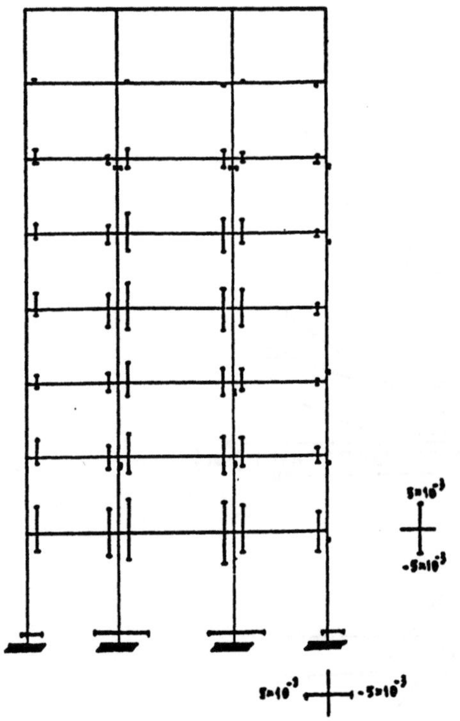

Fig. 4.23 Plastic rotation demands for $q=4$

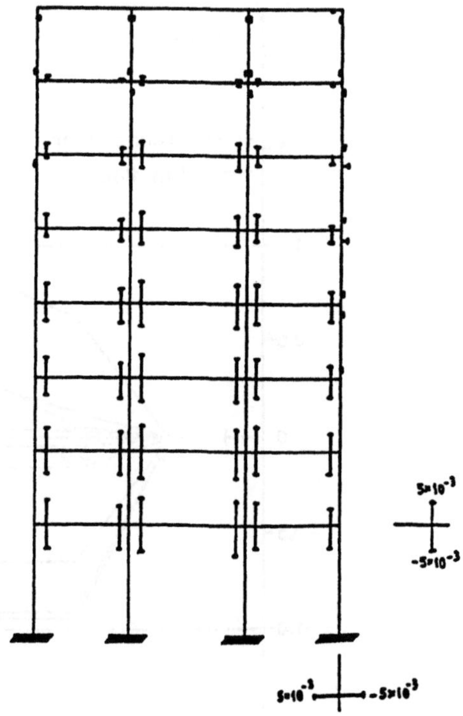

Fig. 4.24 Plastic rotation demands for $q=3$

Details of the Structure. The structure used in this investigation is a symmetric plane, eight-story reinforced concrete frame with three bays, 36 joints, 24 beams, 3 columns, and a total height of 25.5 m, Fig. 4.13. The material data are:

concrete : C20
concrete modulus of elasticity : 29 GPa
steel (longitudinal reinforcement): S400
steel (transverse reinforcement) : S220

The distributed loads are:

dead load: G = 40 kN/m
live load: Q = 10 kN/m

For a vertical load analysis, dead and live loads were concentrated in nodal loads and the effect of the distributed load was taken into account by introducing fixed-en forces and moments at the beam ends.

For the dynamic analysis, the dead load was lumped in nodal masses at th joints. The accelerogram recorded in Corinth during the Alcyonides Earthquake Greece, 1981, was used, which has the smoothed response spectrum (damping rati

$\xi = 0.05$) shown in Fig. 4.14.

Dimensioning of the Structure. The dimensions chosen for the beams and the square columns are given in Fig. 4.15. Moments, axial and shear forces due to dead and live load were determined with program SAP IV [4.6]. A linear dynamic response spectrum analysis was also carried out using the same program, applying the smoothed response spectrum of Fig. 4.14. The design action effects were determined by adding the static effects to the seismic action effects divided by $q = 4$, the first estimated behavior factor. The structure was dimensioned according to the requirements of the new Greek Building Code and the Eurocode 8. The dimensioning process was repeated with the value $q = 3$, because the initial estimate of $q = 4$ was found to be too high. The increase in design loads required larger cross sections.

Elastoplastic Analysis. To test the structure dimensioned for a specific value of q, it was numerically subjected to earthquake excitation. The inelastic time-history integration was carried out using program DRAIN-2D [4.7]. This program characterizes the elastoplastic behavior of beam elements (for which the axial force can be considered negligible) by the maximum and minimum moments, beyond which plastification starts. The elastoplastic behavior of column elements depends on the level of axial loading. DRAIN-2D permits the introduction of simplified M-N-interaction curves defined by four points, Fig. 4.16, which separate the elastic and plastic domains. Maximum and minimum moments were computed for all beams, and simplified interaction curves were determined for all column elements. For the application of the algorithm, a certain strain-hardening ratio is needed, which represents the ratio between the inelastic and initial elastic member stiffnesses. This ratio was chosen to be 0.1 for both types of members.

The program employs the Newmark integration method, i.e. constant acceleration within a time step [4.8]. Rayleigh damping was used, i.e., the damping matrix was formed as a combination of mass matrix and stiffness matrix such as to obtain a global damping ratio of approximately $\xi = 0.05$.

Results. The distribution of plastic hinges at different instances during the earthquake excitation is shown in Fig. 4.17 for the structure with $q = 4$ and in Fig. 4.18 for the structure with $q = 3$. Plastic hinges appear first in the beams of the lower and upper floors and then spread to the intermediate floors. After about half the earthquake duration, plastic hinges also appear in the columns. In the case of $q = 3$, the hinges appear in intermediate and upper floors; in the case of $q = 4$, the column hinges appear in the lower floors, especially at the column bases of the ground floor.

The maximum horizontal floor displacements are depicted in Fig. 4.19 for both q-values. As can be seen, displacements increase linearly with height and remain below the Code allowance.

Based on previous research [4.9] a computer program was written to compute the moment-curvature diagrams for columns, taking into account the increase in concrete strength and ability to sustain larger strains due to lateral reinforcement confinement. Fig. 4.20 shows such a moment-curvature diagram for one of the columns. The maximum available curvature is assumed to be that corresponding to 0.85 times the design moment M_d. Multiplying this curvature by a fictitious plastic hinge length of (conservatively) $l_p = d/4$, and dividing by a safety factor 1.5, the available plastic rotation θ_{av} is determined, which is taken as the maximum value.

Fig. 4.21 shows the ratio θ/θ_{av}, i.e. the rotation demand versus plastic rotation capacity in critical columns during the earthquake, for the case $q = 4$. It can be easily noticed that the normalized deformations of all ground floor columns exceed the chosen rotation capacity limits, i.e. $\theta/\theta_{av} > 1$. This result prompted the redimensioning of the frame with the smaller behavior factor, $q = 3$. The normalized plastic rotations for this case are plotted in Fig. 4.22. Now, the normalized plastic rotations remain sufficiently below their respective capacities, i.e. $\theta/\theta_{av} < 1$, for all columns.

Plastic Rotation Demands. The maximum plastic rotations for the different structural members are shown in Fig. 4.23 for the case $q = 4$ and in Fig. 4.24 for the structure designed for $q = 3$. The indicated values are the maximum positive and negative plastic rotations occurring during the duration of the earthquake for both column or beam ends. The members must be able to sustain these rotations to prevent failure.

According to Fig. 4.23, for $q = 4$ a high rotation capacity is required for the bottom of the ground floor columns; all other columns have relatively small plastic rotation demands. The required plastic rotation capacities of the beams are of them same order as for the ground floor columns, but they decrease continually towards the roof. For $q = 3$, Fig. 4.24, the rotation capacity demands for the beams are similar to those of case $q = 4$, i.e. starting with large values at the ground floor and decreasing towards the top. For the columns, on the other hand, only relatively small rotation capacities are demanded in some intermediate floors.

Conclusions. The elastoplastic analysis of a frame that had been dimensioned for two different behavior factors (q= 3 and 4) revealed the formation of plastic hinges in the columns for both q-values, even though a moderate capacity design criterion (γ_c= 1.5) had been applied. The comparison of the observed and estimated available plastic rotations can constitute a criterion for the reliability of the structure and led to the selection of the final behavior factor q= 3.5, which seems to reflect safely the entire design and detailing philosophy used. The method proved to be fruitful in several other respects, as it revealed the distribution of plastic rotation demands, the sequence of plastic hinge appearances, pragmatic interstory drifts, etc.

4.4 Seismic Design of RC Wall for a Given Behavior Factor

4.4.1 Introduction

The beneficial role of well-designed structural RC walls in the seismic behavior of buildings cannot be overemphasized. Nevertheless, a basic prerequisite has to be always observed: the ductility of such walls should be compatible with the behavior factor initially assumed for the seismic design of the building. For this purpose, a design method is sought for the rational dimensioning of the critical wall region such that the desired extent of inelastic behavior is accounted for explicitly, as reflected in the behavior factor, q. National building codes take care of this problem by means of a set of rules such as,

- a minimum length of confined wall ends;
- a maximum vertical reinforcement ratio;
- a minimum amount of closed stirrups provided in the boundary regions;
- a maximum value of the neutral axis depth relative to the wall length.

Qualitatively correct as they may be, such Code provisions are not connected directly with the selected overall ductility level and do not account for the actual amount of confinement reinforcement provided in the end regions.

The design method proposed here is an attempt to establish a numerical relationship between the q-factor used for the design of the building and the confinement conditions of the wall's boundary regions, i.e., the required confined length, l_c and the volumetric mechanical reinforcement percentage, ω_{wd}, of the closed stirrups or cross ties available. The method follows essentially the procedure proposed by Paulay and Uzumeri [4.10]. The overall behavior factor q is translated roughly into a displacement ductility factor, μ_d, which subsequently is expressed in terms of the local curvature ductility factor, $\mu_{1/r}$. It remains to assure that this local ductility demand is available in the critical wall region, thanks to its appropriate dimensions and detailing.

The nature of the related multiparametric and complex post-yield phenomena justifies a set of approximations introduced for the analysis. Thus, a rather low level of predictive precision is claimed. However, it is believed that the insight offered by this approach is helpful for a rational seismic design, as compared with the semi-empirical Code provisions.

4.4.2 Working Assumptions

To serve the purpose described above, the following simplified assumptions are made.

a) How to relate q to μ_d.

Due to the reduced natural vibration periods of the structural systems considered, the equal energy principle is adapted in assessing their elastoplastic behavior. Thus,

$$q = \sqrt{2\mu_d - 1} \quad \text{or} \quad \mu_d = 0.5(q^2 + 1) \tag{4.48}$$

At this point, it is appropriate to recognize the rather formalistic nature of the q-values imposed by the Codes. The direct numerical manipulation of such values in what follows might therefore be considered questionable.

b) How to relate μ_d to $\mu_{1/r}$.

The RC walls under design are considered as cantilevers fixed at the base. Their interaction with the RC frames in dual systems is taken into account by means of appropriate seismic load distributions over their heights. The following relationship can be derived with the help of Fig. 4.25,

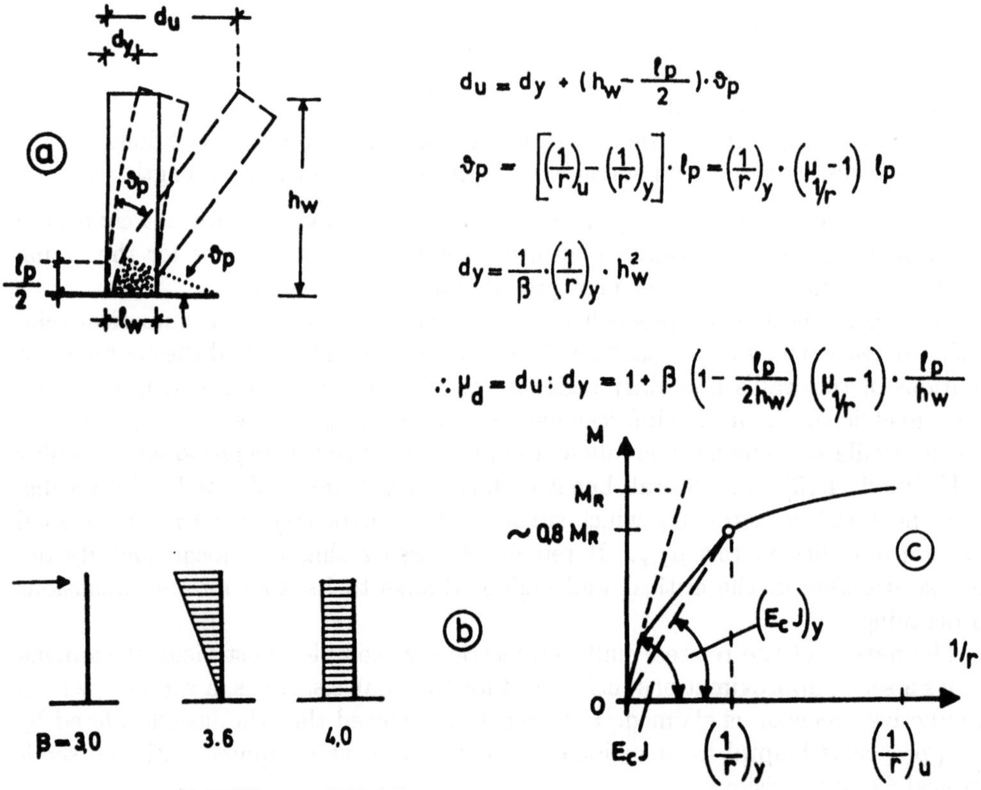

Fig. 4.25 Relationship between curvature and displacement ductility factors, μ_d and $\mu_{1/r}$

$$\mu_{1/r} \quad = \quad 1 + \frac{1}{\beta}\frac{c(\mu_d - 1)}{\lambda_p(1 - 0.5\lambda_p)} \tag{4.49}$$

where β = coefficient reflecting the distribution of seismic load over the wall height (Fig. 4.25a); c = coefficient accounting for possible foundation flexibility, such as introduced by Priestley and Park [4.11]; μ_d = roof displacement ductility factor; $\lambda_p = l_p/h_w$ = normalized height of the critical region or plastic hinge length of the wall of height h_w. Based on suggestions by Steidle and Schäfer [4.12], Wohlfahrt and Koch [4.13], and Priestley and Park [4.11], this plastic hinge length is here taken as,

$$\lambda_p \quad = \quad 0.6\frac{l_w}{h_w} \tag{4.50}$$

c) Approximate expression for base curvature at yield

With the notations of Fig. 4.25c one obtains,

$$\left(\frac{1}{r}\right)_y \quad = \quad \frac{M_y}{(E_cJ)_y} \quad = \quad \frac{0.8M_R}{0.6E_cJ} \quad = \quad \frac{4}{3}\frac{M_R}{E_cJ} \tag{4.51}$$

For a better approximation, see Sect. 4.4.3..

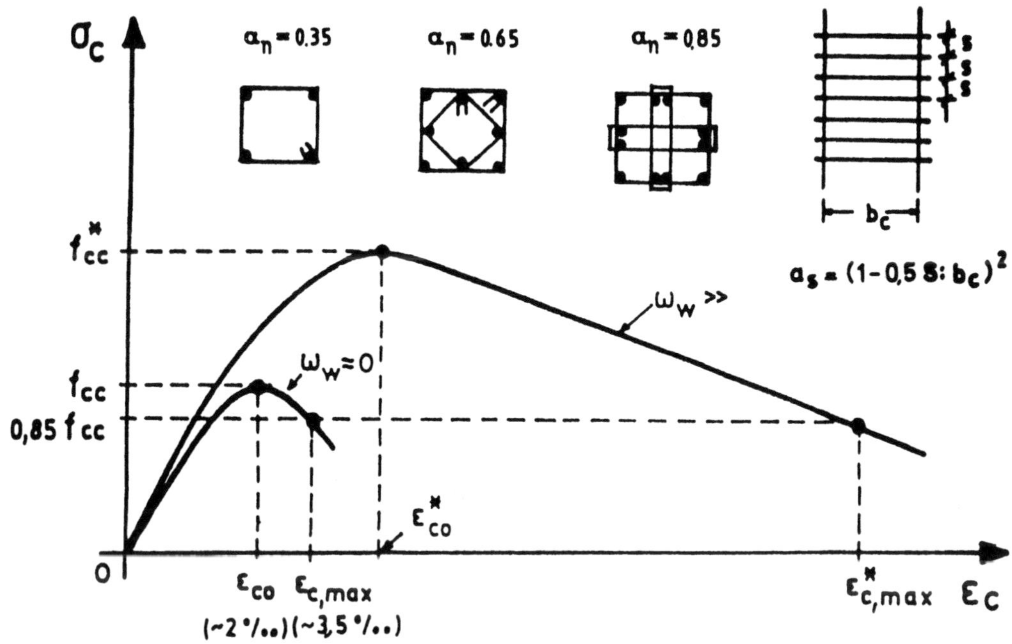

d) Constitutive law for confined concrete

The stress-strain relationship for concrete within the confined edge zones is approximated by the following equations, Fig. 4.26 [4.9]. The strength of confined concrete,

$$f^*_{cc} = f_{cc}(1.0 + 2.5\alpha\omega_{wd}) \quad \text{for} \quad \alpha\omega_{wd} < 0.10$$
$$= f_{cc}(1.125 + 1.25\alpha\omega_{wd}) \quad \text{for} \quad \alpha\omega_{wd} > 0.10 \tag{4.52}$$

Peak concrete strain,

$$\epsilon^*_{co} = \epsilon_{co}\left(\frac{f^*_{cc}}{f_{cc}}\right)^2 \tag{4.53}$$

Concrete strain at $0.85f_{cc}$-level,

$$\epsilon^*_{c,max} = 0.0035 + 0.1\alpha\omega_{wd} \tag{4.54}$$

where f_{cc} = unconfined concrete strength; $\alpha = \alpha_s\alpha_n$ = effectiveness of lateral confinement (Fig. 4.26); for the simple configuration of lateral reinforcement feasible in walls and for stirrups or cross-ties spaced at about $1/3b_c$, $\alpha = 0.25$ is assumed here; ω_{wd} = design value of the volumetric mechanical ratio of confinement reinforcement = (volume of confinement steel $\times f_{yd}$)/ (volume of concrete core $\times f_{cd}$).

4.4.3 Ductility Design of Critical Wall Region

Using the conventional definition of curvature ductility, Fig. 4.25a,

$$\mu_{1/r} = \frac{(l_w/r)_u}{(l_w/r)_y} = \frac{\epsilon_{su}+\epsilon_{cu}}{(l_w/r)_y} \tag{4.55}$$

the following expression can be derived with the help of Eqs. (4.49) and (4.51),

$$\epsilon_{su}+\epsilon_{cu} = \left[1+\frac{1}{2\beta}\frac{c(q^2-1)}{\lambda_p(1-0.5\lambda_p)}\right]\frac{4}{3}\frac{M_R}{E_c J} \tag{4.56}$$

Thus, the required slope of the cross-section at the wall base is given as a function of all parameters governing the problem, including the overall behavior factor q initially selected for the seismic analysis of the building, Fig. 4.27.

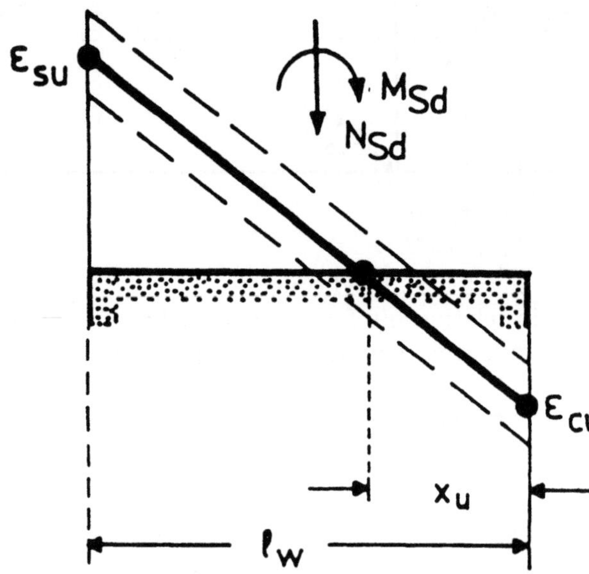

Fig. 4.27 Parallel positions of rotated wall base, observing overall ductility requirement (Eq. 4.56)

The following procedure is suggested for the wall ductility design. First, the critical wall region is dimensioned for the minimum axial load N_{min} and corresponding bending moment M_{s1}. Vertical reinforcement is distributed over the width $l'_c = max(0.2l_w, 2b_w)$ on both edges, for preliminary design purposes.

A moment-curvature diagram is drawn for the above conditions in order to determine a yield curvature which corresponds to the final conditions and N_{max},

$$\left(\frac{l_w}{r}\right)_{y,N_{max}} = \left(\frac{l_w}{r}\right)_{y,N_{min}} \frac{N_{min}}{N_{max}}$$

This expression may be substituted for the more approximate value $\frac{4}{3}\frac{M_R}{E_c J}$ in Eq. (4.56).

The final dimensioning, which should meet the ductility requirement of Eq. (4.56), should be performed for N_{max} and the corresponding moment, M_{sd}. Shear design and verification is controlled first by N_{max} such that flexural failure precedes shear failure. The location of the line of zero strain, x_u (Fig. 4.27), about which the base section rotates, will be determined such that force and moment equlibrium conditions are satisfied,

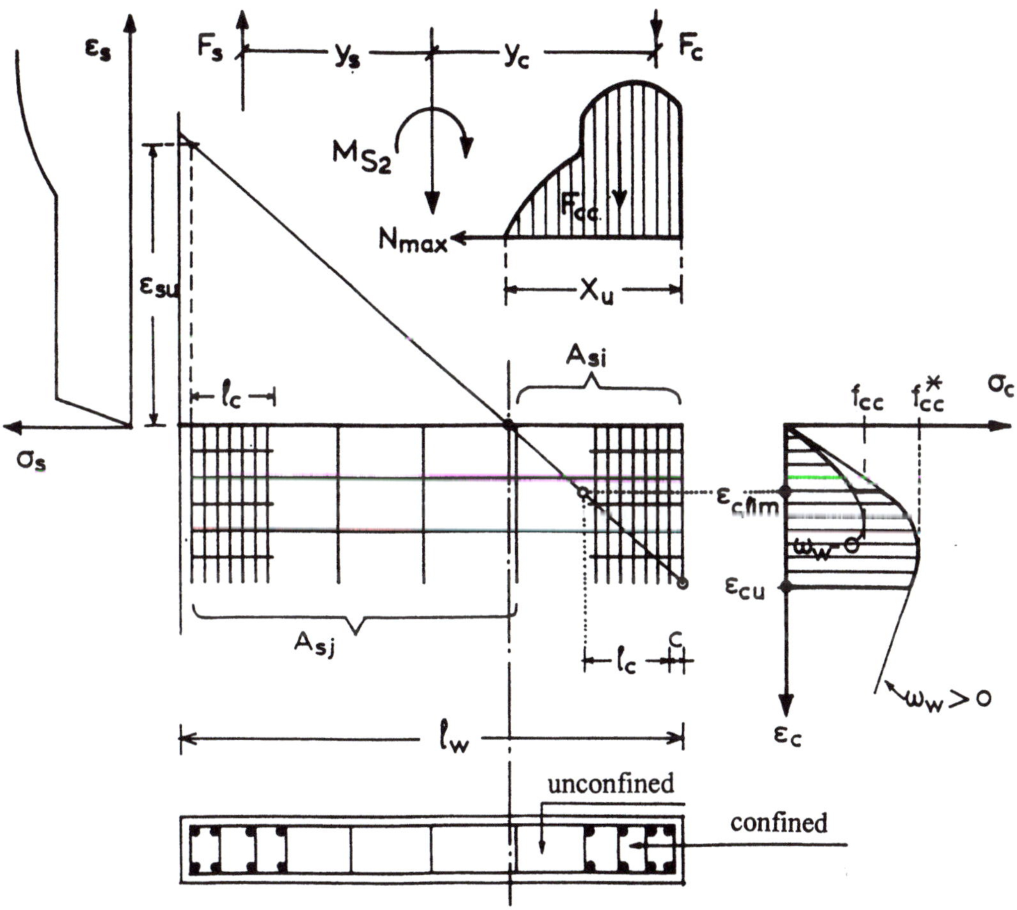

Fig. 4.28 Stress and strain distributions in base section under final seismic load combination

$$N_{max} = F_c - F_s = (F_{cc} + \sum_i A_{si}\sigma_{si}) - \sum_j A_{sj}\sigma_{sj} \qquad (4.57)$$

$$M_{sd} = F_c y_c - F_s y_s \tag{4.58}$$

where the various symbols are defined in Fig. 4.28. The assumption that plane sections remain plane impose the following additional requirements,

$$x_u = (l_w - 2c)\frac{\epsilon_{cu}}{\epsilon_{su} + \epsilon_{cu}} \tag{4.59}$$

$$l_c = x_u \frac{\epsilon_{cu} - \epsilon_{c,lim}}{\epsilon_{cu}} \tag{4.60}$$

where $\epsilon_{c,lim} = 1.5 \times 10^{-3}$ is the concrete strain under which the material may be considered as elastically stable and does not need to be confined.

The five unknowns of the problem $(\epsilon_{su}, \epsilon_{cu}, x_u, l_c, \omega_{wd})$ may now be determined by means of the five Eqs. (4.56) through (4.60). A trial-and-error method may be followed, starting with some initial value for ω_{wd}, possibly using Eq. (4.61) below as a guide. Obviously, the l_c-value will change during iteration from the initial l'_c-value. Consequently, the bearing capacity is expected to change as well.

4.4.4 Parametric Studies

A series of numerical studies has been carried out with the following parameters:

- three q-values (2.5, 3.5, and, to a limited extent, 4.5);
- three different numbers of stories (3, 5, and 7);
- two different total floor areas (250 m^2 and 500 m^2).

The walls were considered as fully fixed at the base. In all cases, an effective bedrock acceleration of $\alpha_g = 0.3g$ was considered, leading to a base shear coefficient of $2.5\alpha_g/gq$.

A static analysis was carried out for a seismic load distributed over an inverted triangle. Appropriate magnification factors were introduced to account for both static torsion and higher modes effects, as suggested in seismic Codes. In each of the fifty cases examined, the wall length was selected to be either 1.5 m, 2.0 m, 3.0 m, or 4.0 m, based on engineering judgment. The wall thickness was 250 mm in each case.; materials were C20 and S400. Because of the numerical procedure followed, the achieved degree of precision differed from case to case.

The results of the investigation are summarized in Fig. 4.29. The length l_c of the confined edge areas and the mechanical confinement reinforcing ratio ω_{wd} provided at distance $s = 1/3b_c$ are plotted as functions of $(\overline{\nu_d} + \overline{\mu_d})$, where $\overline{\nu_d} = N_{s,max}/b_w l_w f_{cd}$ is the normalized axial compression force and $\overline{\mu_d} = M_{sd}/b_w l_w^2 f_{cd}$ is the corresponding normalized bending moment.

On the basis of this rather limited numerical investigation, the following empirical (and somehow conservative) expressions may be proposed. These could be useful for estimating the relevant quantities before the final verification analysis:

$$\omega_{wd} = \left(\frac{q}{2.5}\right)^2 \left|(\overline{\nu_d} + \overline{\mu_d}) - 0.05(4.0 - q)\right| \tag{4.61}$$

$$l_c = 0.10 + 0.45(\overline{\nu_d} + \overline{\mu_d}) \leq 0.50 \tag{4.62}$$

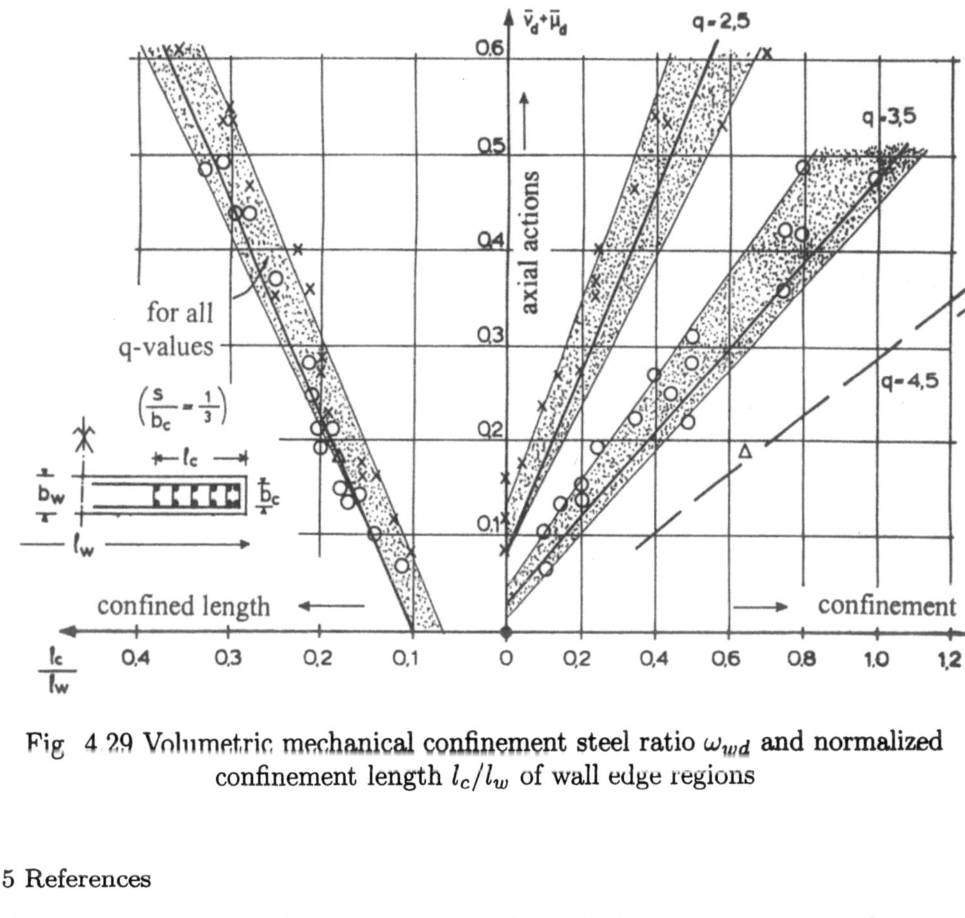

Fig. 4.29 Volumetric mechanical confinement steel ratio ω_{wd} and normalized confinement length l_c/l_w of wall edge regions

4.5 References

[4.1] Commission of the European Communities, 1993. Eurocode No. 8, Structures in Seismic Regions - Design. Part 1 - General and Building.

[4.2] CEB-FIP Model Code, 1990. Thomas Telford, London, 437, 1993.

[4.3] New Zealand Standard, 1982. Code of Practice for the Design of Concrete Structures, NZS3101, Part 1.

[4.4] Paulay, T. and Priestley, M.J.N., 1992. Seismic Design of Reinforced Concrete and Masonry Buildings, John Wiley and Sons, Inc., New York.

[4.5] Eurocode 2, Design of Concrete Structures, Part 1.1.

[4.6] Bathe, K.J., Wilson, E.L., Petersson, F.E., 1973. SAP IV, A Structural Analysis Program for Static and Dynamic Response of Linear Systems, Report EERC 73-11, Berkeley, California.

[4.7] Kanaan, A.E. and Powell, G.H., 1973. DRAIN-2D, A General Purpose Computer Program for Dynamic Analysis of Inelastic Plane Structures, Report EERC 73-6 and 73-22, Berkeley, California.

[4.8] Clough, R.W. and Penzien, J., 1975. Dynamics of Structures, McGraw-Hill, Inc., New York.
[4.9] Tassios, T.P. and Lefas, I.D., 1984. Concrete Ductility Due to Lateral Confinement, Scientific Papers, National Technical University, Vol. 8, No. 1-4, Athens.
[4.10] Paulay, T. and Uzumeri, S., 1975. A critical review of the seismic design provisions for ductile shear walls of the Canadian Code and Commentary. Canadian Journal of Civil Eng., 2.
[4.11] Priestley, M.J.N. and Park, R., 1984. Strength and ductility of bridge substructures. Bridge Design and Research Seminar, Auckland.
[4.12] Steidle, P. and Schäfer, K., 1986. Trag- und Verformungsfähigkeit von Stützen bei großen Zwangsverschiebungen. Deutscher Ausschuß für Stahlbeton, Heft 376.
[4.13] Wohlfahrt, R. and Koch, R., 1986. Versuche an Stützen mit Normalkraft und Zwangsverschiebungen. Deutscher Ausschuß für Stahlbeton, Heft 376.

APPROXIMATE ANALYSIS AND DESIGN TOOLS

P. Gergely†
Cornell University, Ithaca, NY, USA

C. Meyer
Columbia University, New York, NY, USA

5.1 The Role of Approximate Analyses

Dynamic analysis generally implies the calculation of the time-history response (usually displacements and member forces). The response spectrum method of analysis is sometimes also used as a dynamic analysis method, because the response is computed in terms of mode shapes, although that is not dynamic analysis in a strict sense. Dynamic time history analysis of linear systems is now a practical and reliable analysis procedure. However, the analysis of nonlinear systems involves many difficulties and uncertainties and therefore is not yet recommended for routine use. The hysteretic properties of RC elements are functions of many different variables and difficult to reproduce analytically, as discussed in Chapter 2. The response of nonlinear systems can be very sensitive to the properties of the ground motion input. That means that in practice several possible inputs need to be selected and a nonlinear analysis performed for each. The unknown characteristics of future earthquakes adds to the uncertainties met in doing nonlinear analyses.

An indication of the uncertainties in nonlinear dynamic analysis is the fact that true, i.e. blind, predictions of experimental response have been rather poor. In cases where computed responses compared well with experimental results, the experimental data were typically available to the analyst, and important input parameters could be adjusted for optimum agreement. Such analyses are properly referred to as "post-analyses". For example, as was pointed out in Sect. 2.5, the agreement hinges on the initial period of the structure, which normally cannot be calculated accurately for actual buildings.

As a consequence of these uncertainties with nonlinear dynamic analysis, alternate methods have been developed to help designers predict maximum response levels. These methods are the subject of this chapter. They are not true dynamic analysis methods but are useful for estimating nonlinear response levels without calculating responses as functions of time.

The performance of buildings is best judged in terms of story displacements (drifts) if the response is inelastic. Of the five methods summarized in this chapter, four involve displacement as the key parameter and are especially useful in design. They are less useful for estimating the expected response of existing buildings which may have poor details and require more accurate methods of analysis.

5.2 The Pushover Curve

The plot of the total lateral force (base shear) versus roof displacement is called the *pushover curve.* The generation of such a curve is relatively foolproof since it is quite easy to estimate the story shear capacities and thus get an idea of the peak load level. Few currently available computer codes can perform such nonlinear analyses automatically, but it is possible to use standard analysis programs to perform the analysis in a step-by-step procedure. We start with a linear elastic analysis of the frame structure for an arbitrary level of lateral load. By comparing for each frame member the bending moment with its yield capacity, it is straightforward to determine the load level that should have been applied to cause the formation of the first plastic hinge. Let α_1 be the ratio between the computed moment at that first hinge location and its yield capacity, then all bending moments of the first analysis are to be divided by factor α_1. Next, a hinge with zero or near-zero rotational stiffness is inserted at the location where the yield moment was first reached, and the frame with this reduced stiffness is analyzed a second time for an arbitrary load increment. Again, all frame elements are scanned for the largest ratio between computed moment and yield moment reserve, i.e., the actual yield capacity minus the moment obtained in the first run. This ratio, α_2, will be used to scale the results of the second analysis. The moments are added to those of the first analysis and a second hinge is inserted. The analysis progresses in a similar fashion step by step, until a mechanism is formed that cannot resist any further load.

Such a nonlinear static analysis can be tedious, but linear elastic analysis codes can be modified with relative ease to automate the process. It may not be necessary to modify the stiffness each time a new hinge is formed, but rather after a set of new hinges has formed. Generic pushover curves are shown in Fig. 5.1. For frame structures, the curve tends to be close to an elastoplastic (bilinear) curve, because a story mechanism develops soon after the first flexural hinge forms. On the other hand, for structural systems with walls, the force level increases considerably beyond that at which the first hinge forms.

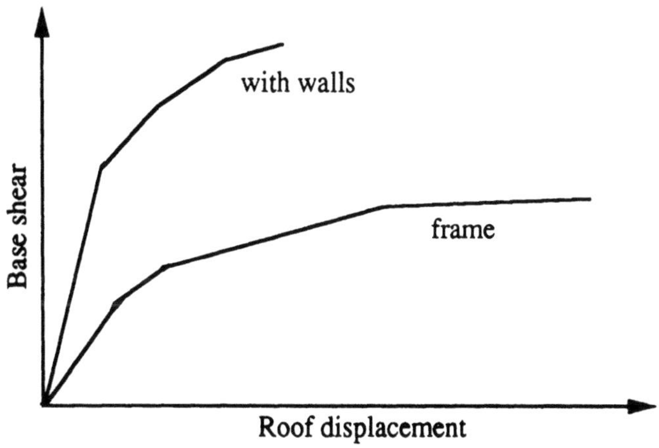

Fig. 5.1 Typical pushover curves

The pushover curve reveals significant information about the characteristics of a structure and is, therefore, important even if no further use is made of it. It permits conclusions to be drawn about the progressive damage in the structure, displays concentrations of hinges, and indicates whether there is a soft story, i.e. whether a story mechanism can develop before significant hinging occurs elsewhere in the structure. It also helps trace the load transfer among the various load-resisting structural systems in the nonlinear range. This is important because many failures have occurred during earthquakes as a result of poor load transfer.

In the simplest approach, the pushover analysis is performed for an elastic force distribution, for example, for code forces which are functions of the floor masses and heights. Alternatively, the lateral force distribution is obtained from a modal analysis, although the modal forces at each floor level cannot be combined in a rational manner. In both approaches the relative force distribution remains constant and does not reflect the stiffness changes that can be significant, e.g., in buildings with a soft story upon formation of the mechanism.

The actual response of flexural members is cyclic rather than monotonic. As a result, the tangent stiffness, greatly reduced by the hinging, is lower than the actual or effective stiffness. To account for the higher effective member stiffness, one may substitute a secant stiffness for members with nonlinearities that corresponds to an assumed member deformation level, Fig. 5.2, and use this stiffness in the next step of the analysis. If the resulting member deformation is different from the assumed one, the secant stiffness can be adjusted accordingly. This iterative analysis converges rapidly.

For irregular structures it may be necessary to improve the force distribution for the pushover analysis. For example, a load vector may be generated as the product of the current displacements and masses. The new displacements resulting from this load application define an improved inertia force vector. An iteration would converge

vector. An iteration would converge to the first mode shape for the current stiffness matrix, and we could also get the natural frequency from the Rayleigh quotient. Again, this analysis can be performed by an ordinary static analysis program with relative ease.

The pushover curve is a property of the structure; thus, we need to estimate the seismic demand. Several approximate ways have been proposed to achieve this.

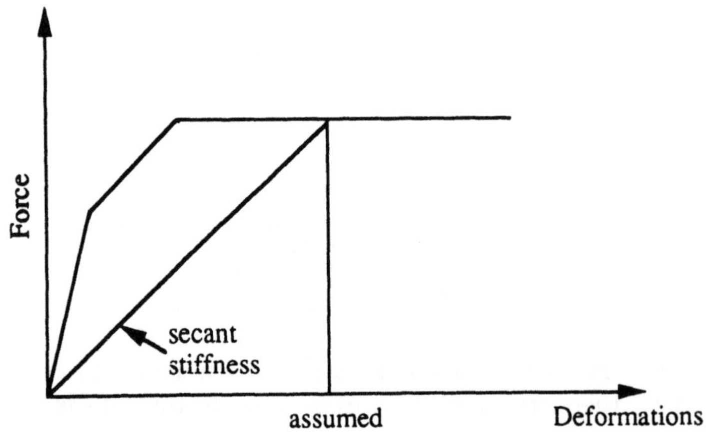

Fig. 5.2 Secant stiffness for iterative approach

5.3 Capacity Spectrum Method

The capacity spectrum method, first proposed by Freeman [5.1], takes a graphical representation of the global force-displacement capacity curve of the structure (pushover curve) and compares it to the response spectrum representation of the earthquake demand. The pushover curve is determined as described above. It is assumed that the structure can resist a number of cycles along the capacity curve and behave in a hysteretic manner so that the effective stiffness is reduced to an equivalent global secant modulus measured to the maximum excursion along the capacity curve for each cycle of motion. The roof displacement and base shear coordinates are converted to spectral displacements, S_d, and spectral accelerations, S_a, Fig. 5.3, respectively, by use of modal participation factors and effective modal weights as determined from the structure's fundamental mode. These values change as the displaced shape changes. An equivalent inelastic period of vibration, T_i, is calculated at various points along the capacity curve using the secant modulus, $T_i = 2\pi\sqrt{S_{di}/S_{ai}\, g}$. Now the capacity spectrum curve can be plotted with the same coordinates as a response spectrum, i.e. S_a vs T, Fig. 5.4.

The spectral accelerations and displacements corresponding to key static lateral force and displacement levels (capacity) are calculated and compared with the demand. The base shear coefficient is $C_B = V/W$, where V = base shear, and W = total weight of the structure. The ratio of roof displacement to spectral displacement is

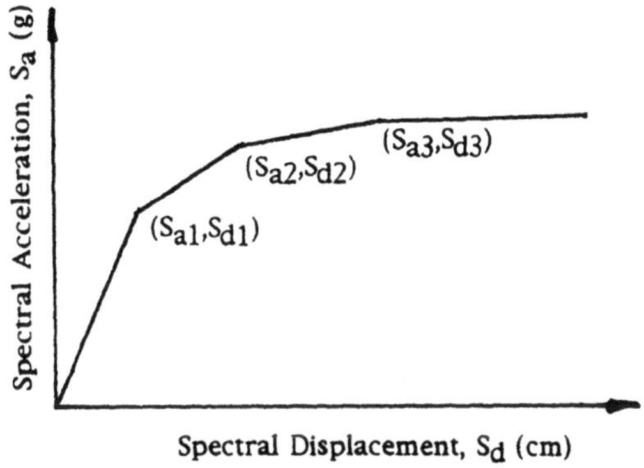

Fig. 5.3 Spectral acceleration S_a vs spectral displacement S_d

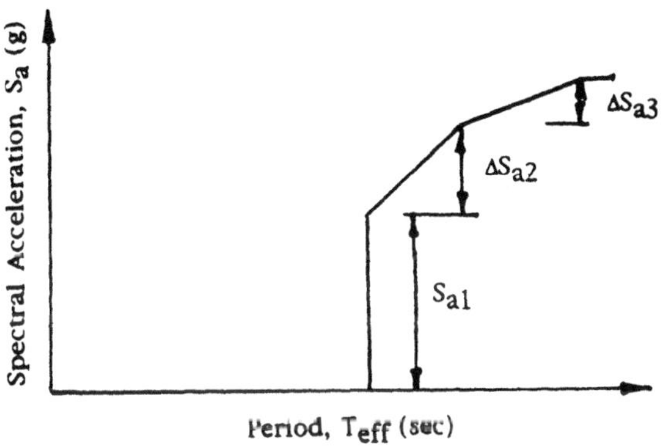

Fig. 5.4 Spectral acceleration S_a vs effective period T_{eff}

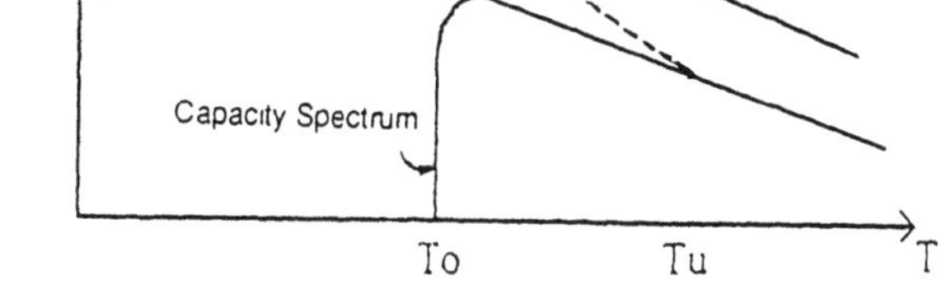

Fig. 5.5 Capacity spectrum method

$$\frac{d_R}{S_d} = \Gamma \phi_R \tag{5.1}$$

where Γ is the participation factor (see Appendix), and ϕ_R is the roof displacement component of the eigenvector. The lateral force vector and the participation factor are

$$\{P\} = [M]\{\phi\}\Gamma S_a \tag{5.2}$$

and

$$\Gamma = \frac{\sum m_i \phi_i}{m_i \phi_i^2} \tag{5.3}$$

respectively. Introducing the nondimensional factor, $\alpha_a = S_a/g$, the spectral acceleration in terms of gravity, the following ratio is readily calculated as

$$\frac{C_B}{\alpha_a} = \frac{V}{W\alpha_a} = \frac{\Gamma S_a \sum m_i \phi_i}{g \sum m_i \alpha_a} = \frac{\Gamma \sum m_i \phi_i}{\sum m_i} \tag{5.4}$$

The spectral displacement S_d corresponding to each load stage is the ratio of d_R and d_R/S_d. The period corresponding to this acceleration and displacement is

$$T = 2\pi \sqrt{\frac{S_d}{\alpha_a g}} \tag{5.5}$$

Thus, we may plot the capacity spectrum defined at each significant stage of the nonlinear load-displacement curve. A capacity spectrum is shown in Fig. 5.5, together with demand spectrum curves for two values of damping, one for the initial structure, and one for the structure near failure. The intersection of the capacity spectrum with the demand spectrum for the appropriate damping value defines the inelastic displacement level of the structure for the particular earthquake. Damping interpolation is necessary because the level of damping depends on the level of displacement. If the capacity curve can extend through the envelope of the demand curve, the building survives the earthquake. The effect of cycling is not reflected in this method, although we could modify the capacity spectrum depending on the estimated effects of cycling. The great advantage of this method is that most engineers are more comfortable with nonlinear static analysis than with time history dynamic analysis, which requires several ground accelerograms, as discussed above. Also, it is very easy to superpose the capacity spectrum on a number of different earthquake spectra.

5.4 Reserve Energy Technique

This method, proposed by Blume in 1960 [5.2], also requires the estimation of the nonlinear load-displacement curve of each story. It is actually quite similar to the capacity spectrum method, as Fig. 5.6 illustrates. It attempts to incorporate more general modes of failure than merely those defined by excessive deformation.

To be realistic, it should include all structural and nonstructural elements. The input energy for a particular mode is calculated as $\overline{m}S_v^2/2$, where S_v is the spectral velocity. The area under the load-deformation curve is the energy absorbed, and the area required to balance the input energy defines the displacement and load level of the structure. For multistory buildings the energy in the various stories must be added approximately, for example by using the first mode shape.

In principle, the method can deal with deteriorating systems. However, there is no apparent quantitative way to measure the rate or extent of deterioration under cyclic loading. In fact, the damage accumulation mentioned in earlier chapters, which is a function of the number of load cycles, appears nowhere in the formulation as originally presented.

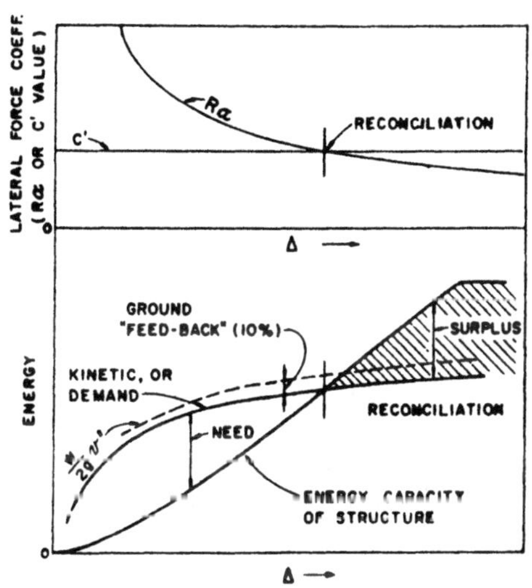

Fig. 5.6 Reserve energy technique

5.5 The Substitue Structure Method

The two main reasons for the reduction of forces due to nonlinear response are the hysteretic energy absorption and the increase in period. The substitute structure method, developed in the early 1970s [5.3], employs linear elastic analysis of a modified version of the actual structure. The substitute structure is defined by replacing the actual member stiffness EI_a by a reduced stiffness,

$$EI_s = \frac{EI_a}{\mu} \tag{5.6}$$

where subscripts s and a refer, respectively, to the substitute and actual structure, and μ is the acceptable damage ratio, which is comparable to but not identical with the ductility ratio. It measures the reduction of the rotational member stiffness (averaged according to the modal flexural strain energy for the building) and is related to the displacement ductility ratio. Modes and frequencies for the undamped substitute structure are obtained using linear dynamic analysis. Equivalent modal damping ratios are computed as

$$\xi_s = 0.02 + 0.2(1 - \frac{1}{\sqrt{\mu}}) \tag{5.7}$$

Thus, the stiffness is reduced (leading to an increase in period) and the damping is increased. Studies [5.4] have shown that this change in stiffness and damping leads to approximately the same response as the hysteresis damping of the inelastic structure. Design forces may be computed using a linear elastic response spectrum analysis. This method has been shown to give results that are sufficiently accurate for preliminary design purposes [5.3]. When using it for final design, it is important to recognize the limitations of this method, which is after all a highly approximate method. In particular, one should remember that the method cannot reliably predict the ductility demands of individual members in the structure.

5.6 Displacement Based Approaches

Design procedures can be developed in which the structural performance is related directly to displacements, rather than indirectly as in the force approach [5.5]. Most common types of damage are easier relate to displacements (mostly inter-story drifts) rather than force or ductility levels, and there is no need to compute forces first. The basic idea is that the maximum inelastic spectral displacements are estimated, namely to be equal to the elastic values in the long-period range and modified in the short-period range. Approximate displacement spectra can be constructed, such as the one in Fig. 5.7. The maximum displacement profile of a building is estimated from the spectral value based on the estimated deformed shape. For example, the actual roof displacement is about 1.2 times greater for a frame and about 1.5 times greater for a wall structure. These values can be refined, based on the accumulated experience of the engineer. From the deformation profile the inelastic story drifts and the corresponding member demands can be calculated. This procedure is relatively simple and rational for the design of buildings. However, when analyzing existing structures, local weaknesses could invalidate the approach or would require special modifications.

5.7 Capacity Design Methods

There is a tendency in the design community towards a two-level design procedure, the service load level (elastic behavior for frequent earthquakes) and a collapse level (for extremely strong earthquakes). The collapse load is determined using the plastic mechanism approach [5.6, 5.7]. This method may be as simple as assuming a beam sway mechanism for the building, where the hinges occur only in the beams and at the bottoms of the columns. The procedure can be complicated for complex geometries. Like the other methods, this one is more suitable for design because we can assure the location of the hinges and the necessary ductility by proper detailing.

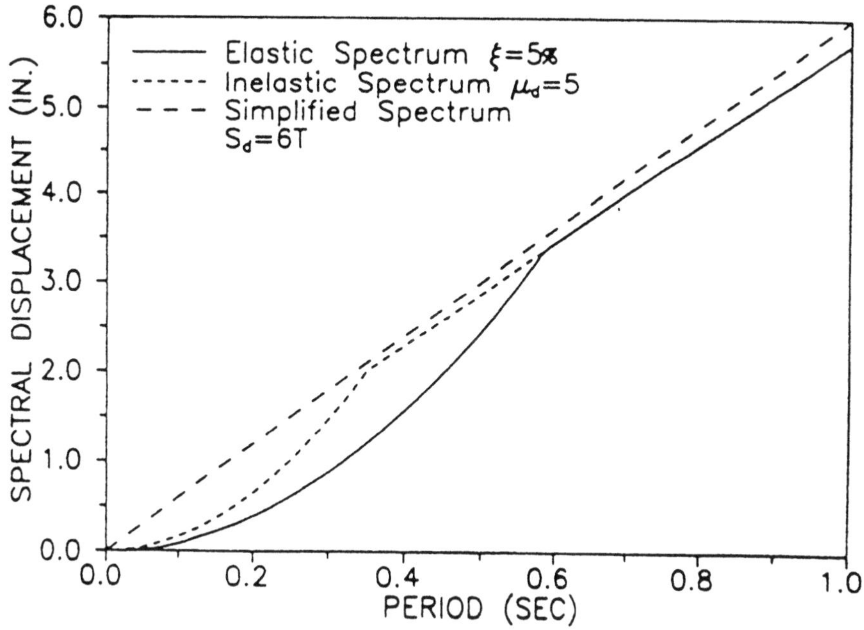

Fig. 5.7 Idealized displacement spectra

5.8 Conclusions

Although a properly done nonlinear dynamic time history analysis would be the most accurate method, its general use in practice is most likely to be inappropriate because a number of major uncertainties involved. The results depend on many factors, and the engineer must be intimately familiar with the analysis program and with structural behavior as a necessary condition for avoiding errors. The simple and straightforward methods sketched out in this chapter are less error-prone, and they deserve to be considered as useful tools for routine analyses. Of course, if detailed information about the motion of the building is needed, for example for the generation of reliable floor spectra, recourse to the more elaborate time history analysis methods may be necessary.

5.9 References

[5.1] Freeman, S.A.: The capacity spectrum method for determining the demand displacement, ACI 1994 Spring Convention, San Francisco, CA, March 1994.

[5.2] Blume, J.A.: A reserve energy technique for the earthquake design and rating of structures in the inelastic range, Proceedings, Vol. II, 2nd World Conference on Earthquake Engineering, Tokyo, 1960.

[5.3] Shibata, A. and M.A. Sozen: The substitute-structure method for seismic design in reinforced concrete, ASCE, Journal of the Structural Division, 102 (1976), no. ST1, 1-18.

[5.4] Gülkan, P. and M.A. Sozen: Response and Energy Dissipation of Reinforced

Concrete Frames Subjected to Strong Base Motions, Structural Research Series No. 377, University of Illinois, Urbana, IL, May 1971.
[5.5] Moehle, J.P.: Displacement-based design of RC structures subjected to earthquakes, Earthquake Spectra, 8 (1992), no. 3.
[5.6] Priestley, M.J.N. and Calvi, G.M.: Towards a capacity-design assessment procedure for reinforced concrete frames, Earthquake Spectra, 7 (1991), no. 3.
[5.7] Park, R.: Ductile design approach for reinforced concrete frames, Earthquake Spectra, 2 (1986), no. 3.

EXAMPLES OF IMPLEMENTATION

J. Blaauwendraad and C. van der Veen
Delft University of Technology, Delft, The Netherlands

C. Meyer
Columbia University, New York, NY, USA

6.1 Analysis Example by DIANA

6.1.1 Introduction

A cylindrical wall of structural concrete protects an inner steel tank for mass LNG storage. The concrete wall is loaded by a passing pressure wave created by a gas cloud explosion at some distance. The dynamic study first assumes linear elastic and thereafter nonlinear material behavior due to cracking and/or plasticity.

The circular cylindrical wall is a shell in which both membrane forces (in-plane) and bending actions (perpendicular to plane of shell) occur. These shell actions will be determined.

The worked-out example focusses on two different aspects, namely to show how to perform a dynamic transient analysis and to stress the importance of validating the output results. A complex dynamic analysis of a structure of such size can easily become costly and time-consuming. Therefore, analysts must continuously look for simple techniques that are suitable for verifying whether or not the output of the complex dynamic analysis is reliable.

6.1.2 Input Data

Structural Data. Geometrical data will be expressed in meters (m). Load data and stiffness data are expressed in kilonewtons (kN) and meters (m). For mass data we will use the unit *ton* (equal to 1000 kilograms) and meter (m). The unit for time is second (sec).

The concrete cylindrical shell has the following dimensions:

Radius, $R = 27.2\ m$
Height, $H = 32.4\ m$
Wall thickness, $h = 0.5\ m$

The shell is open at the upper end (no roof) and rests on a pile foundation at the bottom. The distance between piles is small with regard to the shell circumference. Therefore, the ring of piles can be considered as a uniformly distributed line-spring. In the present analysis the springs are linear-elastic for both horizontal and vertical forces. A pin connection is assumed between the shell and the springs, so the shell's bottom edge can rotate freely. The vertical spring stiffness per m of the shell is k_v. Horizontally the stiffness is k_h, both in the radial and tangential direction. Consulting a foundation engineering expert, the following values were selected:

$$k_h = 2000\ kN/m \text{ per } m$$
$$k_v = 50000\ kN/m \text{ per } m$$

In case of a very rigid foundation one can choose large values for k_h and k_v, say 1,000,000 kN/m, with the effect that bottom boundary displacements of the shell are approximately zero.

The Poisson ratio is set to zero. This ratio has only minor influence for this type of structure. Moreover, the zero value is often applicable, unless the concrete is subjected to triaxial compression. For the modulus of elasticity (Young's modulus) a value typical for structural concrete has been chosen:

$$E = 32.5 \times 10^6\ kN/m^2$$
$$\nu = 0$$

For the damping ratio we chose:

$$\xi = 0.05$$

which can be justified for rather heavily loaded concrete. The use of the parameter ξ presumes proportional damping. Possible additional damping due to the foundation is not considered in the present study. If such damping were to be included, concentrated dampers would have to be added to the nodes of the bottom boundary of the shell. The specific mass of structural concrete is:

$$\rho = 2.5\ ton/m^3$$

The shell structure has been somewhat simplified for the purpose of this example. In a real structure additional ring-type stiffening beams may exist at the top and bottom edges of the shell, and the wall may be thickened locally in case of prestressing to allow for proper anchorage of the prestressing cables. Such details are not essential for the purposes of this study, but may require quite a bit of additional work or confuse the issue. Therefore, they are omitted in the present study.

Load Data. The gas cloud explosion (deflagration) causes a pressure wave which propagates with a velocity v in a terrain with obstacles. A good approximation for v is the velocity of sound in air at atmospheric conditions, i.e., $v = 340\ m/s$. Here we consider the "effective pressure" above ambient.

When we measure the effective pressure at the passing of an undisturbed pressure wave in a terrain without obstacles, a pressure time history is recorded. In this example a history is used in which the effective pressure $p(t)$ increases linearly from zero to a value p in some time interval t_d, Fig. 6.1. At time t_d the effective pressure drops abruptly to zero. When the cylindrical shell is being passed, the effective pressure changes due to reflections from the obstacle. In this case study, a reflection coefficient r is used which depends on the cylinder coordinate ϕ. The formula for $r(\phi)$ has been derived by the Prins Maurits Laboratory of TNO in The Netherlands,

$$r(\phi) \quad = \quad 4.333(1 + 0.0368\ cos\phi)^7 - 3.333$$

The resulting effective pressure is always normal to the shell surface and is constant over the height of the shell, but its distribution changes continuously with time in the circumferential direction. To give an impression of the value of $r(\phi)$, it has been calculated for three salient positions along the shell:

$$\begin{array}{ll} \text{front } (\phi = 0): & r(\phi) = 2.25 \\ \text{side } (\phi = \frac{\pi}{2}): & r(\phi) = 1 \\ \text{back } (\phi = \pi): & r(\phi) = 0 \end{array}$$

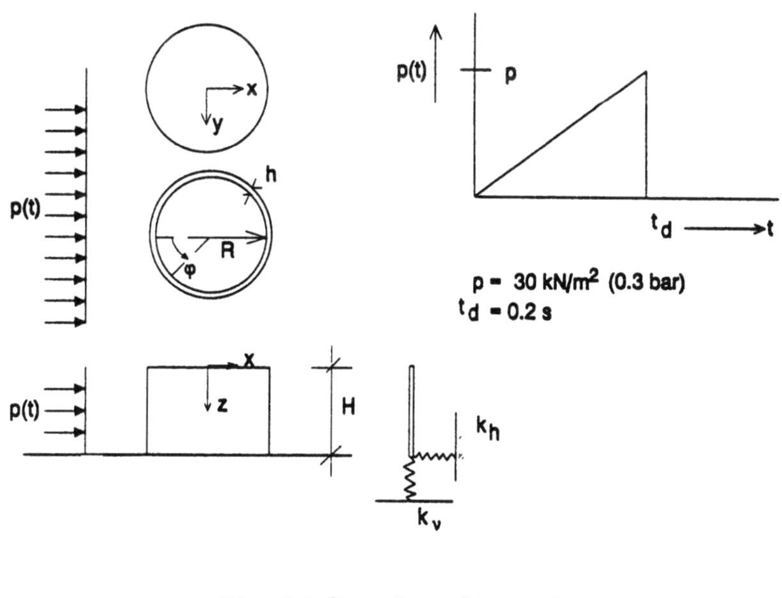

Fig. 6.1 Overview of input data

Summary of input data:

$H = 32.40\ m$
$R = 27.20\ m$
$h = 0.50\ m$
$\rho = 2.5\ ton/m^3$ (specific mass)
$\xi = 0.05$ (damping)
$E = 32.5 \times 10^6\ kN/m^2$
$\nu = 0$
Reflection coefficient: $r(\phi) = 4.333(1 + 0.0368\ cos\phi)^7 - 3.333$

Element Choice. The DIANA package offers flat shell elements and curved shell elements. The flat shell elements can be used for linear static analysis only. So, for nonlinear dynamics only curved shell elements can be considered. In this category, five different elements are at disposal for a combined "linear static" (L), "dynamic" (D) and "nonlinear" (N) analysis (see DIANA User's Manual). All these elements produce stresses $(\sigma_{xx}, \sigma_{yy}, \sigma_{xy}, \sigma_{zx}, \sigma_{zy})$ at the integration points. Only one of the elements (CQ 40S) offers the option to present results also as stress resultants across the thickness $(m_{xx}, m_{yy}, m_{xy}, q_{xz}, q_{yz}, n_{xx}, n_{yy}, n_{xy})$. As such output is preferable for the cracked state, we select element type CQ 40S. The C in this name refers to curved, Q to quadrilateral and S to shell. In the analysis of the cylindrical wall, the element will be used as a curved rectangle with straight vertical edges and curved horizontal edges. The element has 8 nodes (4 at the corners and 4 at mid-side positions), and each node has 5 degrees of freedom. So the number 40 in the element name refers to the number of degrees of freedom per element. The element has been derived from a solid element on the basis of a parabolic displacement field. This means strains vary linearly, and therefore, moments, shear forces and membrane forces likewise vary linearly over the element. The values of these stress resultants are given at the corner nodes.

Global and Local Axes. A global and a local coordinate system are distinguished. The global system is necessary to specify the coordinates of the nodes in the wall structure and to specify the boundary conditions. Its origin is at the center of the circle at the top of the cylinder. The positive x-axis is in the horizontal direction of the pressure wave, the y-axis is also horizontal and the positive z-axis points downwards, Fig. 6.2.

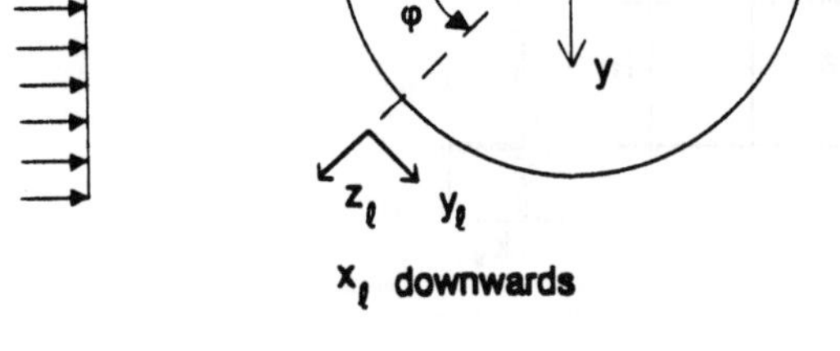

Fig. 6.2 Global and local axes

The local set of axes is defined for each element. The x_ℓ-axis points downwards (like the global z-axis), the y_ℓ-axis points into the circumferential direction and the z_ℓ-axis outward. This local set of axes is used to specify the loading on the structure and is the reference frame for the output stress resultants. The equations of motion are derived in the global directions. Therefore, the boundary constraints along the lines of symmetry must also be specified in those directions. The spring supports may be specified in global directions, too. The horizonal stiffnesses are equal in radial and tangential directions, so the same values can be used in the x- and y-direction.

6.1.3 Dynamic Analysis Considerations

Exact Dynamic Analysis. The dynamic analysis for a pressure wave passing the shell structure can be done in two ways:

1. Modal analysis, which profits from eigenmodes. This approach is only applicable for linear elastic analyses.
2. Direct integration of the equations of motion. This approach is suitable for both linear and nonlinear analysis.

In both approaches we can speak of a real dynamic analysis. Except for approximations of the finite element method, both approaches produce "exact" answers. In the case of linear elastic analyses one can use either method. Method 1 is cheaper than method 2 if a relatively small number of eigenmodes are sufficient. If the load effect is highly localized, one must consider the contribution of a large number of eigenmodes in method 1. In that case method 2 can be preferable. The explosion loading in the present study can be considered to be reasonably well distributed in the circumferential direction, without large gradients. Therefore method 1 could be chosen. However, we do not only want to perform a linear analysis, but also continue with a nonlinear analysis. In that case only method 2 can be considered. In any case, we do not have much of a choice when using DIANA, even for linear analyses, because the SHOCK module of this FEM package offers only the direct integration method for transient analysis. DIANA does have a modal analysis module (MODAL), but for harmonic loads only. It has not yet been implemented for shock type loads.

Approximate Dynamic Analysis. In many situations a complex exact dynamic analysis is not needed. For example, in a design phase, it may be possible to accept a simple approximation with less demand on time and money. For a preliminary design not much precision is needed, but rather speed and insight. Another reason for using a simpler method of analysis is the need to validate the results of a complex exact dynamic analysis. The more complex the analysis, the higher the risk for errors. It is then of great value if the order of magnitude of the results can be estimated independently. An approximate analysis is based on the idea that a complex structure can be seen as a simple one-degree-of-freedom mass-spring-system. An approximate analysis consists of three steps:

- a linear static analysis;
- computation of the period T for the lowest eigenmode;
- determination of the dynamic load factor (DLF), given the load duration t_d and

the period T of the structure;
- multiplication of the static analysis results by DLF.

Representation of Mass and Damping. DIANA offers in principle two options for the input of mass, "Consistent" and "Lumped". For the present study a transient analysis is performed, using DIANA/SHOCK, which accepts only Rayleigh damping:

$$M\ddot{u} + C\dot{u} + Ku = p(t)$$

in which

$$C = aM + bK$$

In this case a consistent mass matrix is obligatory in DIANA, so such mass representation will be used. The coefficients a and b must be chosen appropriately. In a modal analysis, a damping ratio ξ would normally be selected. For heavily loaded or cracked structures, a value 0.05 (5%) might be applied to all modes. If we know the first two frequencies ω_1 and ω_2, the coefficients a and b are determined as follows. With $\xi_1 = \xi_2 = 0.05$, let

$$\alpha = \frac{\omega_1}{\omega_2}, \quad \beta = \frac{\xi - \alpha\xi_2}{\omega_2 - \alpha\omega_1}$$

Then,

$$a = 2\omega_1\omega_2\beta, \quad b = 2\beta$$

and for the first two modes with frequencies ω_1 and ω_2, the damping will be $\xi_1 = \xi_2 = 0.05$. For higher modes with frequencies ω_i, $i \geq 3$, the damping values ξ_i will be higher than 0.05 and increase with mode number i.

Analyses to be Performed. The final goal in the present example, a nonlinear dynamic analysis, will be approached in five distinct steps:

1. Linear static analysis
2. Determination of eigenfrequencies and eigenmodes
3. Determination of dynamic load factor (DLF)
4. Linear dynamic analysis by direct integration
5. Nonlinear dynamic analysis by direct integration

The linear static analysis is done for a pressure loading which causes the maximum horizontal resultant in the x-direction. If $t = 0$ is the time at which the pressure wave reaches the cylindrical shell, the maximum loading will occur at time $t = 0.2$ *sec* with a maximum effective pressure of 30 kN/m^2 at the front vertical line ($\phi = 0$; $x = -R$). The results of the static analysis give a first impression of the behavior of the structure. The absolute values do not have any meaning for the dynamic analysis, but the distribution of forces is already revealed more or less. In addition, the finite element mesh and several input data can be checked adequately. Finally, the displacement field of the static run can be used in the Rayleigh method to estimate the lowest frequency.

The determination of eigenfrequencies and eigenmodes is most meaningful if a modal analysis is planned. In that case one can see the contribution of each mode to a selected design variable (e.g. a bending moment), and it becomes clear which higher modes do not contribute noticeably. If a non-modal analysis is done, i.e. a direct integration, the computation of eigenfrequencies and eigenmodes is still not completely meaningless. At least the lowest frequency is worth knowing, to determine the dynamic load factor (DLF). In the following, each of the five analyses will be discussed in the framework of separate tasks and their respective results.

6.1.4 Linear Static Analysis

Task 1: Loading

- Indicate when the loading generates the largest horizontal resultant force.
- Determine the reflected effective pressure at the following five positions in the circumferential direction: $\phi = 0,\ \frac{\pi}{4},\ \frac{\pi}{2},\ \frac{3\pi}{4},\ \pi$.
- Plot the pressure distribution.

Results of Task 1

For the dimensions of the protection wall (see structural data) and pressure wave velocity v (see load data), the passing time is $t = 0.16$ *sec*. Consequently, the loading position is determined which generates the largest horizontal force, Fig. 6.3.

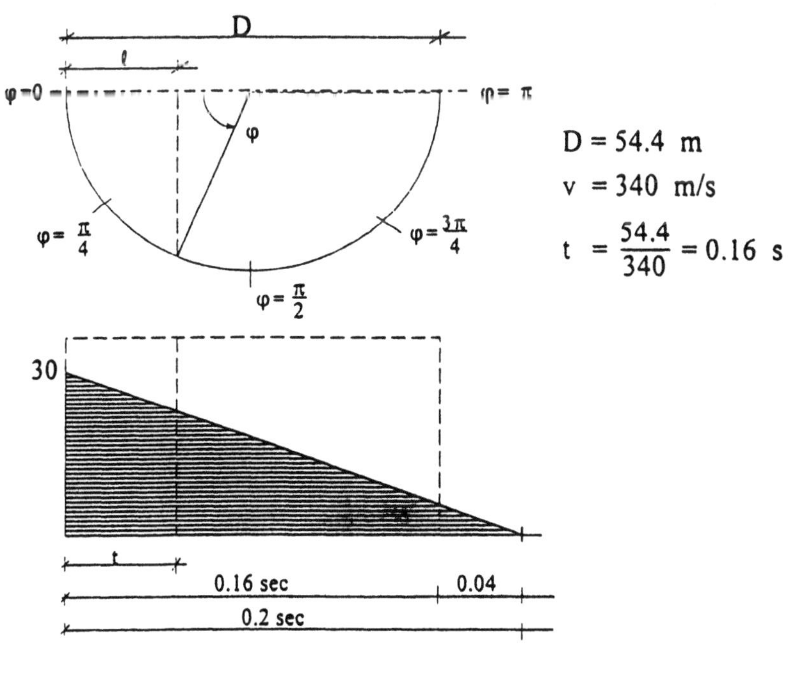

Fig. 6.3 Loading position which generates the largest horizontal resultant force

The pressure p, reflection coefficient r and effective pressure pr is calculated for the five positions in the circumferential direction. Typical values are listed in Table 6.1. The distribution along the circumferential direction is shown in Fig. 6.4.

ϕ	ℓ	t	p	r	pr
0	0.00	0.000	30.0	2.25	67.5
$\pi/4$	8.00	0.023	26.5	1.85	49.0
$\pi/2$	27.20	0.080	18.0	1.00	18.0
$3\pi/4$	46.43	0.137	9.5	0.27	2.6
π	54.40	0.160	6.0	0.00	0.0

Table 6.1 Typical values to calculate the effective pressure

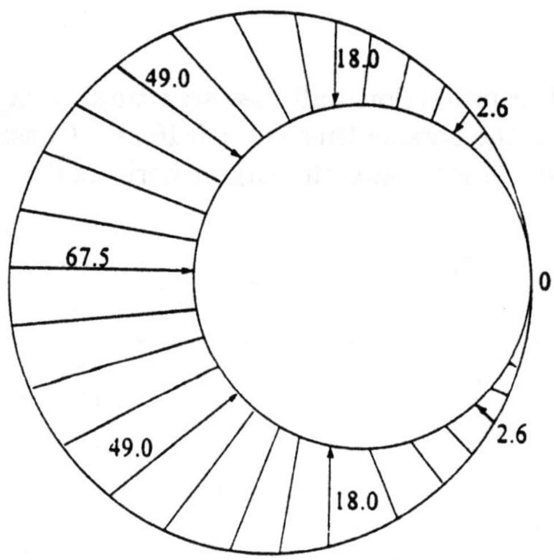

Fig. 6.4 Typical distribution of effective pressure along circumference

Task 2: Expectation of shell behavior

Indicate your expectation about the distribution of the following quantities:
- horizontal displacements along the upper shell edge;
- membrane forces n_{xx} in the vertical and n_{yy} in the circumferential direction;
- moments m_{xx} in the vertical and m_{yy} in the circumferential direction.

Results of Task 2

The shell can be considered as a short clamped beam of circular cross section. Then its behavior can be expected as shown in Fig. 6.5.

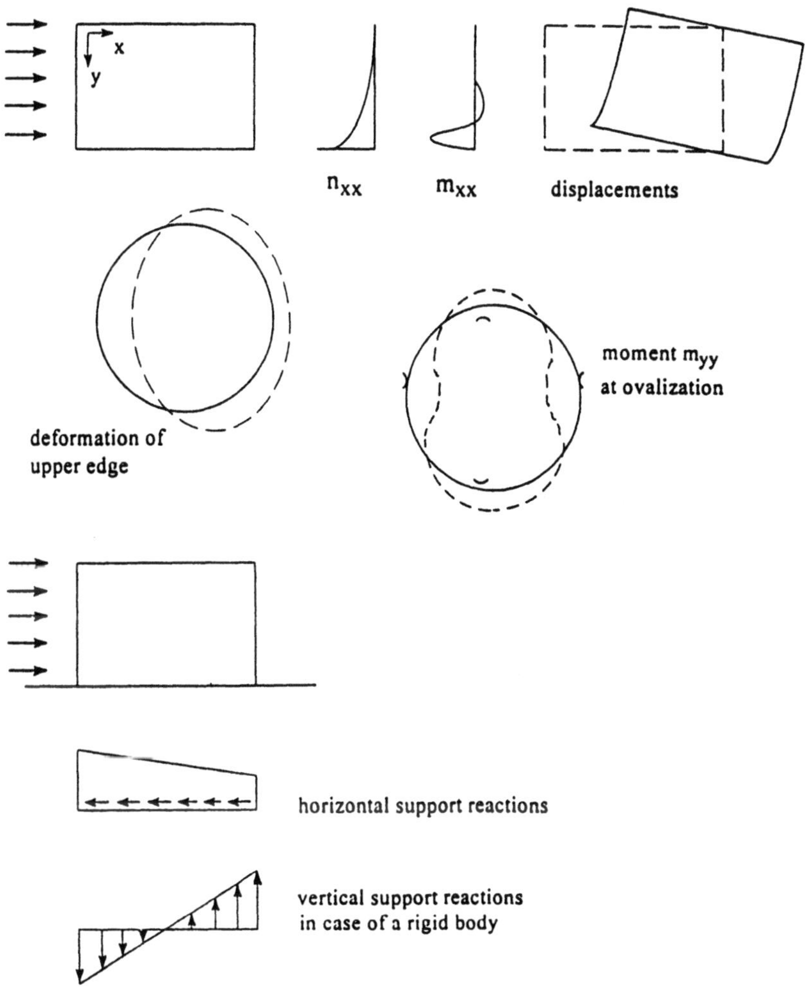

Fig. 6.5 Expected shell behavior

The effect of the edge disturbance zone on the moment distribution m_{xx} can be calculated with the aid of the theory of plates on an elastic foundation. In this way we find for the characteristic length,

$$\ell_{char} = \sqrt[4]{\frac{4EI}{k}}$$

With

$$EI = \frac{E}{12}h^3 \quad \text{and} \quad k = \frac{Eh}{R^2} \quad (\text{per } m)$$

we obtain,

$$\ell_{char} = \sqrt[4]{\frac{R^2h^2}{3}} = 0.76\sqrt{Rh} = 0.76\sqrt{27.2 \times 0.50} = 2.80\ m$$

Task 3: Choice of element mesh

Decide on the element mesh to be chosen, taking into account the expectation about the stress resultants as found in Task 2:

- plot the mesh, with size indications;
- number the nodes;
- number the elements;
- specify the boundary conditions.

Results of Task 3

In the circumferential direction, 15 elements are sufficient for the half circle. Considering n_{xx}, an equidistant mesh of 8 to 10 elements would be fine. A finer mesh is needed for the bending disturbance edge zone in the lower region of the shell. The moment m_{xx} is of the kind,

$$m_{xx} = F\ell_{char}\ sin\frac{x}{\ell_{char}}\ exp\left(-\frac{x}{\ell_{char}}\right)$$

where F is the horizontal support reaction.

Four elements are needed per half sine wave of the function. So, the first four elements cover a height of

$$\pi\ell_{char} = 8.8\ m$$

$$x_{max} = \frac{\pi}{4}\ell_{char}$$

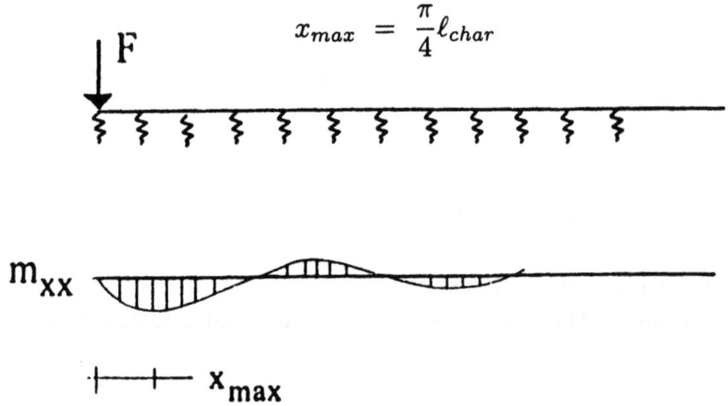

Fig. 6.6 Element mesh and boundary conditions

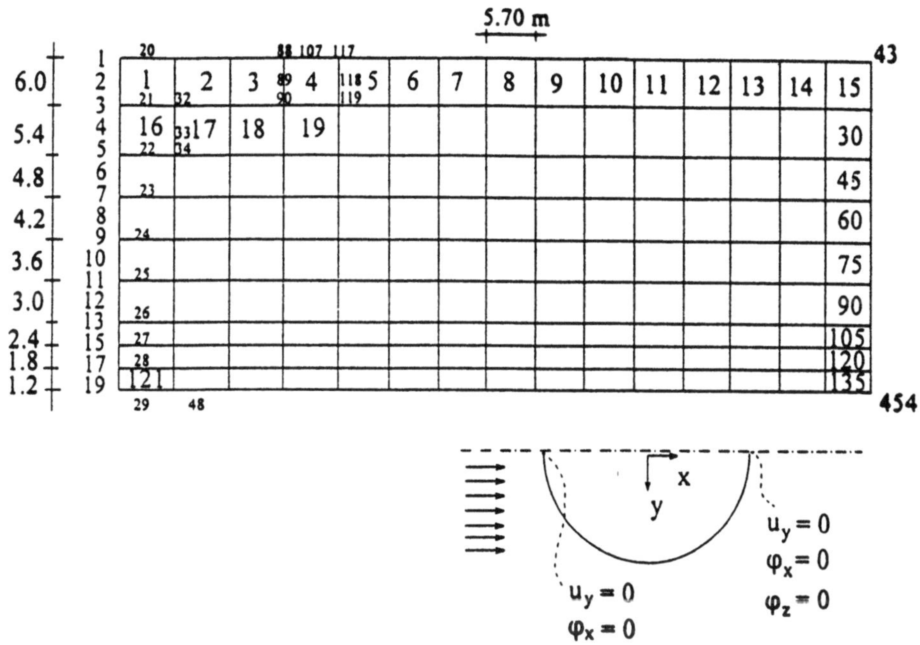

Fig. 6.6 (Cont'd) Element mesh and boundary conditions

Task 4: Conversion of pressure into nodal forces

The loading of Task 1 must be converted into lumped forces at the nodes of the element mesh. In a vertical series of elements the same pressure is applied to each element. Hereafter we refer to such a vertical series as a "column". Within an element of such a column the pressure is constant in the vertical direction, but does change in the circumferential direction.

- Indicate how the distributed pressure is converted into nodal forces (which nodal forces result in case of constant pressure over the area of an element?)

Results of Task 4

Results for Column 4 (for example), with

$$\Delta p = 27.7\ kN/m^2, \quad r = 1.99, \quad \pi R/15 = 5.70\ m,$$

El.	Area A $\Delta p \cdot r \cdot A$	Node	Nodal Load
4	$6.0 \times 5.7 = 34.2\ m^2$	88	-1/12 × 1880 = -157 kN
	1880 kN	107	1/3 × 1880 = 627
		117	-1/12 × 1880 = -157
		89	1/3 × 1880 = 627
		118	1/3 × 1880 = 627
		90	-1/12 × 1880 = -157
		108	1/3 × 1880 = 627
		119	-1/12 × 1880 = -157
19	$5.4 \times 5.7 = 30.8\ m^2$	90	-1/12 × 1700 = -142
	1700 kN	108	1/3 × 1700 = 567
		119	-1/12 × 1700 = -142
		91	1/3 × 1700 = 567
		120	1/3 × 1700 = 567
		92	-1/12 × 1700 = -142
		109	1/3 × 1700 = 567
		121	-1/12 × 1700 = -142

Etc.

Task 5: Spring constants of pile foundation

The pile foundation is represented by elastic springs at the bottom nodes. Determine these lumped spring stiffnesses. At each node three springs must be used, two in the horizontal direction and one in the vertical direction. The stiffnesses are specified for the global axes.

Results of Task 5

Element edge -

$$\begin{aligned}\text{x}-,\text{y}-\text{direction}: K_1 &= 1.9\ kN/mm = 1{,}900\ kN/m\\ 2K_1 &= 3{,}800\ kN/m\\ K_2 &= 7{,}600\ kN/m\\ \text{z}-\text{direction}: K_1 &= 47{,}500\ kN/m\\ 2K_1 &= 95{,}000\ kN/m\\ K_2 &= 190{,}000\ kN/m\end{aligned}$$

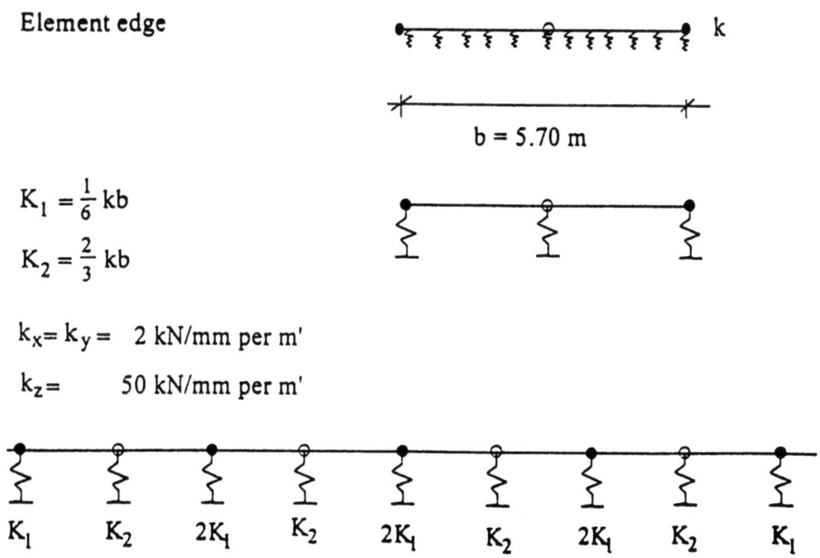

Fig. 6.7 Foundation springs

Task 6: Displaying results

The most important design quantities along the most important lines/cross sections of the shell are plotted. Compare these results with the expectations in Task 2.

Results of Task 6

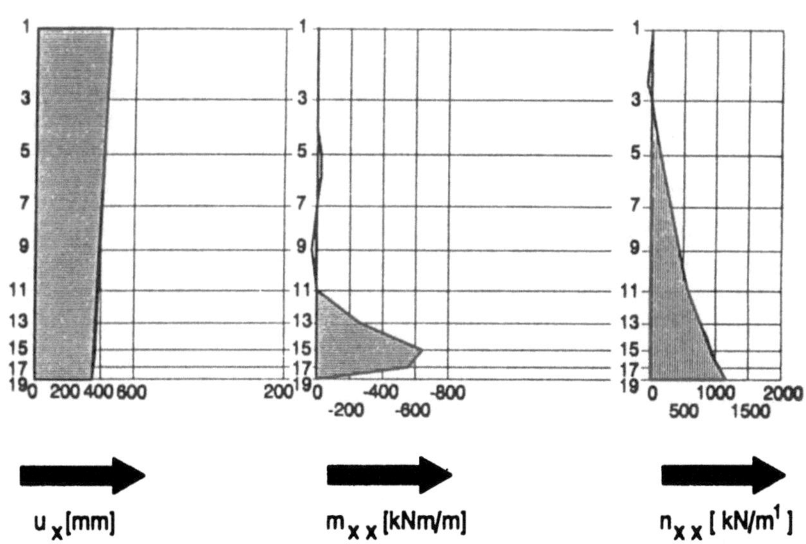

Fig. 6.8 Static analysis results

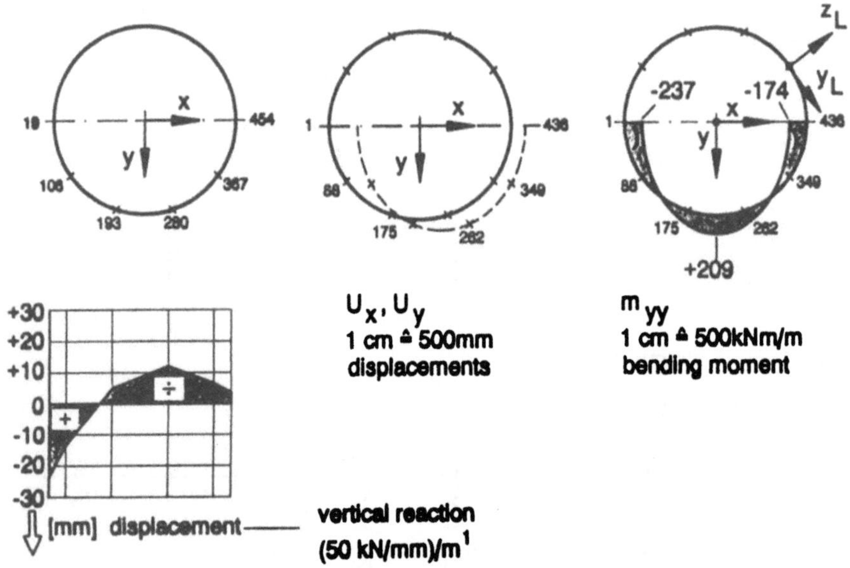

Fig. 6.8 (Cont'd) Static analysis results

6.1.5 Approximate Dynamic Response

Eigenfrequencies and Eigenmodes. Compared with the static analysis, a number of changes or additions must be made now. First of all, mass must be specified. In DIANA, one can specify specific mass, and the program generates the mass matrix. As mentioned in Sect 6.1.3, DIANA/SHOCK and Rayleigh damping require a consistent mass matrix.

The element mesh for a dynamic analysis is not necessarily the same as that for a static analysis. However, without good cause one will not change the mesh. A part of the input exists then already and has been tested (which is even more important). In the present study no clear reason exists to modify the mesh. It is expected that sufficient mass degrees of freedom occur per wave length in the deflection mode.

The next decision regards the choice of the method to compute the eigenfrequencies and the eigenmodes (if the software offers more than one option). The considered explosion loading is in fact a distributed loading which does not change rapidly in the circumferential direction. This justifies the use of a restricted number of (lowest) eigenfrequencies and eigenmodes. For this study 15 modes are sufficient.

The Housholder tri-diagonalisation is very time-consuming and only has advantages if one wants to know all eigenvalues and eigenvectors. In the present study this is not the case. The inverse power iteration lends itself well for the computation of a few eigenvalues of large systems. That is needed in this study, so this method can be chosen. The subspace iteration is suitable for the solution of very large eigenvalue problems and has some advantages over the inverse power iteration. It can be hard to select a proper convergence criterion, both for the inverse power iteration and the subspace iteration. In the present study both methods can be considered equally well. In DIANA the subspace iteration is available.

Application of Eigenvalue Analysis. The computed lowest eigenvalue can be used to estimate the dynamic load factor (DLF) of the shell. We first consider a simple mass-spring-system with one degree of freedom. The system has a period T and is loaded by a force which increases linearly in t_d seconds from zero to F. The maximum dynamic displacement $u_{max.d}$ is related to the maximum static displacement $u_{max.s}$ for a static load F by the dynamic load factor,

$$DLF = \frac{u_{max.d}}{u_{max.s}}$$

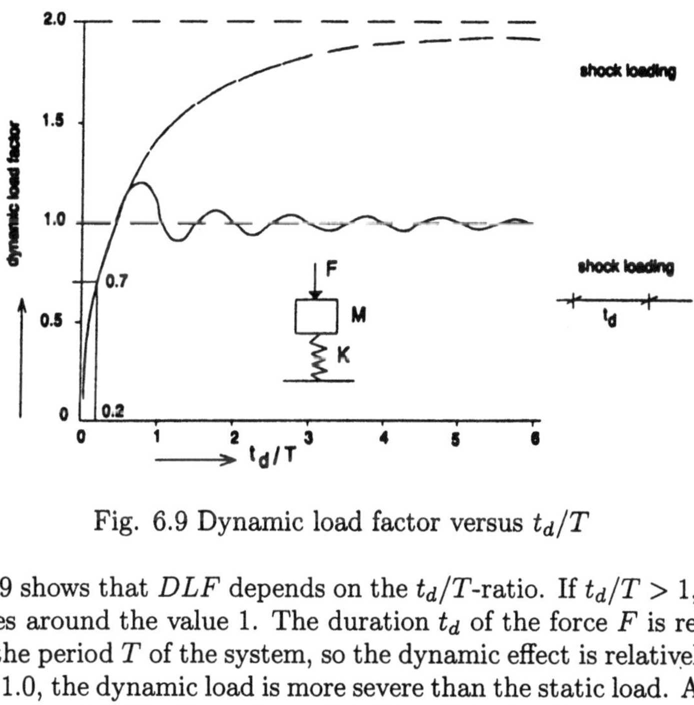

Fig. 6.9 Dynamic load factor versus t_d/T

Figure 6.9 shows that DLF depends on the t_d/T-ratio. If $t_d/T > 1$, the value of DLF oscillates around the value 1. The duration t_d of the force F is relatively long compared to the period T of the system, so the dynamic effect is relatively small. For $0.4 < t_d/T < 1.0$, the dynamic load is more severe than the static load. At maximum, the increase is about 25%. For values of $t_d/T < 0.4$, the dynamic response is less than the static behavior. Now the displacement is reduced because $DLF < 1$.

Task 7: Dynamic load factor

- Determine the t_d/T-ratio for the shell structure and use Fig. 6.9 to determine the value of DLF. The period of the first mode was computed to be $T_1 = 1$ *sec.*
- Apply DLF to the static analysis results of Task 6. This yields a design estimate for the dynamic behavior of the structure.

Results of Task 7

$$t_d = 0.2\ sec$$
$$T = 1.0\ sec$$
$$t_d/T = 0.2$$

From Fig. 6.9 we read, $DFL = 0.7$. This value is applied to the static analysis results to yield a design estimate for the dynamic behavior of the structure. A dynamic approximation is made for displacement u_x, moment m and normal force n at node 1 and the moment and normal force at node 15. Furthermore, a dynamic approximation is made for the horizontal and vertical reaction forces.

Dynamic approximations:

$$
\begin{aligned}
u_{x1} &= 0.7 \times 0.45 = 0.32\ m\\
m_{yy1} &= 0.7 \times -237 = -166\ kN\\
n_{yy1} &= 0.7 \times -1633 = -1143\ kN/m\\
m_{xx15} &= 0.7 \times -577 = -404\ kN\\
n_{xx15} &= 0.7 \times 300 = 210\ kN/m\\
R_{hor,19} &= 0.7 \times 675 = 473\ kN\\
R_{vert,19} &= 0.7 \times 1121 = 784\ kN
\end{aligned}
$$

It is expected that this estimate holds rather well for the displacements, the support reactions and membrane forces. The expectations are not high for the bending moments. Particularly the moments m_{yy} in the circumferential direction may be quite different when the pressure wave files past the structure.

Estimate of Eigenfrequency. Instead of making a DIANA run for the computation of the eigenfrequencies, one can approximate the lowest frequency by the Rayleigh method. We reduce the shell structure to a simple one-degree-of-freedom mass-spring-system. This is done by assuming a displacement field with one control displacement as fundamental unknown. We choose the displacements for the static run and adopt the horizontal displacement u_{x1} of node 1 at the upper edge ($\phi = 0$; $x = -R$) as the control displacement. The vertical displacements appear to be very small if compared to the horizontal displacements. Therefore, only the horizontal displacements are considered. These are defined in the global x,y,z coordinate system. Along the shell surface, a coordinate s is defined in the circumferential direction ($0 < s < \pi R$). If we denote the maximum horizontal displacement u_{x1} simply as u, the displacement field can be written as follows:

$$u_x(z,s) = \psi_x(z,s) \cdot u$$

$$u_y(z,s) = \psi_y(z,s) \cdot u$$

The shape functions $\psi_x(z,s)$ and $\psi_y(z,s)$ are considered to be:

$$\psi_x(z,s) = f(z) \cdot g_x(s)$$

$$\psi_y(z,s) = f(z) \cdot g_y(s)$$

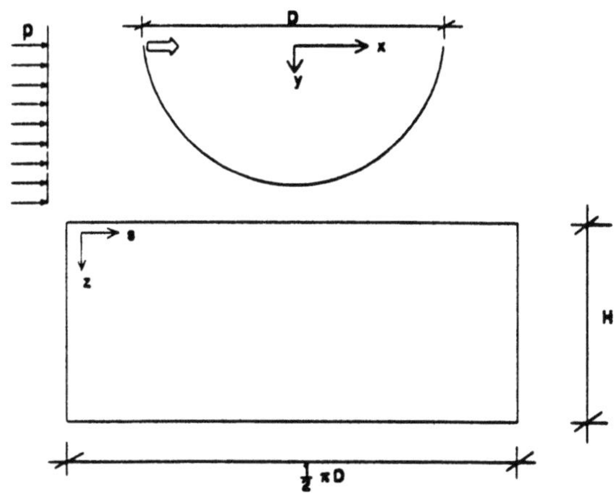

Fig. 6.10 Front view of shell surface

This means that the distribution over the height is the same for each position s along the shell and also is the same for displacements in the x- and y-direction. The vertical distribution through node 1 is chosen for $f(z)$.

We do not introduce large errors in this way. The approach can be physically interpreted such that we first concentrate mass in the vertical direction at the upper edge and after that the distributed mass along the upper edge is replaced by an equivalent mass M at node 1. In this weighting procedure we have to apply squares of ψ_x and ψ_y, because the kinetic energy must be represented correctly in the equivalent mass-spring-system. For the approximate analysis not much precision is necessary. We employ 6 nodes along the upper edge, so the distance between 2 nodes consists of 3 elements.

The stiffness K of the mass-spring-system must be approximated as well. This is done by computing the force F which is equivalent to the distributed pressure. For this purpose ψ_x and ψ_y must be used, because the potential energy of the loads must be represented well. The stiffness is then computed from,

$$K = \frac{F}{u}$$

in which the value of u_{x1} is used for u.

Task 8: Approximate eigenfrequency

- Find the static values of u_x and u_y
- Determine ψ_x and ψ_y at the 6 nodes of the upper shell edge and plot the distributions along the edge. These plots represent $g_x(s)$ and $g_y(s)$.
- Plot ψ_x along the vertical line through node 1 and normalize the distribution. This plot represents $f(z)$.

- Apply $f^2(z)$ to determine which mass M_i is assigned to each of the 6 nodes at the upper shell edge.
- Also determine the equivalent forces F_i at these 6 nodes due to the pressure, applying $f(z)$.
- Determine the equivalent mass M by:
  $M = \sum_{i=1}^{6} (g_{xi}^2 + g_{yi}^2) M_i$
- Determine the equivalent force F:
  $F = \sum_{i=1}^{6} (g_{xi} F_{xi} + g_{yi} F_{yi})$
- Compute the equivalent stiffness K from:
  $K = \frac{F}{u}$
- Compute the eigenfrequency:
  $\omega^2 = \frac{K}{M}$, and from this the period T.
- Compare this period with the period $T = 1.002$ *sec*, obtained from a DIANA eigenvalue analysis.

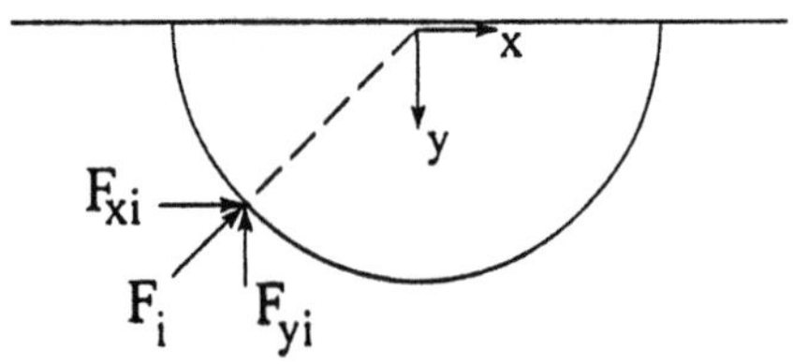

Results of Task 8

nodes	1	88	175	262	349	436
	û					
u_x	**0.452**	0.373	0.295	0.286	0.214	0.154
u_y	0.000	0.032	0.131	0.121	0.026	0.000
g_x	1.00	0.83	0.65	0.63	0.47	0.34
g_y	0.00	0.07	0.29	0.27	0.06	0.00
g_x^2	1.00	0.69	0.42	0.40	0.22	0.12
g_y^2	0.00	0.00	0.08	0.07	0.03	0.00
$g_x^2+g_y^2$	1.00	0.69	0.50	0.47	0.25	0.12
M_i	277020	554040	554040	554040	554040	277020
$M_i(g_x^2+g_y^2)$	277020	382288	277020	260399	138510	33242
	$M_{eq} = \Sigma M_i(g_x^2+g_y^2) = 1368478$ kg					
g_y	0.00	0.07	0.29	0.27	0.06	0.00
F_{yi}	0.00	-15975*	-13839*	-4493*	-409*	0
g_x	1.00	0.83	0.65	0.63	0.47	0.34
F_{xi}	16642	21988	4497	-1460*	-562*	0
$F_{xi}g_x+F_{yi}g_y$	16642	17132	-1090	-2133	-289	0
	$F_{eq} = \Sigma F_{xi}g_x+F_{yi}g_y = 30262$ kN					

* the force and displacement have an opposite sign.

$$K_{eq} = \frac{F_{eq}}{\hat{u}} = \frac{30262}{0.452} = 66951\ kN/m$$

$$M_{eq} = 1368478\ kg = 1368.5\ ton$$

$$\omega^2 = \frac{66951}{1368.5} = 48.9\ s^{-2}$$

$$\omega = 7.01\ s^{-1}$$

$$f = \frac{\omega}{2\pi} = 1.11\ s \rightarrow T = 0.90\ s \text{ (the computer calculation gives T = 1.0 s)}$$

$$\frac{t_d}{T} = \frac{0.2}{0.90} = 0.22 \rightarrow \text{dyn. factor} \approx 0.76 \quad \zeta = 0.00$$

$$0.71 \quad \zeta = 0.05$$

node	u_x	$f(z)$	$f(z)^2$
1	0.452	1.00	1.00
3	0.438	0.97	0.94
5	0.426	0.94	0.89
7	0.415	0.92	0.84
9	0.405	0.90	0.80
11	0.398	0.90	0.78
13	0.391	0.88	0.75
15	0.380	0.87	0.71
17	0.367	0.84	0.66
19	0.356	0.81	0.62

mean value $f(z)$ = 0.89 $\rho = 2500\ kg/m^3$

mean value $f(z)^2$ = 0.89 $H = 32.4\ m$ $\frac{\pi R}{5}$. $17.1\ m$ h = 0.50 m

$M_0 = 0.80\ \{17.1 * 32.4 * 0.50 * 2500\} = 554040\ kg$
$M_1 = M_6 +$ $1/2\ M_0$ $= 277020\ kg$
$M_2 = M_3 = M_4 = M_5 = M_0$ $= 554040\ kg$

			α	$F_{ix}=F_i \cos\alpha$	$F_{iy}=F_i \sin\alpha$
$p*r =$ 67.50 kN/m^2	F_1= 67.50*½ A^*=	16642 kN	0°	16642	0
= 55.15	F_2= 55.12 * A^* =	27179 kN	36°	21988	15975
= 29.51	F_3= 2.51 * A^*=	14551 kN	72°	4497	13839
= 9.58	F_4= 9.58 * A^*=	4724 kN	108°	- 1460	4493
= 1.41	F_5= 1.21 * A^*=	695 kN	144°	- 562	409
= 0.00	F_6= 0.00*½A^*=	0 kN	180°	0	0

$$A^* = \frac{\pi R}{5} \cdot H \cdot 0.89 = 493\ m^2$$

6.1.6 Linear Dynamic Analysis

The new load in the linear dynamic analysis is the explosion loading. This load is specified per column (vertical series) of elements. The element size (5.7 m) is small compared with the length of the pressure wave ($t_d \cdot v_{sound} = 68\ m$). Therefore we can compute the pressure in the middle of the row as a function of time and assign this function to all nodes in the row.

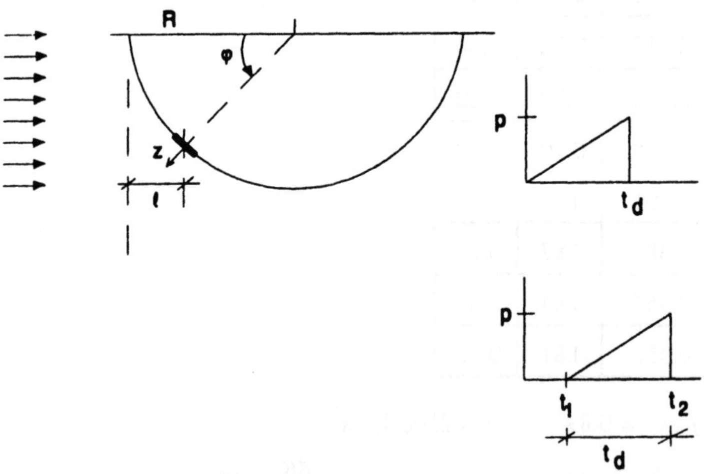

Fig. 6.11 Load time histories for each column of elements

The pressure wave reaches the middle of a row at time t_1:

$$t_1 = \frac{\ell}{v_{sound}}$$

and has completely passed the middle at time t_2:

$$t_2 = t_1 + t_d$$

The length ℓ can be calculated from

$$\ell = R(1 - cos\phi)$$

in which ϕ is the angle which defines the middle of the column under consideration.

For the rest we use:

$$\begin{aligned} R &= 27.20\ m \\ t_d &= 0.2\ sec \\ v_{sound} &= 340\ m/sec \\ p &= \text{-}30\ kN/m^2 \end{aligned}$$

The pressure is negative, because the z-axis of the local element set of axes is directed outward. The relevant reflection coefficients are applied per row. These coefficients differ slightly for the nodes in an element.

The DIANA input has been organized as follows:

- The elements have been defined in 15 groups (thc 15 columns) as in the static run.
- A load case is specified per group, and this load consists of the reflection coefficients (load case 1 to 15).
- The 15 load cases are combined into one load case 18. In this combination the load-time history of each group is applied to the load case (with reflection coefficients) which refers to this group.

Task 9: Load-time history

- Determine the load time history for all 15 groups. (If t_2 is exactly equal to a multiple of the chosen time step, it must be slightly increased to make sure that the correct load is applied; e.g. 0.2001 instead of 0.2).

Results of Task 9

We have to calculate the start-time (t_1) and the end-time t_2 of the local interval. Some values are calculated and presented in Table 6.2.

Column	ϕ (Rad)	ℓ (m)	t_1 (sec)	t_2 (sec)
1	$\pi/30$	0.149	0.0004	0.2004
2	$3\pi/30$	1.331	0.0039	0.2039
3	$5\pi/30$	3.644	0.0171	0.2171
4	$7\pi/30$	6.986	0.0205	0.2205
...	...	...	...	...
15	$29\pi/30$	54.251	0.1596	0.3596

Table 6.2 Start- and end-times of load time history

Choice of time step. The choice of the time step is influenced by two considerations. The first one regards the load time history and the second one the correct description of the system response. As for the load time history, the time step Δt must be sufficiently small in comparison with the load duration time t_d,

$$\Delta t \ll t_d \quad .$$

A more precise criterion for Δt is the condition that not more than one row of elements is unloaded during one time step. The choice $\Delta t = 0.02$ *sec* satisfies the condition that Δt is small compared with t_d, but does not prevent the first three rows from unloading suddenly in the eleventh step ($0.2 < t < 2.2$). During this time step, t_2 of the load time history is reached for the first three groups (0.2004, 0.2039, 0.2171, respectively). This means that one quarter of the shell gets unloaded during one time step. We cannot expect to compute membrane forces and bending moments correctly in this way.

For a smooth unloading of the shell surface each difference Δt_2 must comprise a sufficient number of time steps Δt,

$$\Delta t \ll \Delta t_{2,min}$$

The correct reproduction of the system response poses another condition on the time step. For this purpose we need to know which frequencies and modes of the system are to be activated. We want to make sure that the contribution of the high-frequency modes is computed correctly. Preferably 20 time steps must be applied during the (small) period T_i of the highest mode (using as a criterion that the period be computed with less than 5% error).

$$\Delta t \leq \frac{T_i}{20}$$

To decide on the highest frequency to be taken into account, one must determine which frequencies are dominant in the load time history. This requires an expansion of the load into a Fourier series. If a modal analysis approach is used, such an expansion is not needed. In that case one can see which system frequencies contribute to the total response and to what extent. For the present shell structure, only the first 15 modes are of importance. For the rest it will be seen that the loading criterion is highly dominant in the present study.

Number of time steps. To decide on the number of time steps we can consider the following. If one wants to see some repetition in the global response, the first eigenmode must be repeated two times. So the total time at which the computation is ended (t_e) must be larger that $2T_1$, where T_1 is the period of the first mode.

$$t_e > 2T_1$$

Above that the load must be released completely at all elements of the shell, which further requires that, with shell diameter D,

$$t_e \geq \frac{D}{v_{sound}} + t_d$$

Task 10: Time steps

- Select the time step Δt.
- Select the total time t_e.

Results of Task 10

Only one row of elements is allowed to unload in one time step Δt. It follows from the applied load time history that the minimum difference between t_2 of two unloaded columns amounts to $\Delta t_{2,min} = 0.0035$ *sec*. The duration $t_d = 0.25$ *sec* of the force and the period of the 15th eigenmode, $T_{15} = 0.1$ *sec* are relatively long compared to $\Delta t_{2,min}$. Consequently, a time step $\Delta t = 0.002$ *sec* has been used.

The total time t_e must be larger than

$$t_e \; > \; 2T_1 \; = \; 2 \times 1 \; = \; 2 \; sec$$

and

$$t_e \; > \; \frac{D}{v_{sound}} + t_d \; = \; \frac{54.40}{340} + 0.2 \; = \; 0.36 \; sec$$

Thus, $t_e = 2$ *sec*.

Task 11: Linear dynamic response

The following design quantities are plotted versus time:

- displacement u_x of node 1
- membrane force n_{xx} in element 106, node 15
- membrane force n_{yy} in element 1, node 1
- moment m_{xx} in element 106, node 15
- moment m_{yy} in element 1, node 1
- horizontal support reaction R_x at node 19

- vertical support reaction R_z at node 19

Compare these results with those of the approximate dynamic analysis.

Results of Task 11

The dynamic load factor for the approximate dynamic analysis is 0.76 ($\xi = 0$). The results of the dynamic analysis are compared with those of the static analysis in Table 6.3.

A dynamic load factor in the range of 0.51 to 0.77 was computed. Thus the DLF corresponds reasonably well with the value of the DLF calculated with the approximate analysis.

	dynamic	static	DLF
u_x node 1	0.25	0.45	0.56
u_x node 19	0.19	0.36	0.53
m_{yy} node 1	-120	-237	0.51
n_{yy} node 1	-1250	-1633	0.77
m_{xx} node 15	-310	-577	0.54
n_{xx} node 15	697	900	0.77
n_{xx} node 19	818	1123	0.73
vert. reaction node 19	818	-1121	0.73
horz. reaction node 19	351	675	0.52

Table 6.3 Comparison of static and dynamic response

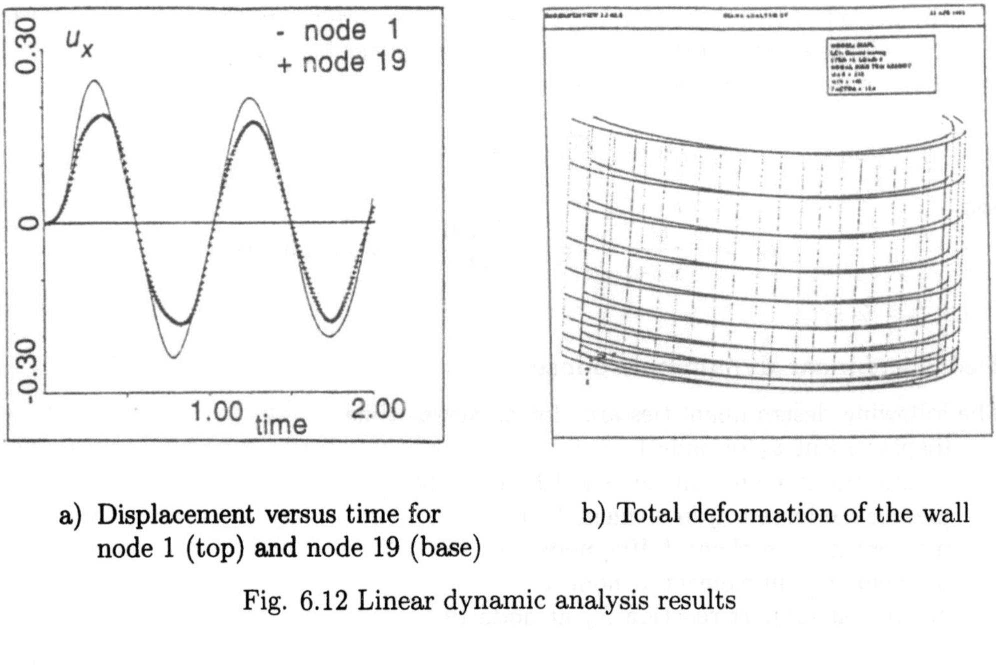

a) Displacement versus time for node 1 (top) and node 19 (base)

b) Total deformation of the wall

Fig. 6.12 Linear dynamic analysis results

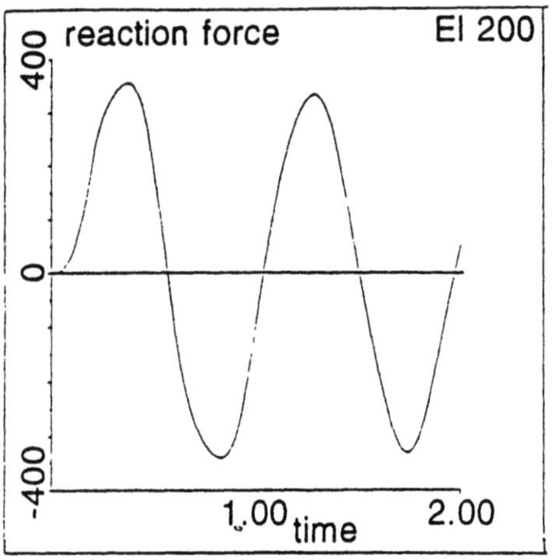

c) Horizontal reaction at node 19

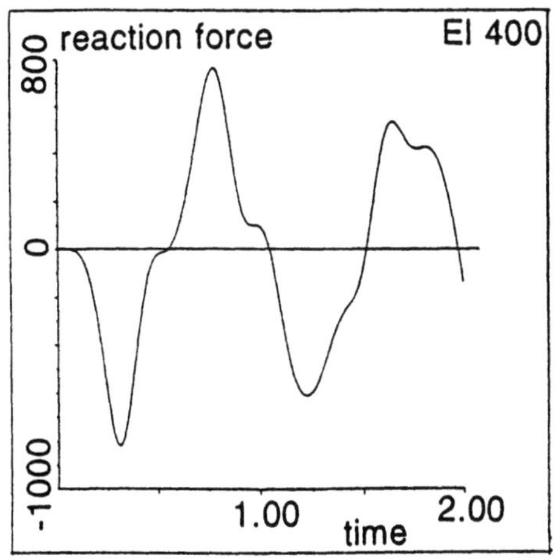

d) Vertical reaction at node 19

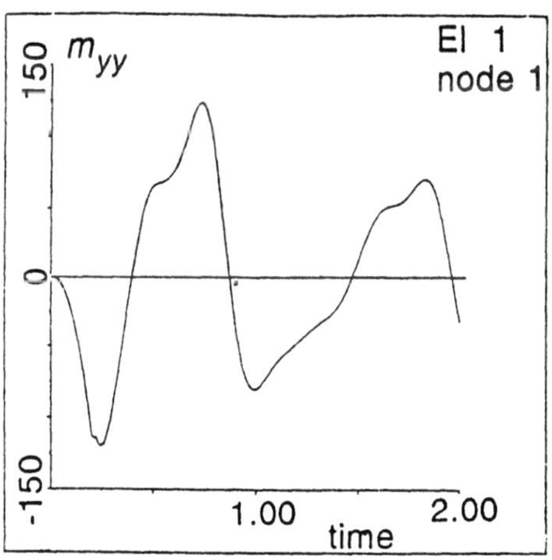

e) Moment m_{yy}, element 1, node 1

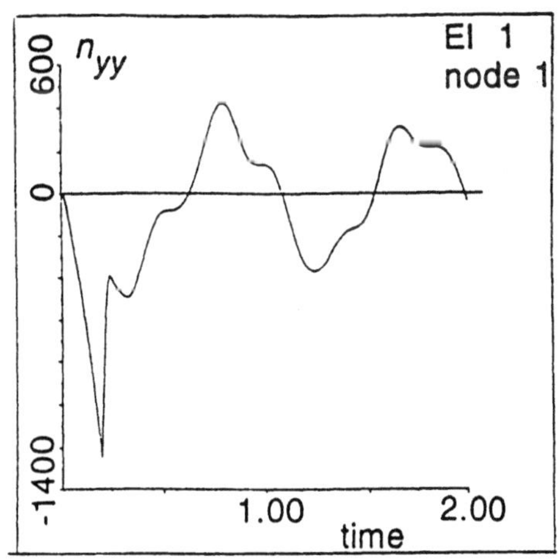

f) Normal force n_{yy}, element 1, node 1

Fig. 6.12 (Cont'd) Linear dynamic analysis results

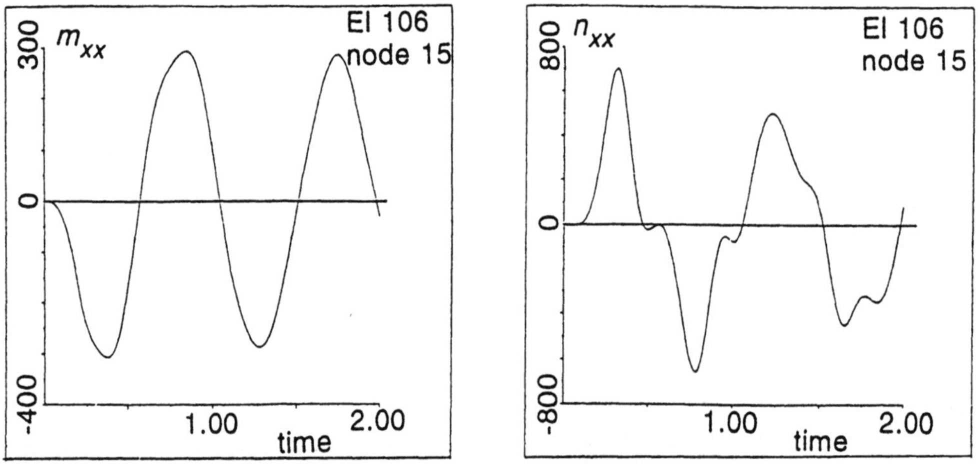

g) Moment m_{xx}, element 106, node 15 h) Normal force n_{xx}, element 106, node 15

Fig. 6.12 (Cont'd) Linear dynamic analysis results

Task 12: Displaying results

The most important design quantities are plotted along important lines/cuts of the shell as in Task 6 for the static analysis. The selected time is that at which these quantities are maximum.

Results of Task 12

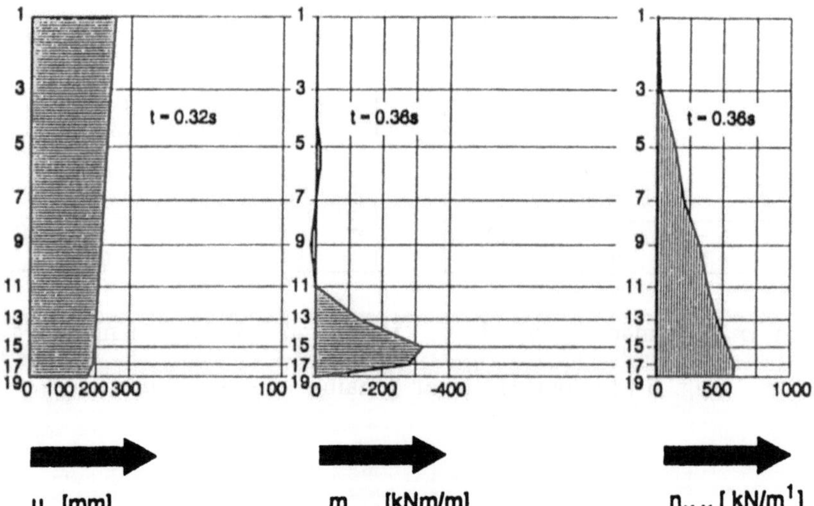

Fig. 6.13 Maximum linear dynamic response values

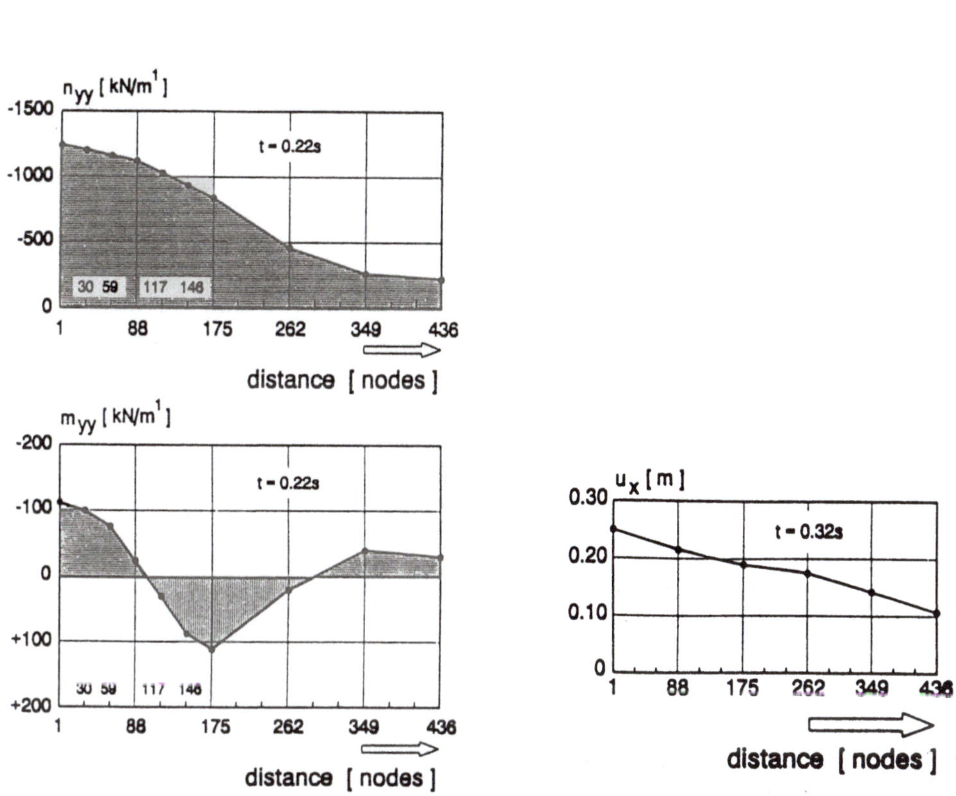

Fig. 6.13 (Cont'd) Maximum linear dynamic response values

6.1.7 Nonlinear Dynamic Analysis

The nonlinear dynamic analysis differs from the linear dynamic analysis for the additional material data and loading cases. No reason exists to modify the element mesh or the time step.

Definition of load. The non-linear elastic dynamic analysis of the shell structure takes into account three different load cases:

- dead weight
- prestressing
- explosion loading

These load cases are applied in two steps. The dynamic analysis has to be preceded by a static analysis for dead weight and prestressing. This is the first step. The displacements and stresses resulting from a combination of these two load cases form the initial conditions at $t = 0$ for the second step of the dynamic analysis for the explosion loading. In the present version of DIANA one has to reenter the dead weight in the second step as a permanent load which is constant in time. The prestressing loads need not be repeated, because this load case is an internal loading.

Integration schemes. For the linear elastic dynamic analysis a 2 x 2 x 2 integration scheme is used. Thus, in the ζ, η and ϕ directions, two integration points are used, see Fig. 6.14. In this case the Gauss integration through the thickness is performed and all integration points are located inside the element volume. However, for the nonlinear dynamic analysis more integration points through the thickness are necessary in order to simulate nonlinear material behavior through the thickness, e.g. cracking with softening and plasticity. In this example, 7 integration points are used through the thickness, together with a Simpson integration. Some integration points are located in the upper and lower faces or in the mid-plane of the element. This may facilitate interpretation of strains and stresses.

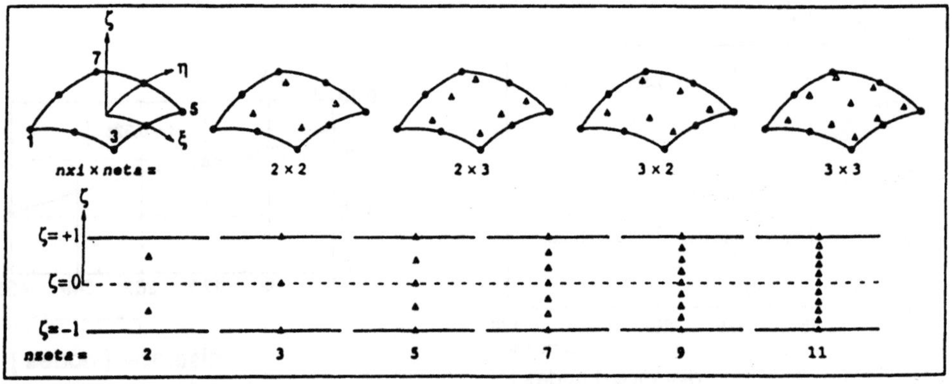

Fig. 6.14 Element CQ40S - Integration schemes

Material data. The concrete is modelled as follows:

- Mohr-Coulomb yield criterion
- no hardening
- linear tension-softening
- tension cut-off (constant stress)

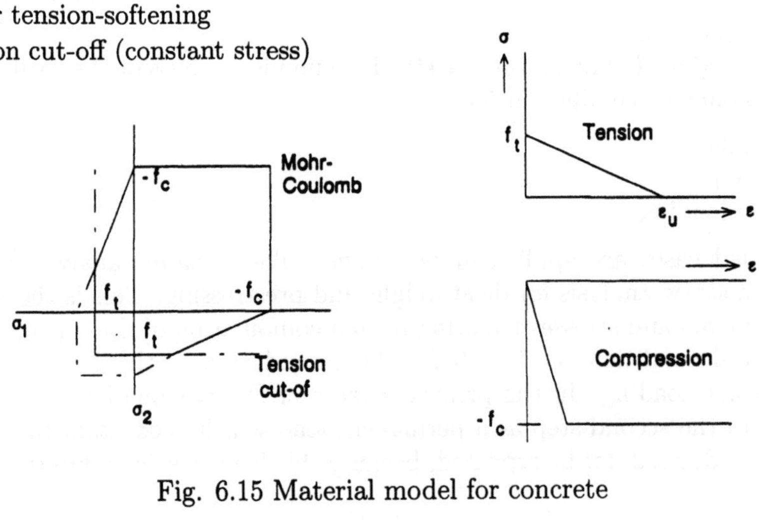

Fig. 6.15 Material model for concrete

The mild steel and the prestressing steel are modelled by the Von Mises criterion:

- Von Mises yield criterion
- no hardening

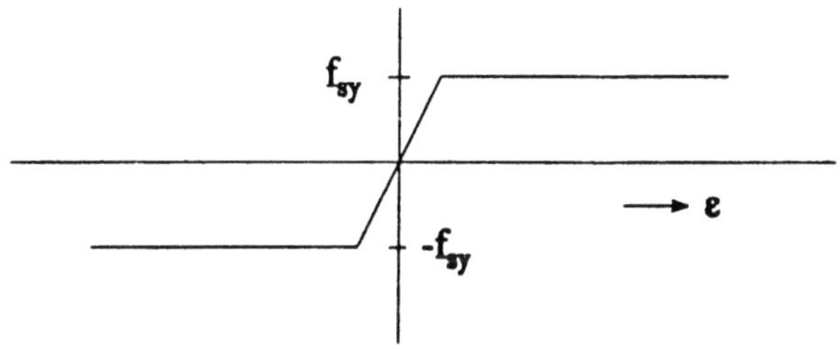

Fig. 6.16 Material model for steel

Task 13

Determine the material data for concrete (Type C35), the reinforcing steel (yield strength 500 N/mm^2) and prestressing steel (1675 N/mm^2).

Results of Task 13

The cylinder compressive strength for concrete is 0.85 x 35 = 29.75 N/mm^2, and the friction angle ϕ is taken as 30°. The cohesion then follows from

$$c = f_c \frac{1 - sin\phi}{2cos\phi} = 8.59 \ N/mm^2$$

Furthermore, the angle of dilatancy ψ is taken equal to ϕ for simplicity. Otherwise we get non-associated plasticity ($\psi \neq \phi$), which gives a non-symmetric tangential stiffness matrix. However, for plane stress situations, the value of ψ is not essential.

A mean value for the concrete tensile strength is used according to the Dutch Code and amounts to $f_{ctm} = 2.8 \ N/mm^2$. In the tensile regime, a constant stress cut-off criterion is used: a crack arises if the major principal tensile stress exceeds f_{ctm}. Linear tension-softening is used after cracking. In fact, we model the tension-stiffening of the reinforced concrete with the aid of tension softening. The ultimate strain ϵ_u equals

$$\epsilon_u = \frac{\sigma_{yield}}{E_s} \times 0.3$$

which yields $\epsilon_u = 0.714 \times 10^{-3}$. Note: If only membrane forces are active, $\epsilon_u = \sigma_{yield}/E_s$. In case of pure bending, use $\epsilon_u = 0.3 \times \sigma_{yield}/E_s$. In this case we expect cracking near the base. As a result of the edge effect, bending will dominate.

Furthermore, a constant shear retention factor $\beta = 0.2$ is used. For the reinforcement and prestressing steel, yield strengths of 500 and 1675 N/mm^2 are used, respectively.

Task 14

Compare the displayed results with the results of Tasks 11 and 12 and draw conclusions regarding the effects of cracking.

Results of Task 14

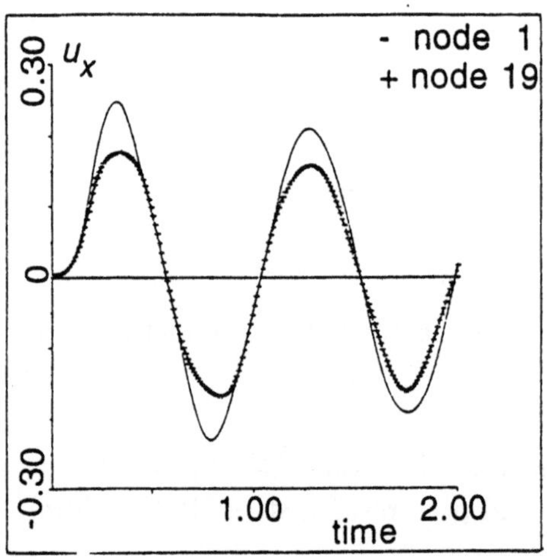

a) Displacements versus time for node 1 (top) and node 19 (base)

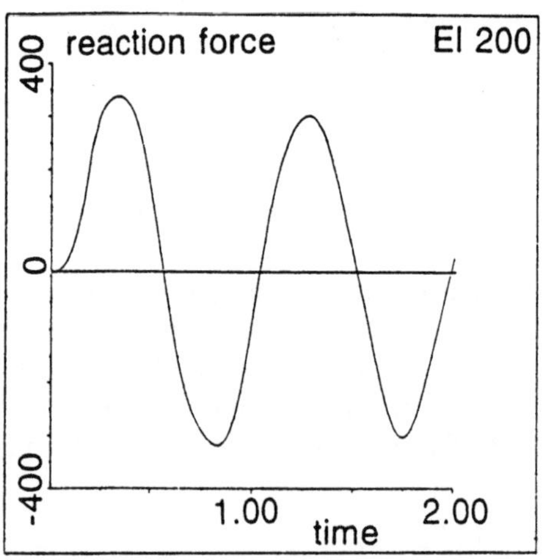

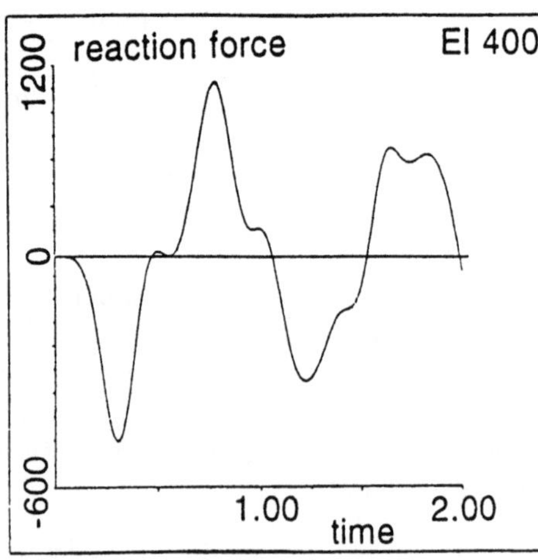

b) Horizontal reaction at node 19 c) Vertical reaction at node 19

Fig. 6.17 Overview of nonlinear dynamic analysis results

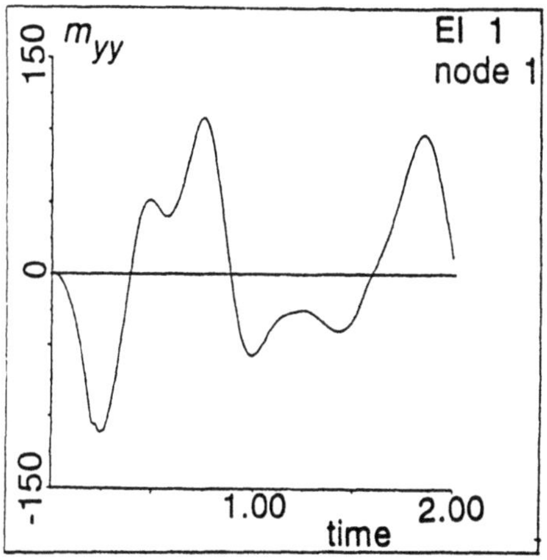

d) Moment m_{yy}, element 1, node 1

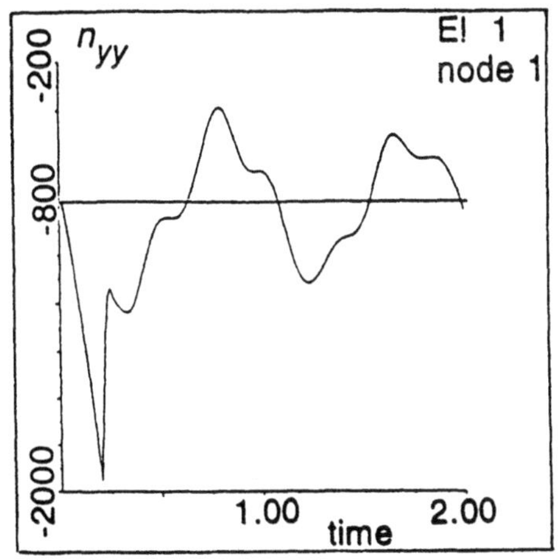

e) Normal force n_{yy}, element 1, node 1

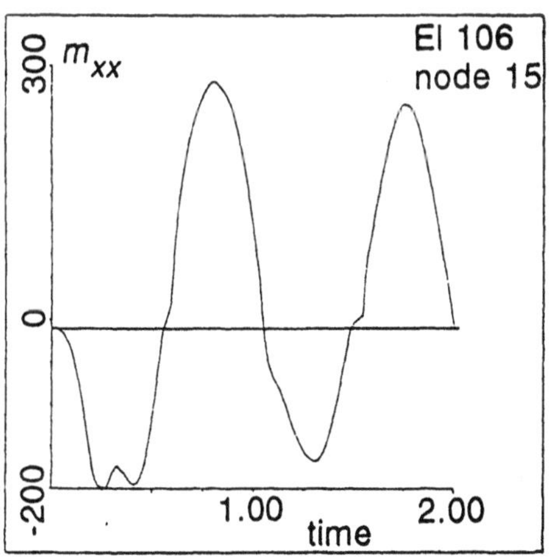

f) Moment m_{xx}, element 106, node 15

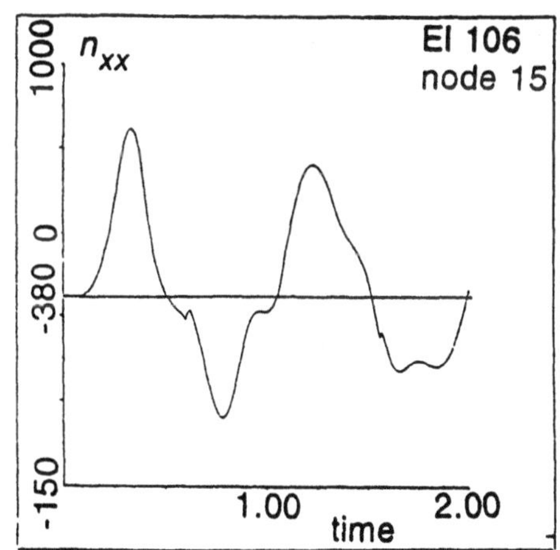

g) Normal force n_{xx}, element 106, node 15

Fig. 6.17 (Cont'd) Overview of nonlinear dynamic analysis results

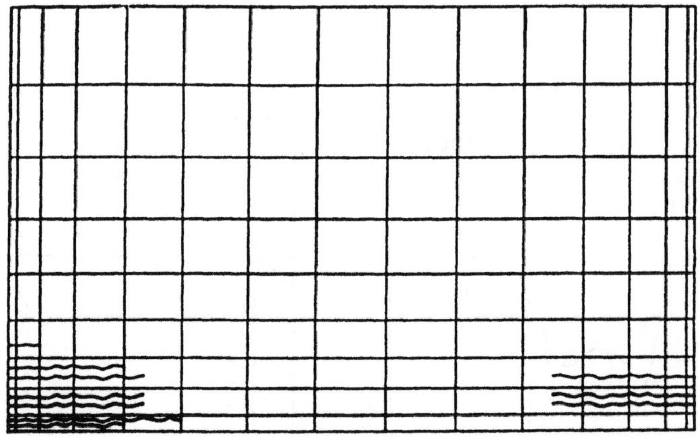

h) Shell cracking

Fig. 6.17 (Cont'd) Overview of nonlinear dynamic analysis results

Comparable results are obtained with the nonlinear dynamic analysis. Differences are found for the moment m_{xx} and vertical normal force n_{xx} at node 15, when we look more in detail. At the instant of cracking the moment and normal force reduce significantly, see Fig. 6.18. Cracks occur in the wall only at the front (inside) and rear side of the wall. Note that at the front side (where the pressure wave hits the wall for the first time) five elements which are connected to the foundation are cracked completely through the thickness of the wall. This phenomenon affects very much the nonlinear response of the normal force, see Fig. 6.18.

It was found that no plasticity occurs in the reinforcement and prestressing steel. However, we have to bear in mind that the DFL was smaller than one (0.7 to 0.5).

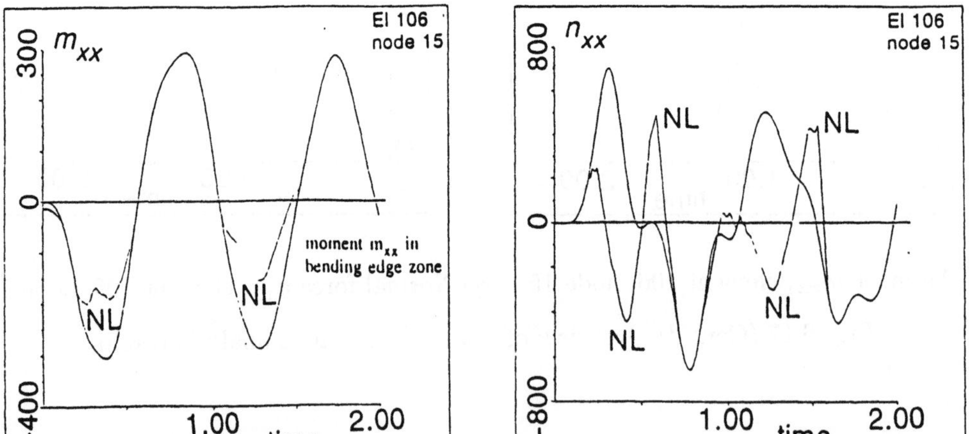

Fig. 6.18 Comparison of linear and nonlinear dynamic analysis results

Conclusion. The general conclusion may be drawn that the global behavior prediction of the wall is not significantly affected by the nonlinear dynamic analysis. There is, however, a distinct difference in the local behavior as a result of excessive cracking.

6.2 Dynamic Analysis with SARCF

6.2.1 Comments on Program SARCF

Almost all computer programs that are available to perform nonlinear dynamic analysis of reinforced concrete buildings are based on the empirical models described in Chapter 2. Program SARCF (Seismic Analysis of Reinforced Concrete Frames), developed at Columbia University [6.1-6.6], is basically not much different from the other programs (e.g. IDARC [6.7]), except that the frame element model incorporates the spread elasticity model (see Sect. 2.3.2). Also, by considering strength and stiffness degradation under cyclic loading as a low-cycle fatigue phenomenon, the modelling of damage and failure can be considered to be more "rational" than others, although it would still not satisfy the rigorous standards of a mechanics-based model.

Earlier versions of the program, SARCF and SARCF-II, have been documented in Refs. [6.5, 6.8]. A detailed documentation of an updated version, SARCF-III, is currently in preparation [6.9], complete with theoretical manual, programmer's manual and Fortran source listing. It is available from the National Center for Earthquake Engineering Research in Buffalo. Before practitioners attempt to use the program, they should be reminded, however, that sophisticated programs require continuous support, development, and maintenance.

The program is based on the well-known structural analysis program DRAIN-2D [6.10], developed at the University of California at Berkeley. The major modifications and enhancements can be summarized as follows.

1. The frame element developed by Roufaiel [6.1], with enhancements due to Chung [6.2, 6.3] was incorporated to replace the element models supplied with the standard DRAIN-2D version.
2. At any time t during a dynamic time history analysis it is possible to interrupt the time step integration to compute mode shapes and natural frequencies, using the current stiffness of the frame at time t.
3. At each time step during the dynamic time history analysis, a damage index is computed for both ends of each element in the frame.
4. Whereas DRAIN-2D expects the user to supply a ground motion time history and performs only one time history analysis per run, SARCF is programmed to also generate upon user request an arbitrary number of artificial ground motion time histories using the Monte Carlo simulation technique.
5. The program can perform a static analysis for gravity loads before entering the dynamic time history analysis. Thus the problem of nonlinear superposition is solved correctly.
6. Following the analysis of a frame structure for a specified number of ground motions, the program computes mean values and standard deviations of all element

damage indices.

7. The program has a special automatic design option. Upon user request it will rely on a number of built-in design rules to modify the input frame such that the damage distribution becomes as uniform and as low as specified by the user.

This last program feature, i.e., the automatic design capability, merits further comment, as it illustrates the potential of nonlinear analysis software such as SARCF.

Probably the most difficult task in designing reinforced concrete buildings to withstand strong earthquake ground motions is to devise an acceptable mechanism for dissipating the large amounts of energy imparted upon the building by the earthquake. Short of resorting to unusual technologies such as base isolation, frictional dampers, active or passive control mechanisms, current aseismic design philosophy typically relies on energy dissipation in structural components undergoing large inelastic deformations. There are basically two different approaches to achieve this goal. In the first approach, a number of structural members are deliberately selected to act as "fuses", i.e. weak spots in the structure assigned to develop plastic hinges and to dissipate energy under tightly controlled conditions. The designer has to detail the selected structural elements very carefully to assure that the energy dissipation demand can be met without premature failure. In the other approach, all or most structural elements are called upon to share equally in the task of energy dissipation so that the resulting damage is distributed uniformly over the entire frame, thereby assuring that the average damage level is a minimum. In the design of ductile moment-resistant frames it is accepted practice to use strong columns and weak beams. Thus the energy dissipation duty is limited to the beams.

In the automatic design procedure of SARCF, a preliminary design is modified iteratively until the damage has reached a preselected uniform distribution. This means that the chosen design approach utilizes the second of the two options just mentioned. The main justification for this choice is the reduction of detailing requirements for plastic hinge regions, which can be achieved by lowering the expected overall damage level for a given earthquake intensity. To achieve the design objective, the following steps are taken, Fig. 6.19.

1. Perform a preliminary design of a frame, for example by satisfying the static lateral load requirements of the Uniform Building Code. This task the user is expected to perform prior to using SARCF.
2. Perform a nonlinear dynamic analysis of the frame for seismic ground shaking of specified or assumed intensity, duration and spectral content, to be selected in consultation with an expert familiar with the seismicity of the building site, or by Monte Carlo simulation of a random process.
3. Compute for both ends of each frame element a damage index for each of the separate runs, as well as representative mean values and variances.
4. Evaluate the damage distribution by comparing it with an acceptable distribution as specified by the user.
5. If the damage is unacceptable, automatically introduce certain design changes on the basis of design rules that are incorporated in the program.

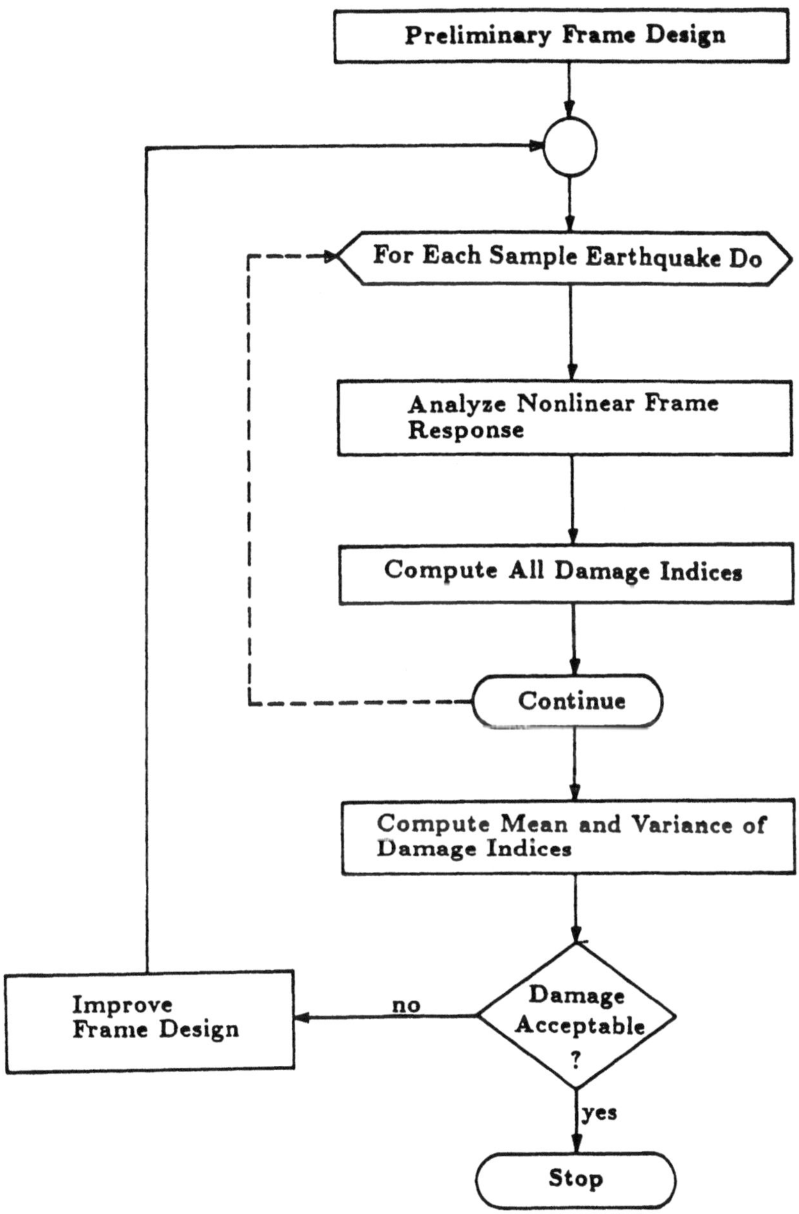

Fig. 6.19 Automatic design flowchart

6. Repeat steps 2 through 5 until the level of the frame damage is acceptable for the specified intensity of ground motion.

An elaborate set of numerical experiments on a four-story, three-bay building frame was conducted to derive a few basic design rules that were then incorporated into the program [6.4, 6.6]. The major conclusions of these studies were as follows:

1. The amount of flexural reinforcement appears to be the most effective and consistent tool for influencing the damage distribution in a frame.
2. The amount of confinement steel in plastic regions does not appear to influence the damage distribution to any useful extent. (This conclusion was contrary to expectation and may be due to insufficiently accurate mathematical modeling of the confinement effect).
3. The effects of changing member depths are often inconclusive. Aside from the common architectural constraints, members depths therefore do not lend themselves as an effective variable in an automatic design method.

6.2.2 Example

A typical office building frame with four stories and three bays is shown in Fig. 6.20, together with the flexural reinforcement areas for beams and columns. After subjecting this frame to ten different sample ground motions, the various frame elements suffered varying degrees of damage, as measured by the mean damage values indicated in Fig. 6.21. Damage appears to be heavily concentrated in the beams of the lower stories, indicating that these contribute overproportionally to the energy dissipation of the frame. The results given in Fig. 6.22 are those for the same frame and the same ten ground motion histories, but after the frame has undergone twelve automatic design iterations. The program was provided with a target mean damage value of 0.1 for the beams, together with a tolerance allowance of 0.05, and a maximum deviation of 0.05. As can be seen, all damage acceptance criteria are satisfied by the modified frame: no column damage index exceeds 0.01, except at the base, because of the fixed-end boundary conditions, which are usually unrealistic anyway; the mean value of all beam damage indices is 0.115, which is less than 0.15, i.e. the target value plus tolerance; no single beam damage index deviates from the mean value by more than the allowed amount of 0.05.

The design rules incorporated into the program were derived from studies involving a single concrete frame building and one earthquake random process. They are in need of refinements and modifications as additional information becomes available. This example was presented here not to demonstrate a finished automatic design procedure, but rather to illustrate an application concept for a powerful nonlinear analysis program such as SARCF. Although considerable additional work is needed before it can be incorporated into practice, the potential of this nonlinear analysis and design methodology for a design office is obvious.

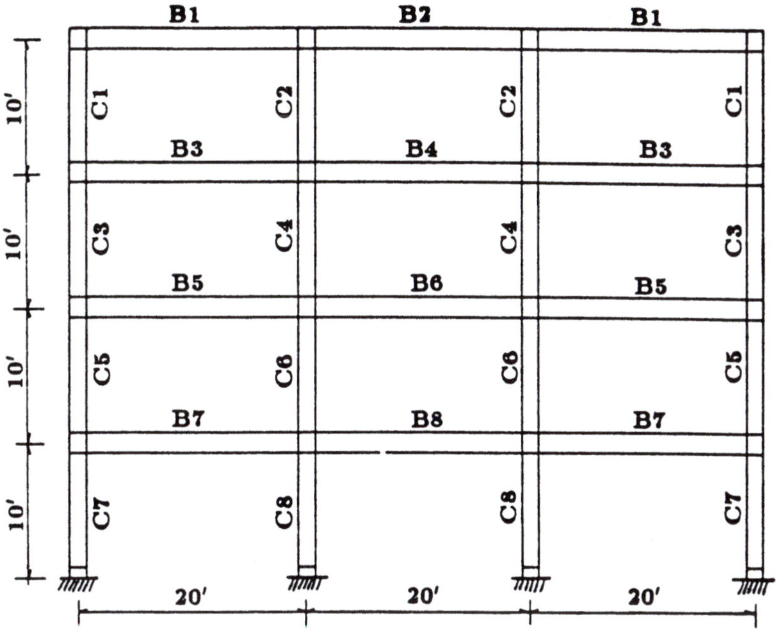

ELEVATION
(Frame Spacing 20')

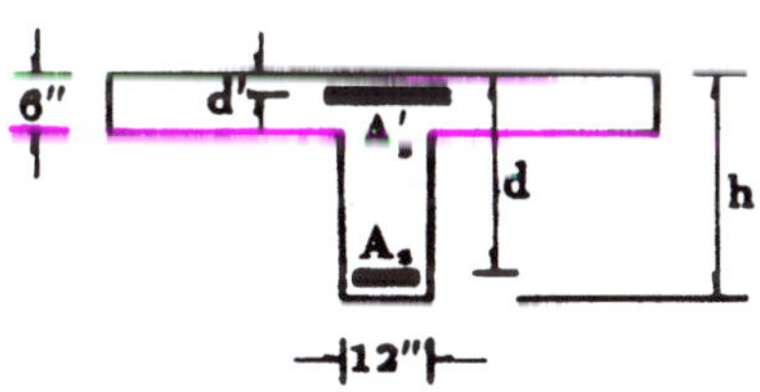

BEAM

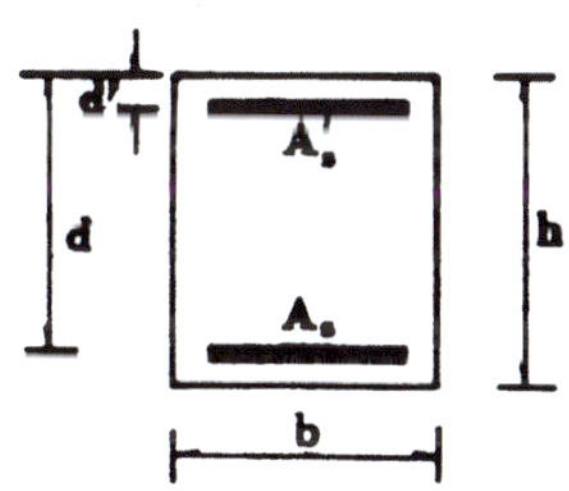

COLUMN

	d' (in)	A_s (in^2)	A'_s (in^2)	h (in)
B1,B2	2.0	2.280	2.280	20.0
B3,B4	2.0	2.622	2.622	20.0
B5,B6	2.0	2.622	2.622	22.0
B7,B8	2.0	2.736	2.736	22.0

	d' (in)	A_s (in^2)	A'_s (in^2)	b (in)	h (in)
C1,C2,C3	1.500	1.800	1.800	12.0	15.0
C4,C5	1.500	2.850	2.850	12.0	15.0
C6,C7	1.875	2.850	2.850	15.0	18.0
C8	1.875	2.964	2.964	15.0	18.0

Fig. 6.20 Example office building frame

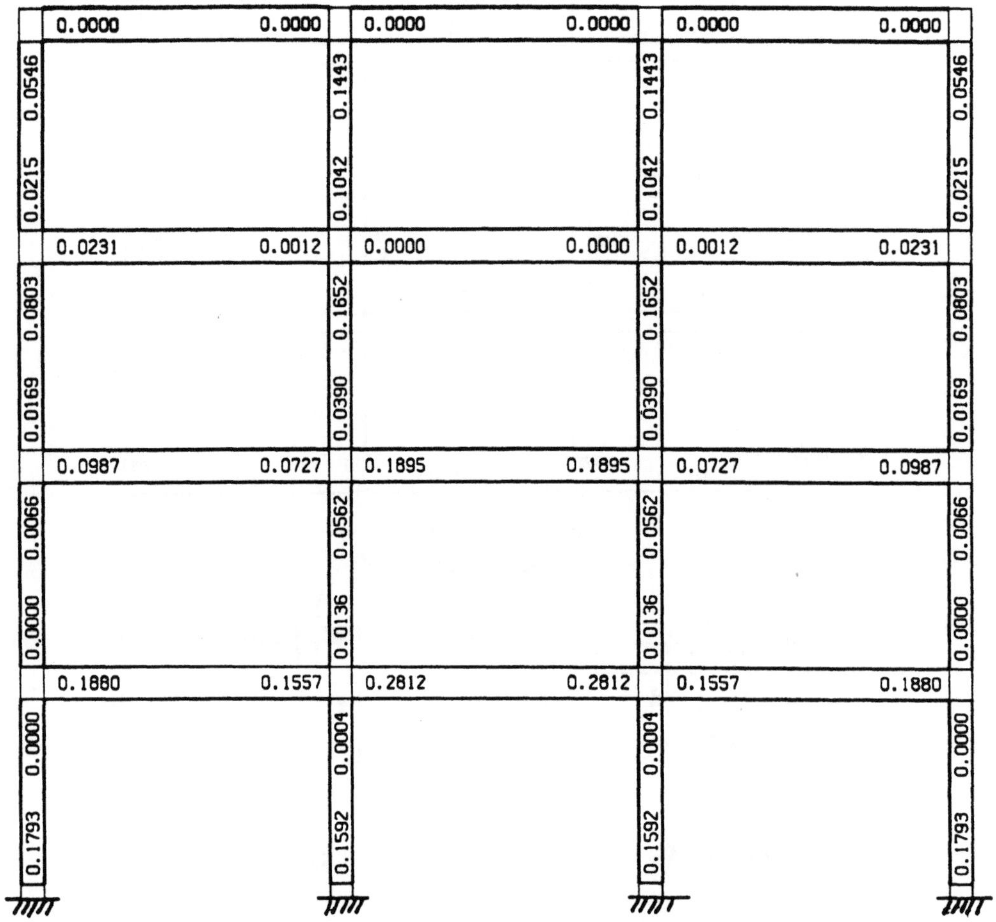

Mean Damage of All Beams : $\overline{D} = 0.084$

Standard Deviation of All Beams : $\sigma_D = 0.094$

Fig. 6.21 Mean damage indices for example building frame

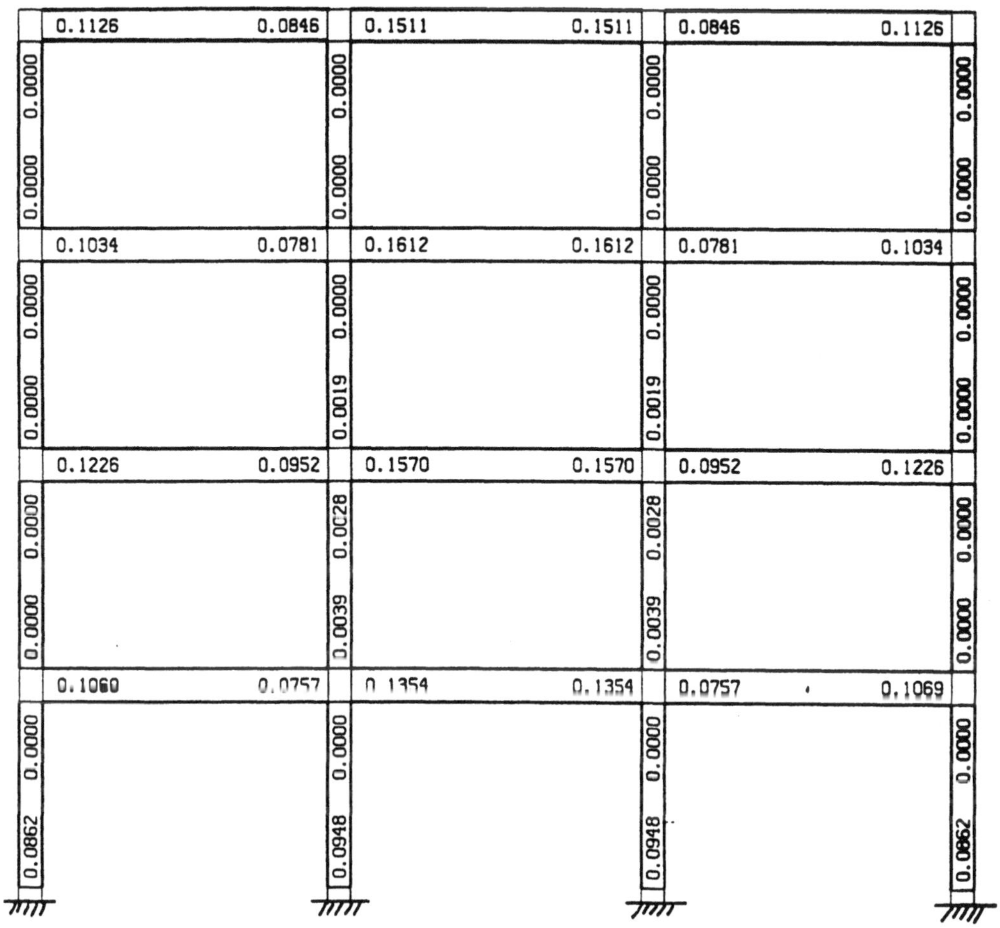

Mean Damage of All Beams : $\overline{D} = 0.115$

Standard Deviation of All Beams : $\sigma_D = 0.029$

Fig. 6.22 Mean damage indices for example building frame after 12 design iterations

6.3 References

6.1 Roufaiel, M.S.L. and C. Meyer: Analytical modeling of hysteretic behaviour of R/C Frames, ASCE J. Str. Eng. 113, No. 3 (1987).
6.2 Chung, Y.S., C. Meyer, and M. Shinozuka: Seismic Damage Assessment of Reinforced Concrete Members, Report No. NCEER-87-0022, National Center for Earthquake Engineering Research, Buffalo, New York, Oct. 1987.
6.3 Chung, Y.S., C. Meyer and M. Shinozuka: Modeling of seismic damage, ACI Structural Journal, May-June 1989.
6.4 Chung, Y.S., C. Meyer and M. Shinozuka: Automated Seismic Design of Reinforced Concrete Buildings, Technical Report NCEER-88-24, National Center for Earthquake Engineering Research, State University of New York at Buffalo, 1988.
6.5 Chung, Y.S., C. Meyer and M. Shinozuka: SARCF - Seismic Analysis of Reinforced Concrete Frames, Technical Report NCEER-88-44, National Center for Earthquake Engineering Research, State University of New York at Buffalo, 1988.
6.6 Chung, Y.S., C. Meyer and M. Shinozuka: Automatic seismic design of RC building frames, ACI Structural Journal, May-June 1990.
6.7 Park, Y.J., A.M. Reinhorn and S.K. Kunnath: IDARC - Inelastic Damage Analysis of Reinforced Concrete Frame - Shear-Wall Structures, Technical Report NCEER-87-8, National Center for Earthquake Engineering Research, State University of New York at Buffalo, 1987.
6.8 Rodriguez-Gomez, S., Y.S. Chung, and C. Meyer: SARCF-II User's Guide - Seismic Analysis of Reinforced Concrete Frames, Technical Report NCEER-90-27, National Center for Earthquake Engineering Research, State University of New York at Buffalo, 1990.
6.9 Meyer, C. and J. Yuan: SARCF-III - Seismic Analysis of Reinforced Concrete Frames, Technical Report, National Center for Earthquake Engineering Research, State University of New York at Buffalo in print.
6.10 Kanaan, A.E. and G.H. Powell: General Purpose Computer Program for Inelastic Dynamic Response of Plane Structures, Report No. EERC-73-6, University of California at Berkeley, Cal., 1982.

APPENDIX A

REVIEW OF DYNAMIC ANALYSIS METHODS

C. Meyer
Columbia University, New York, NY, USA

A.1 Purpose and Objectives

Many excellent textbooks are available on the subject of dynamic analysis [A.1-A.3]. The purpose of this appendix is therefore not a futile attempt to duplicate any of that valuable information by condensing it into a few pages. It is rather to present a brief summary review. Yet, this task is burdened with the dilemma common for such reviews: there is the danger that it is either too short or superficial to provide anything worthwhile to the reader unfamiliar with the subject, or it is too long for those readers already knowledgeable of the material. For this reason it was decided to place this review in an appendix to the lecture notes. It may serve as a reference for readers new to the field, without really being prerequisite for an understanding of the preceding chapters. It is included here solely for the purpose of summarizing those methods that are needed to analyze reinforced concrete structures for dynamic loads.

A.2 Equation of Motion for Single-Degree-of-Freedom Oscillator

A structure that can vibrate only in one particular mode can be represented as a single-degree-of-freedom (SDOF) system. This means, once the geometric shape of the deformed shape of the system is known or prescribed, only the time-dependent amplitude is needed to define the position of each point in the structure in space and time, and it can be idealized by the schematic system of Fig. A.1, which consists of a mass m, supported by a spring with the constant k (not necessarily linear elastic) and subject to damping characterized by a dashpot with damping coefficient c. Isolating the mass as a free body, all applied forces must be in dynamic equilibrium. First, there is the inertia force, $m\ddot{x}_t$, where

$$\ddot{x}_t = \ddot{x} + \ddot{x}_g \tag{A.1}$$

is the absolute acceleration, i.e., the acceleration relative to ground plus ground acceleration. Then there is the velocity-proportional damping force, $c\dot{x}$; the spring or restoring force, kx; and possibly some external forcing function $p(t)$, i.e.,

$$m\ddot{x}_t + c\dot{x} + kx = p(t) \tag{A.2}$$

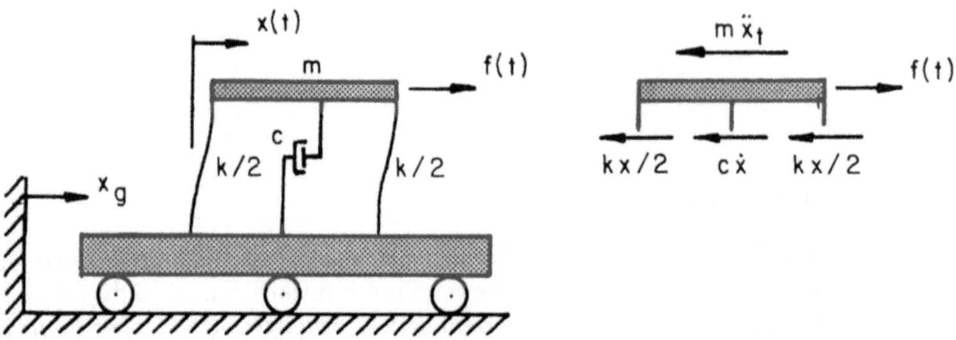

Fig. A.1 Single-degree-of-freedom oscillator

Substituting Eq. A.1 into A.2 gives

$$m\ddot{x} + c\dot{x} + kx = p(t) - m\ddot{x}_g \tag{A.3}$$

Ignoring the external forcing function, i.e. setting $p(t) = 0$, Eq. A.3 can also be written as

$$\ddot{x} + 2\xi\omega\dot{x} + \omega^2 x = -\ddot{x}_g \tag{A.4}$$

where

$$\omega = \sqrt{\frac{k}{m}} \tag{A.5}$$

is the natural frequency of the system, and

$$\xi = \frac{c}{2\omega m} = \frac{c}{c_{crit}} \tag{A.6}$$

is the damping ratio expressed as a fraction of critical damping.

A.3 Solution of Equation of Motion in Time Domain

If there is neither ground acceleration $\ddot{x}_g$ nor damping, Eq. A.4 reduces to,

$$\ddot{x} + \omega^2 x = 0 \tag{A.7}$$

which has the general solution,

$$x = A\sin\omega t + B\cos\omega t \tag{A.8}$$

where the vibration amplitudes A and B depend on the initial conditions. For example, if $x(0) = x_o$, then $B = x_o$, and if $\dot{x}(0) = \dot{x}_o$, then $A = \dot{x}_o/\omega$, so that

$$x = \frac{\dot{x}_o}{\omega}\sin\omega t + x_o\cos\omega t \tag{A.9}$$

(see Fig. A.2). Note that $\omega = \sqrt{k/m}$ is the undamped natural frequency of the system, with units Rad/sec; or, if divided by 2π, with units cycles per second (or Hz). Its inverse is the period of vibration, given in units of seconds,

$$T = \frac{2\pi}{\omega}$$

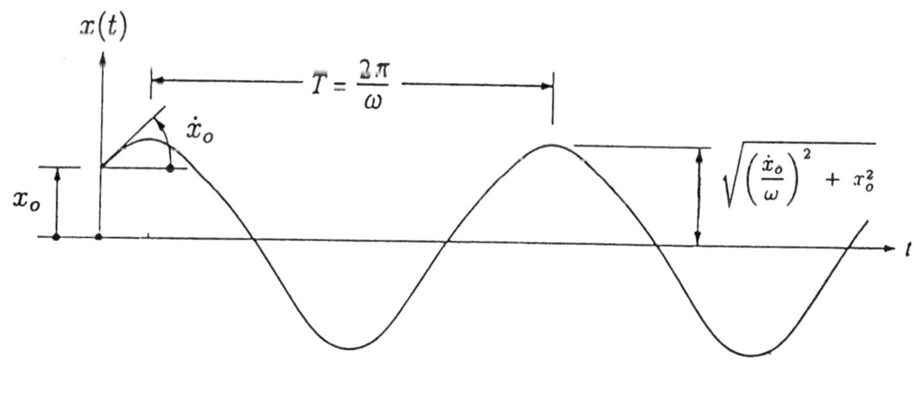

Fig. A.2 Free vibration response

With damping, but still without a forcing function, Eq. A.4 reads,

$$m\ddot{x} + c\dot{x} + kx = 0 \tag{A.10}$$

Substitution of the trial solution, $x = Ae^{st}$, leads to,

$$(ms^2 + cs + k)Ae^{st} = 0$$

For a nontrivial solution, $Ae^{st} \neq 0$, and it follows that with $\omega^2 = k/m$,

$$s^2 + \frac{c}{m}s + \omega^2 = 0$$

which has the two solutions,

$$s = -\frac{c}{2m} \pm \sqrt{(\frac{c}{2m})^2 - \omega^2} \qquad (A.11)$$

This solution implies that there exist three different types of damping,

1. critical damping, for $(\frac{c}{2m})^2 - \omega^2 = 0$, or $c = c_{crit} = 2m\omega$;
2. overcritical damping, for $c > c_{crit}$;
3. undercritical damping, i.e. $c < c_{crit}$, the most common case.

For undercritical damping, the radical in Eq. A.11 is negative. Substituting the damping ratio ξ defined in Eq. A.6 into Eq. A.11, we get

$$\begin{aligned} s &= -\xi\omega \pm \sqrt{(\xi\omega)^2 - \omega^2} \\ &= -\xi\omega \pm i\omega_d \end{aligned} \qquad (A.12)$$

where $i = \sqrt{-1}$, and $\omega_d = \omega\sqrt{1 - \xi^2}$ is the damped frequency, which is less than ω, but for realistic damping values not by much.

We obtain the general solution to the damped vibration problem by inserting the result of Eq. A.12 into our trial solution, $x = Ae^{st}$, i.e.,

$$\begin{aligned} x(t) &= A_1 e^{-\xi\omega t + i\omega_d t} + A_2 e^{-\xi\omega t - i\omega_d t} \\ &= e^{-\xi\omega t}(A_1 e^{i\omega_d t} + A_2 e^{-i\omega_d t}) \\ &= e^{-\xi\omega t}(B_1 sin\omega_d t + B_2 cos\omega_d t) \end{aligned} \qquad (A.13)$$

B_1 and B_2 are determined from the initial conditions. This damped vibration response is shown qualitatively in Fig. A.3. Typical damping values for reinforced concrete structures range from 4 to 7% of critical.

If the SDOF system is subject to a harmonic forcing function with frequency ω_p and amplitude p_o, Eq. A.10 becomes

$$m\ddot{x} + c\dot{x} + kx = p_o sin\omega_p t \qquad (A.14)$$

The solution of this equation consists now of a free-vibration component and a steady-state component. The free-vibration component damps out in time, Fig. A.3, hence we need to consider only the steady-state part of the solution,

$$x(t) = A_1 sin\omega_p t + A_2 cos\omega_p t \qquad (A.15)$$

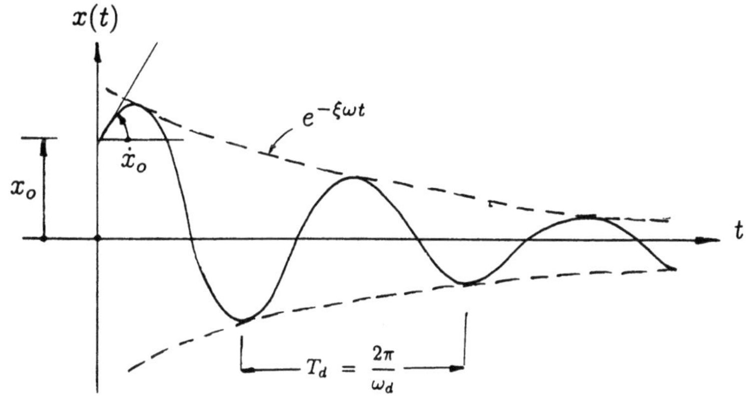

Fig. A.3 Damped vibration response

Substitution of Eq. A.15 into A.14 provides two independent equations,

$$[-A_1\omega_p^2 - A_2\omega_p(2\xi\omega) + A_1\omega^2]sin\omega_p t = \frac{p_o}{m}sin\omega_p t$$
$$[-A_2\omega_p^2 + A_1\omega_p(2\xi\omega) + A_2\omega^2]cos\omega_p t = 0$$

With the frequency ratio $\beta = \omega_p/\omega$, these equations simplify to

$$A_1(1-\beta^2) - A_2(2\xi\beta) = \frac{p_o}{k}$$
$$A_2(1-\beta^2) + A_1(2\xi\beta) = 0$$

Solved for A_1 and A_2,

$$A_1 = \frac{p_o}{k}\frac{1-\beta^2}{(1-\beta^2)^2 + (2\xi\beta)^2}$$
$$A_2 = \frac{p_o}{k}\frac{-2\xi\beta}{(1-\beta^2)^2 + (2\xi\beta)^2}$$

the general solution becomes,

$$x(t) = \frac{p_o}{k}\frac{(1-\beta^2)sin\omega_p t - 2\xi\beta cos\omega_p t}{(1-\beta^2)^2 + (2\xi\beta)^2} \qquad (A.16)$$
$$= \rho sin(\omega_p t - \theta)$$

with

$$\rho = \frac{p_o}{k}[(1-\beta^2)^2 + (2\xi\beta)^2]^{-\frac{1}{2}}$$

and

$$\theta = tan^{-1}\frac{2\xi\beta}{1-\beta^2}$$

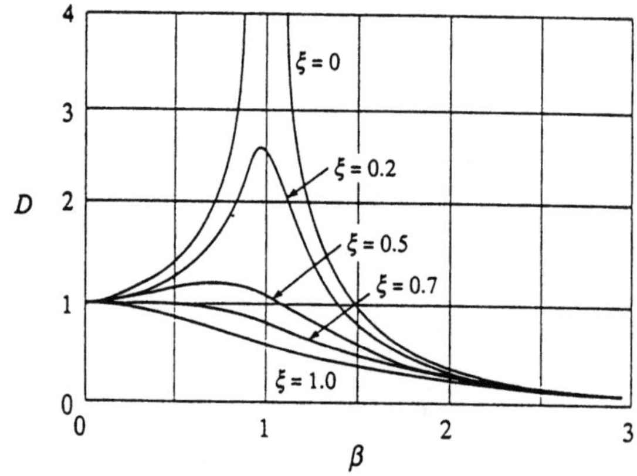

Fig. A.4 Dynamic amplification factor

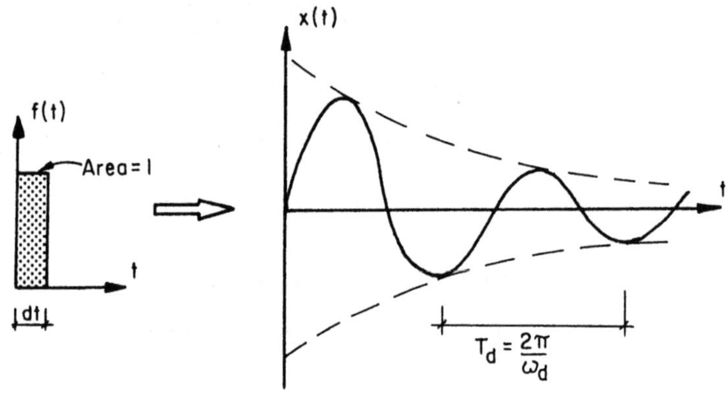

Fig. A.5 Response to unit impulse

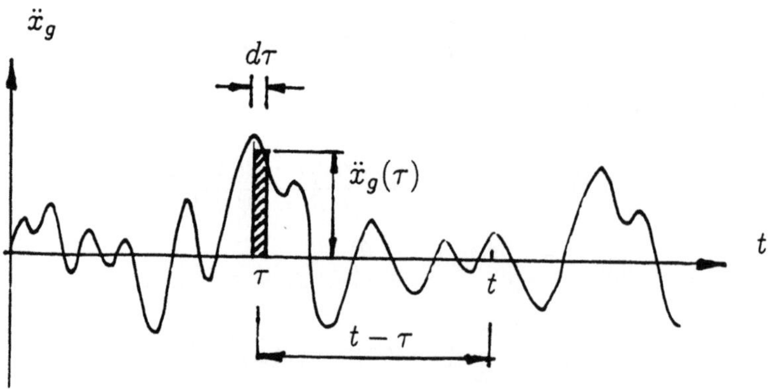

Fig. A.6 Random ground motion history

Note that p_o/k is the static deformation of the system, and

$$D = \frac{\rho}{p_o/k} = \frac{1}{\sqrt{(1-\beta^2)^2 + (2\beta\xi)^2}} \qquad (A.17)$$

is the so-called *dynamic amplification factor*, plotted in Fig. A.4 as a function of frequency ratio β and damping ratio ξ. As can be seen, near resonance, where $\beta \approx 1$, a lowly damped system can experience very large displacements.

The response to a single unit impulse is given by the following solution, Fig. A.5,

$$x(t) = \frac{1}{\omega_d} e^{-\xi\omega t} sin\omega_d t \equiv h(t) \qquad (A.18)$$

where $h(t)$ is referred to, appropriately, as *unit impulse response function.* An arbitrary forcing function such as $\ddot{x}_g$ shown in Fig. A.6, which might be the ground acceleration record of an actual earthquake, can be considered as a series of many unit impulses occurring at different times τ, each one multiplied with the actual load amplitude $\ddot{x}_g(\tau)$ over the differential time increment $d\tau$. The response x at time t to one such general impulse occurring at time τ, follows directly from Eq. A.18, namely,

$$\begin{aligned} x(t) &= \frac{\ddot{x}_g(\tau)}{\omega_d} e^{-\xi\omega(t-\tau)} sin\omega_d(t-\tau)d\tau \\ &= f(\tau)h(t-\tau)d\tau \end{aligned} \qquad (A.19)$$

where the forcing function $\ddot{x}_g$ is now simply referred to as f. If Eq. A.19 represents the response to the one impulse occurring at time τ, then the combined response to all impulses occurring between time 0 and t is found by integrating Eq. A.19 accordingly,

$$\begin{aligned} x(t) &= \frac{1}{\omega_d}\int_o^t \ddot{x}_g(\tau)e^{-\xi\omega(t-\tau)} sin\omega_d(t-\tau)d\tau \\ &= \int_o^t f(\tau)h(t-\tau)d\tau \end{aligned} \qquad (A.20)$$

This is the *Duhamel integral*, also known as *convolution integral.* It represents the response of a SDOF system to an arbitrary forcing function – in our case a specified and possibly recorded ground acceleration history. $\ddot{x}_g$ is not an analytic function, but rather specified or defined at discrete points, Fig. A.7. By assuming a linear variation inbetween those points, we may employ numerical integration to obtain a solution to Eq. A.20. The problem is then reduced to solving Eq. A.4 for the time step from t_i to t_{i+1}, Fig. A.7, where for simplicity the ground acceleration is again referred to by $f(t)$,

$$\begin{aligned} \ddot{x} + 2\xi\omega\dot{x} + \omega^2 x &= -\ddot{x}_g(t) \\ &= -f_i - \frac{\Delta f_i}{\Delta t_i}(t-t_i) \quad t_i \le t \le t_{i+1} \end{aligned} \qquad (A.21)$$

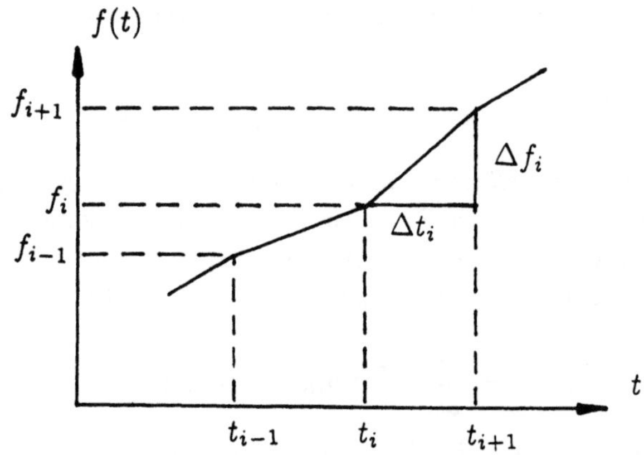

Fig. A.7 Discretized forcing function

where $\Delta f_i = f_{i+1} - f_i$ and $\Delta t_i = t_{i+1} - t_i$. The general solution of Eq. A.21 is given by,

$$\begin{aligned} x &= e^{-\xi\omega(t-t_i)}[C_1 sin\omega\sqrt{1-\xi^2}(t-t_i) + C_2 cos\omega\sqrt{1-\xi^2}(t-t_i)] \\ &- \frac{f_i}{\omega^2} + \frac{2\xi}{\omega^3}\frac{\Delta f_i}{\Delta t_i} - \frac{1}{\omega^2}\frac{\Delta f_i}{\Delta t_i}(t-t_i) \end{aligned} \tag{A.22}$$

The constants C_1 and C_2 can be determined from the initial conditions $x(t_i) = x_i$ and $\dot{x}(t_i) = \dot{x}_i$, whereupon both, displacement x_{i+1} and velocity $\dot{x}_{i+1}$ at time t_{i+1} can be expressed in terms of the initial conditions and the given values for ground acceleration at times t_i and t_{i+1}. Details can be found in Ref. [A.4]. The acceleration $\ddot{x}_{i+1}$ at time t_{i+1} is then determined by solving Eq. A.21 at time t_{i+1}. This completes the solution for the response of a SDOF system to a linearly discretized ground acceleration history.

A.4 Earthquake Response Spectra

The solution of the preceding section can be used directly to compute response spectra for ground motion histories. Let

$$\begin{aligned} S_d(\omega,\xi) &= \max_{i=1,N} [x_i(\omega,\xi)] \\ S_v(\omega,\xi) &= \max_{i=1,N} [\dot{x}_i(\omega,\xi)] \\ S_a(\omega,\xi) &= \max_{i=1,N} [\ddot{x}_i(\omega,\xi)] \end{aligned} \tag{A.23}$$

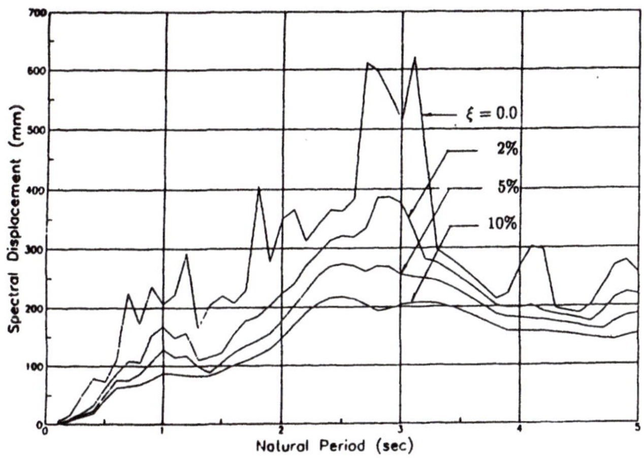

a) Displacement Spectrum, S_d

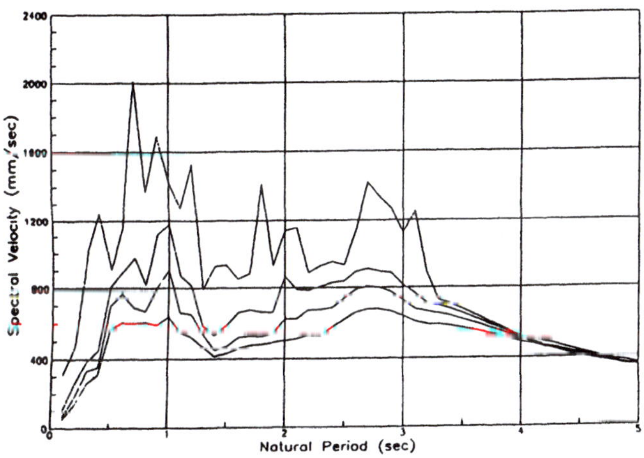

b) Velocity Spectrum, S_v

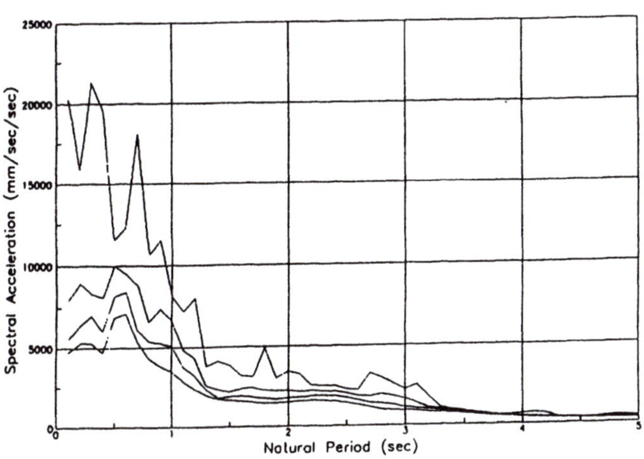

c) Acceleration Spectrum, S_a

Fig. A.8 Response spectra for 1940 El Centro Earthquake

where the maximum values are determined by scanning the responses of all N time steps for which a solution has been determined. These three maximum response values can be plotted as functions of frequency ω (or period $T = 2\pi/\omega$) and damping ratio ξ, Fig. A.8. Choosing an increment for T-values at which to calculate and plot response spectrum values is somewhat arbitrary and depends on the purpose for which the spectrum is being determined. Regarding damping values, spectrum curves are frequently generated for ξ= 0, 1, 2, 5, and 10%.

Note that the *base shear* is equal to the force in the spring, i.e.,

$$V_o(t) = kx(t) = m\omega^2 x(t)$$

and

$$V_{o,max} = m\omega^2 S_d(\omega, \xi)$$

Likewise, the *overturning moment* is

$$M_{o,max} = hV_{o,max}$$

where h is the height of the mass m above ground.

A thorough treatment of response spectrum methods can be found in [A.5].

A.5 Solution of Equation of Motion in Frequency Domain

The equation of motion for a SDOF oscillator subjected to harmonic excitation with frequency ω_p, is given by,

$$m\ddot{x} + c\dot{x} + kx = p_o e^{i\omega_p t} \qquad (A.24)$$

The solution can be assumed to be of the form,

$$x = H(i\omega_p) p_o e^{i\omega_p t} \qquad (A.25)$$

Substitution of Eq. A.25 into A.24 determines the response amplitude H, namely

$$H(i\omega_p) = \frac{1}{-m\omega_p^2 + i\omega_p c + k} \qquad (A.26)$$

With the introduction of the previous abbreviations, $\beta = \omega_p/\omega$, $\omega^2 = k/m$, $c = 2m\omega\xi$, Eq. A.26 can be written as

$$H(i\beta) = \frac{1}{k[(1-\beta^2) + 2i\xi\beta]} \qquad (A.27)$$

This function is referred to as *complex frequency response function.* Rewriting it as

$$H(i\beta) = \frac{1}{k} \frac{(1-\beta^2) - 2i\xi\beta}{[(1-\beta^2) + 2i\xi\beta][(1-\beta^2) - 2i\xi\beta]} = \frac{1}{k} \frac{(1-\beta^2) - 2i\xi\beta}{(1-\beta^2)^2 + 4\xi^2\beta^2} \qquad (A.28)$$

to get the complex conjugate, the response amplitude becomes

$$\begin{aligned} \overline{x}(t) &= p_o|H(i\beta)| = \frac{p_o}{k}\frac{\sqrt{(1-\beta^2)^2+4\xi^2\beta^2}}{(1-\beta^2)^2+4\xi^2\beta^2} \\ &= \frac{p_o}{k}\frac{1}{\sqrt{(1-\beta^2)^2+4\xi^2\beta^2}} \end{aligned} \tag{A.29}$$

where $D = [(1-\beta^2)^2 + 4\xi^2\beta^2]^{-\frac{1}{2}}$ is the dynamic amplification factor encountered earlier, see Eq. A.17.

If for the harmonic forcing function

$$p(t) = C_n e^{in\omega_p t}$$

the steady-state response is

$$x(t) = H(in\omega_p)C_n e^{in\omega_p t}$$

then for the more general forcing function

$$p(t) = \sum_{n=-\infty}^{\infty} C_n e^{in\omega_p t} \tag{A.30}$$

the steady-state response is given by

$$x(t) = \sum_{n=-\infty}^{\infty} H(in\omega_p)C_n e^{in\omega_o t} \tag{A.31}$$

where $H(in\omega_p) = 1/k[(1-n^2\beta^2) + 2in\beta\xi]$ and $\beta = \omega_p/\omega$ is the frequency ratio.

The forcing function $p(t)$ is expressed in Eq. A.31 as a Fourier Series, i.e. as the sum of discrete frequency components. In going from such a discrete set of frequencies to a continuous frequency spectrum, the forcing function is expressed as a Fourier Transform,

$$p(t) = \frac{1}{2\pi}\int_{-\infty}^{\infty} Q(ip)e^{ipt}dp \tag{A.32}$$

where

$$Q(ip) = \int_{-\infty}^{\infty} p(t)e^{-ipt}dt \tag{A.33}$$

For this general forcing function, which in general is not anymore harmonic, the response becomes

$$x(t) = \frac{1}{2\pi}\int_{-\infty}^{\infty} H(ip)Q(ip)e^{ipt}dp \tag{A.34}$$

A.6 Multi-Degree-of-Freedom Systems

This discussion shall be limited to building frame structures composed of beams and columns, Fig. A.9, in which each node has in general 3 degrees of freedom (DOF's), if only planar motion is considered. A building with n stories and m column lines has thus 3nm DOF's. In most cases floor diaphragms can be assumed to be rigid, which reduces the number of DOF's to n(2m+1). If also column axial deformations can be neglected, this reduces further to n(m+1). Significant inertia forces are typically associated only with lateral degrees of freedom. Unless vertical response is specifically called for, it is common to assign zero mass and damping to all rotational and vertical joint DOF's, which can be condensed out from the system equations by static condensation [A.6]. The resulting system equations can then be written in the form

$$[M]\{\ddot{x}\} + [C]\{\dot{x}\} + [K^*]\{x\} = -[M][I]\ddot{x}_g \tag{A.35}$$

where $[M]$ is the (diagonal) mass matrix, containing the various floor masses; $[C]$ is the system damping matrix; $[K^*]$ is the stiffness matrix for the lateral degrees of freedom, Fig. A.10. All vertical and rotational DOF's have been condensed out, i.e. k^*_{ij} is the reaction produced at the i'th floor, if the j'th floor undergoes a unit displacement, with all other floors held in place, but all joints are free to rotate and to move vertically. $\ddot{x}_g$ is the ground acceleration specified as a function of time. If the building has n stories, Eq. A.35 represents a set of n coupled differential equations of second order.

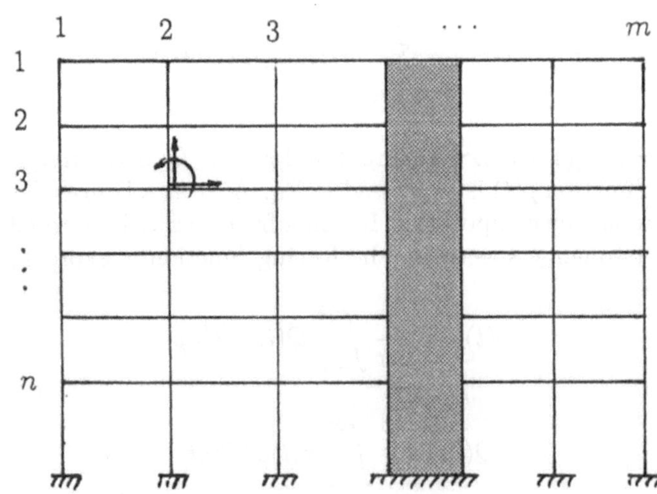

Fig. A.9 General building frame

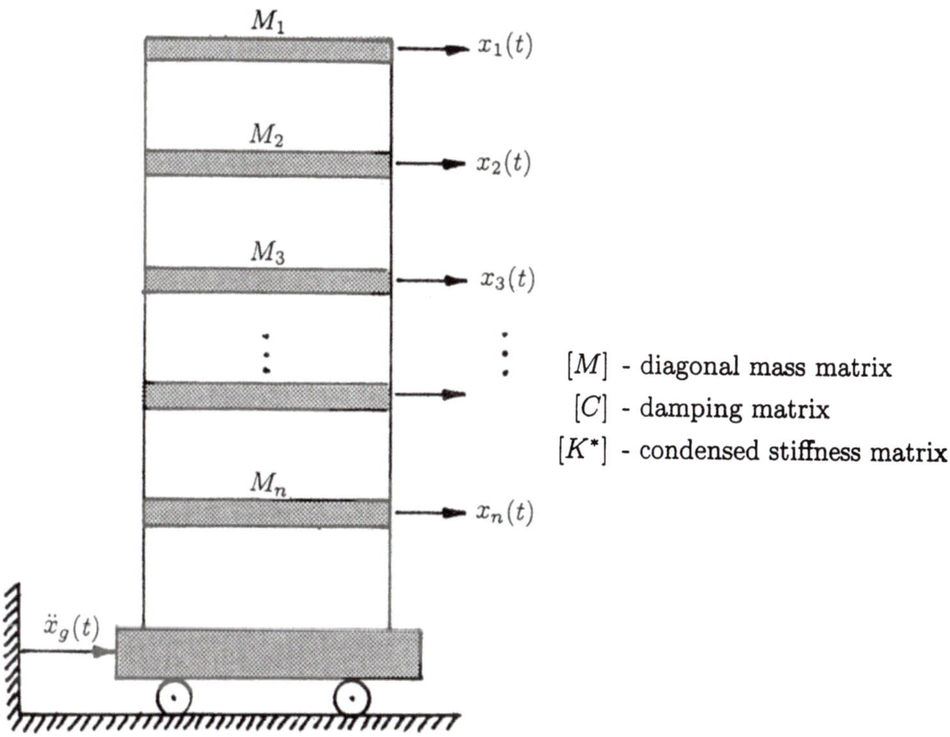

Fig. A.10 n-story building frame with joint rotations and vertical DOF's condensed out

A.7 Eigenvalue Problem

Ignoring the damping and ground motion in Eq. A.35 leads to the free vibration problem,

$$[M]\{\ddot{x}\} + [K]\{x\} = \{0\} \qquad (A.36)$$

where the * superscript for $[K]$ has been dropped for simplicity. Eq. A.36 constitutes an eigenvalue problem, which can be rewritten with $\ddot{x} = -\omega^2 x$ as

$$([K] - \omega^2[M])\{x\} = \{0\} \qquad (A.37)$$

A non-trivial solution of Eq. A.37 exists only if $\det([K] - \omega^2[M]) = 0$. The roots for which this determinant vanishes are the eigenvalues or squared natural frequencies ω_i^2, $i = 1...N$, of the system. Substituting them into Eq. A.37 gives the eigenvectors or mode shapes $\{x_i\} = \{\phi_i\}$.

Example. The two-story frame of Fig. A.11 has equal floor weights $W_1 = W_2 = 500^k$ and the stiffness coefficients, $k_{11} = -k_{12} = 600\ k/in$, $k_{22} = 1000\ k/in$. Find the two natural frequencies and corresponding mode shapes for this structure.

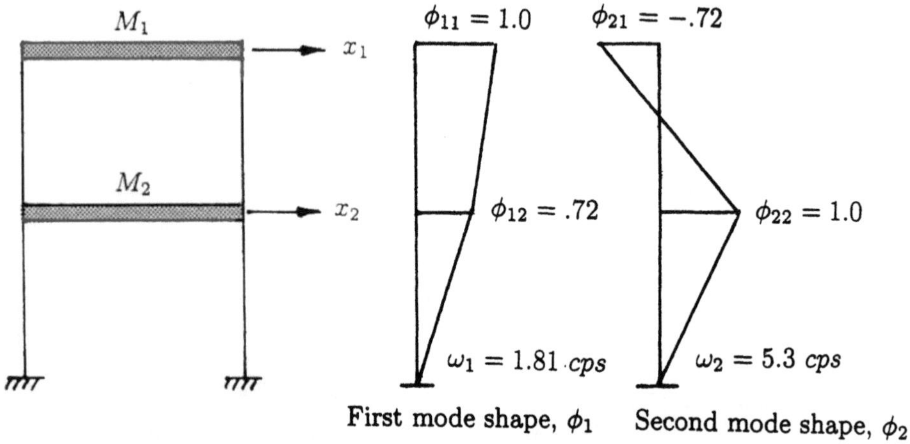

Fig. A.11 Example - Mode shapes of 2-story frame

First converting the floor weights to masses, $M_1 = M_2 = 500/g = 500/386.4 = 1.294\ k - sec^2/in$, the eigenvalue problem becomes,

$$\begin{bmatrix} 1.294 & 0 \\ 0 & 1.294 \end{bmatrix} \begin{Bmatrix} \ddot{x}_1 \\ \ddot{x}_2 \end{Bmatrix} + \begin{bmatrix} 600 & -600 \\ -600 & 1000 \end{bmatrix} \begin{Bmatrix} x_1 \\ x_2 \end{Bmatrix} = \begin{Bmatrix} 0 \\ 0 \end{Bmatrix}$$

The requirement $\det([K] - \omega^2[M]) = 0$ becomes

$$\det \begin{bmatrix} 600 - 1.294\omega^2 & -600 \\ -600 & 1000 - 1.294\omega^2 \end{bmatrix} = 0$$

or

$$(600 - 1.294\omega^2)(1000 - 1.294\omega^2) - 600^2 = 0$$

$$\omega^4 - 1236.5\omega^2 + 143,332 = 0$$

which has the two solutions,

$$\omega_{1,2}^2 = \frac{1236.5}{2} \pm \sqrt{\frac{1236.5^2}{4} - 143,332}$$

i.e., $\omega_1^2 = 129.5$, or $\omega_1 = 11.4\ Rad/sec = 1.81\ cps$ and $\omega_2^2 = 1107$, or $\omega_2 = 333.3\ Rad/sec = 5.3\ cps$.

Substituting $\omega^2 = 129.5$ into Eq. A.37, i.e., with $600 - 1.294\omega_1^2 = 432.4$ and $1000 - 1.294\omega_1^2 = 232.4$, we obtain the two equations

$$\begin{bmatrix} 432.4 & -600 \\ -600 & 832.4 \end{bmatrix} \begin{Bmatrix} \phi_{11} \\ \phi_{12} \end{Bmatrix} = \begin{Bmatrix} 0 \\ 0 \end{Bmatrix}$$

Let $\phi_{11} = 1$, then $(-600)(1) + 832.4\phi_{12} = 0$, i.e. $\phi_{12} = 0.72$. Likewise, substituting $\omega^2 = 1107$ into the system equations, i.e., with $600 - 1.294\omega_2^2 = -832.5$ and $1000 - 1.294\omega_2^2 = -432,5$, we obtain the two equations

$$\begin{bmatrix} -832.5 & -600 \\ -600 & -432.5 \end{bmatrix} \begin{Bmatrix} \phi_{21} \\ \phi_{22} \end{Bmatrix} = \begin{Bmatrix} 0 \\ 0 \end{Bmatrix}$$

Let $\phi_{22} = 1$, then $-832.5\phi_{21} - (600)(1) = 0$, i.e. $\phi_{21} = -0.72$. These results are plotted in Fig. A.11.

A.8 Modal Analysis

Rather than solving the equations of motion,

$$[M]\{\ddot{x}\} + [C]\{\dot{x}\} + [K]\{x\} = -[M][I]\ddot{x}_g \tag{A.38}$$

directly for some ground motion history, it may be advisable to introduce the coordinate transformation,

$$\{x\} = [\phi]\{y\} \tag{A.39}$$

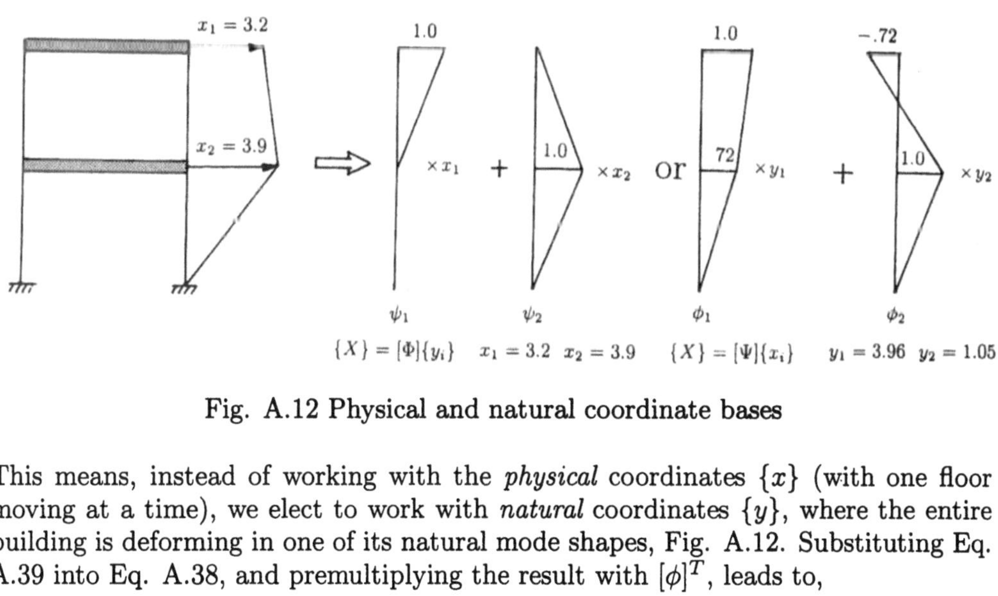

Fig. A.12 Physical and natural coordinate bases

This means, instead of working with the *physical* coordinates $\{x\}$ (with one floor moving at a time), we elect to work with *natural* coordinates $\{y\}$, where the entire building is deforming in one of its natural mode shapes, Fig. A.12. Substituting Eq. A.39 into Eq. A.38, and premultiplying the result with $[\phi]^T$, leads to,

$$[\phi]^T[M][\phi]\{\ddot{y}\} + [\phi]^T[C][\phi]\{\dot{y}\} + [\phi]^T[K][\phi]\{y\} = -[\phi]^T[M][I]\ddot{x}_g \tag{A.40}$$

It can be shown that $[\phi]^T[M][\phi]$ and $[\phi]^T[K][\phi]$ are diagonal matrices. In addition, we *assume* also $[\phi]^T[C][\phi]$ to be diagonal, i.e.,

$$\begin{aligned} [\overline{M}] &= [\phi]^T[M][\phi] \\ [\overline{C}] &= [\phi]^T[C][\phi] \\ [\overline{K}] &= [\phi]^T[K][\phi] \end{aligned} \tag{A.41}$$

These orthogonality properties of the modal matrix $[\phi]$ cause the equations of motion to be uncoupled, i.e. Eq. A.40 can be written as,

$$\overline{M}_i\ddot{y}_i \;+\; \overline{C}_i\dot{y}_i \;+\; \overline{K}_i y_i \;=\; -[\phi]^T[M][I]\ddot{x}_g$$

or

$$\ddot{y}_i \;+\; 2\xi_i\omega_i\dot{y}_i \;+\; \omega_i^2 \;=\; -\Gamma_i\ddot{x}_g(t) \qquad i=1,...,n \tag{A.42}$$

where

$$\Gamma_i = \frac{[\phi]^T[M][I]}{\overline{M}} \tag{A.43}$$

is the participation factor. Eq. A.42 can be solved as if it represented the motion of a single-degree-of-freedom system, with its own frequency $\omega_i^2 = \overline{K}_i/\overline{M}_i$ and damping ratio $\xi_i = \overline{C}_i/2\overline{M}_i\omega_i$. While $\overline{C}_i$-values are not readily determined, it is relatively easy to establish values for the damping ratios ξ_i by exciting the i'th mode and measuring the corresponding damping.

After solving Eq. A.42 for each mode i, one more step is needed. Because the response is formulated in the modal coordinates y_i it needs to be transformed back to physical coordinates x_i, using Eq. A.39. Fig. A.13 is an attempt to graphically illustrate this coordinate transformation.

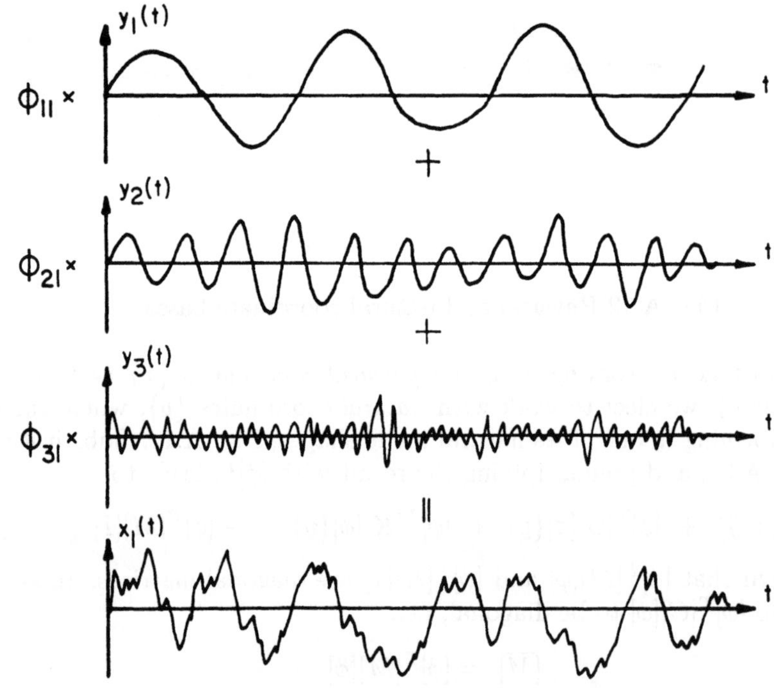

Fig. A.13 Linear superposition of modal responses

Once the floor displacements x_i are known at each time step, rotations and vertical displacements of all joints are determined by backsubstitution. Then member forces (shears, axial forces, bending moments) can be computed for each member, at each time step. Of course, we are typically interested in the maximum value of each such time history. Also significant are maximum relative story displacements, such as the largest displacement difference between the third and fourth floor. Such interstory drifts are an important measure of nonstructural damage.

The main advantage of the modal analysis method sketched here is, that instead of integrating all N equations of a system with N DOF's in Eq. A.38, only n equations of the kind of Eq. A.42 need to be solved, where typically, $n \ll N$. But a fundamental assumption underlying this modal analysis is that the total response can be determined by linear superposition of the various modal responses, i.e., Eq. A.39. This assumption becomes invalid once the structure responds nonlinearly to load. In this case, the equations of motion, Eq. A.38, need to be integrated directly, without first transforming them to natural coordinates.

A.9 Direct Integration

The equations of motion, Eq. A.38, can be written in incremental form,

$$[M]\{\Delta\ddot{x}\} + [C]\{\Delta\dot{x}\} + [K]\{\Delta x\} = \{\Delta F(t)\} \qquad (A.44)$$

For numerical integration, we approximate the velocity and acceleration increments, preferably using an implicit integration scheme, such as Newmark's method [A.7],

$$\begin{aligned} \{\Delta\dot{x}\} &= c_1\{x_{t+\Delta t} - x_t\} + c_2\{\dot{x}_t\} + c_3\{\ddot{x}_t\} \\ \{\Delta\ddot{x}\} &= c_4\{x_{t+\Delta t} - x_t\} + c_5\{\dot{x}_t\} + c_6\{\ddot{x}_t\} \end{aligned} \qquad (A.45)$$

Substitution of Eq. A.45 into A.44 gives,

$$[K^*]\{\Delta x\} = \{\Delta F^*\} \qquad (A.46)$$

where

$$\begin{aligned} [K^*] &= [K] + c_4[M] + c_1[C] \\ \{\Delta F^*\} &= \{\Delta F\} - [M](c_5\{\dot{x}_t\} + c_6\{\ddot{x}_t\}) - [C](c_2\{\dot{x}_t\} + c_3\{\ddot{x}_t\}) \end{aligned}$$

Eq. A.46 can be solved directly for $\{\Delta x\}$, while $\{\Delta\dot{x}\}$ and $\{\Delta\ddot{x}\}$ follow from Eq. A.45. At that point, displacements, velocities, and accelerations are known at time $t+\Delta t$ and can serve as initial conditions for the subsequent time step. In this fashion, the complete response histories for all DOF's of the structure are obtained directly and simultaneously, and from the displacement histories, the member force histories can be computed.

If the structure responds nonlinearly to load, the stiffness matrices $[K]$ changes, i.e. $[K^*]$ changes too. This is generally the case for reinforced concrete frames responding to strong earthquake ground motion. If, on the other hand, the structure remains linear, like during a small or moderate earthquake, $[K^*]$ needs to be factorized only once, and each new time increment can be thought of as merely constituting an additional load case.

A.10 Randomness in Structural Response Analysis

We are confronted with uncertainty and randomness on a daily basis. Certainty is rather the exception than the norm. In structural analysis, there are basically three different sources of uncertainty. First and most significantly, there are the loads, then there are the structural properties, that is, material properties and geometric characteristics, and finally there are our mathematical models and methods of analysis with their own built-in uncertainties.

The extreme randomness of the loads for which we are charged to design our structures and, to a lesser extent, the uncertainties of material properties have a significant impact on structural analysis. One could easily be inclined to resign, because if we know so little about the structures we are analyzing, and nothing about the loads for which we are designing them, why bother with an analysis in the first place! Fortunately, the situation is not quite so bleak. But it is worthwhile to look occasionally at the way we analyze structures and to ask ourselves whether we are employing the appropriate methods. It makes no sense to employ an expensive, highly "accurate" analysis method, if we know little or nothing about the loads, or if the material properties exhibit significant statistical scatter. Conversely, if the expected loads can be specified with some degree of certainty, it is quite logical to use an analysis tool that is compatible with this level of certainty. Yet, even though most important design loads have high uncertainty, it is inappropriate to replace readily available analysis methods with highly approximate ones as standard procedure. By the same logic, we might just as well substitute guesswork. By developing an accurate mathematical model of a structure and utilizing a reliable accurate analysis method, we actually can reduce one of the three sources of uncertainty. This allows us to assess the significance of the other two that remain. In any case it is prudent to use an analysis method of "appropriate" level of accuracy, although that is often not easy to define.

Random vibration analysis recognizes the fact that most dynamic loads are of a random nature. It furnishes estimates of expected structural response to dynamic loading that is specified in statistical terms. It provides the design engineer with a useful quantitative assessment of the expected response to random loading. How to cope with uncertainty is illustrated schematically in Fig. A.14, with earthquake ground motions as the sample random process. A strictly deterministic analysis is of limited value, because there is no guarantee that a structure that responds satisfactorily to one sample input function, does likewise for another one. In the semi-deterministic approach, deterministic analyses are repeated for a set of input functions, which are representative of the random process under consideration, and a statistical evaluation of the individual responses follows such analyses. Finally, there is the probabilistic analysis approach, in which individual deterministic analyses are avoided, and instead, transfer functions are developed, which determine the expected response directly from the provided input information.

It is the objective of random vibration analysis to obtain a rational assessment of the expected response of a structure to loading that is prescribed in statistical terms only and to translate this response information into an assessment of the expected damage. In order to do so, we still need as much information on the expected loading as possible. We also need to know the material (as well as geometric) properties with as much certainty as possible. Then we may proceed with the analysis. Most random vibration analysis techniques are restricted to linear elastic systems and therefore not directly applicable to realistic structures under extreme loading conditions, such as reinforced concrete frames subjected to severe earthquake ground motions. Various "approximate" methods have been devised to overcome this limitation, usually with little success and therefore of limited practical value. A tried and true alternative is an expensive one, namely Monte Carlo type numerical simulations.

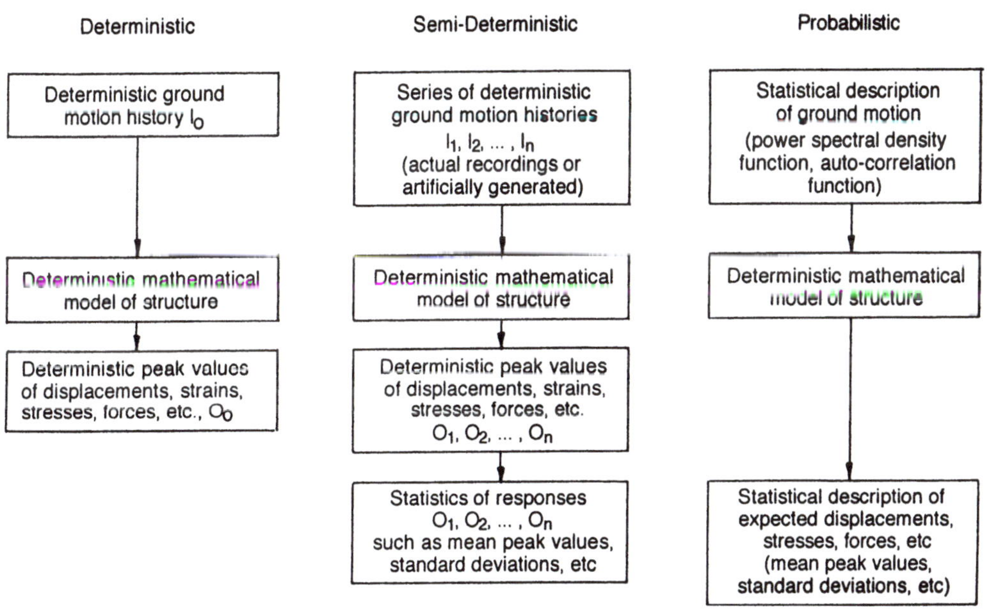

Fig. A.14 Coping with uncertainty

A.11 Random Processes

A random process is a physical process for which the outcome is unpredictable, no matter how carefully all controllable variables are kept constant [A.8]. For example, consider the acceleration history experienced by a piece of equipment traveling by truck from point A to B. Even if we employ always the same driver, use the same truck and route and carefully instruct the driver to "always drive the same", the outcome will be different for each run, i.e., it will be random. Each run will produce a *sample* function $x^{(j)}(t)$. An *ensemble* of five such sample functions is shown in Fig. A.15. Other examples of random processes are wind velocity measurements on the fassade of a building and ground acceleration recordings at a given site due to seismic activity.

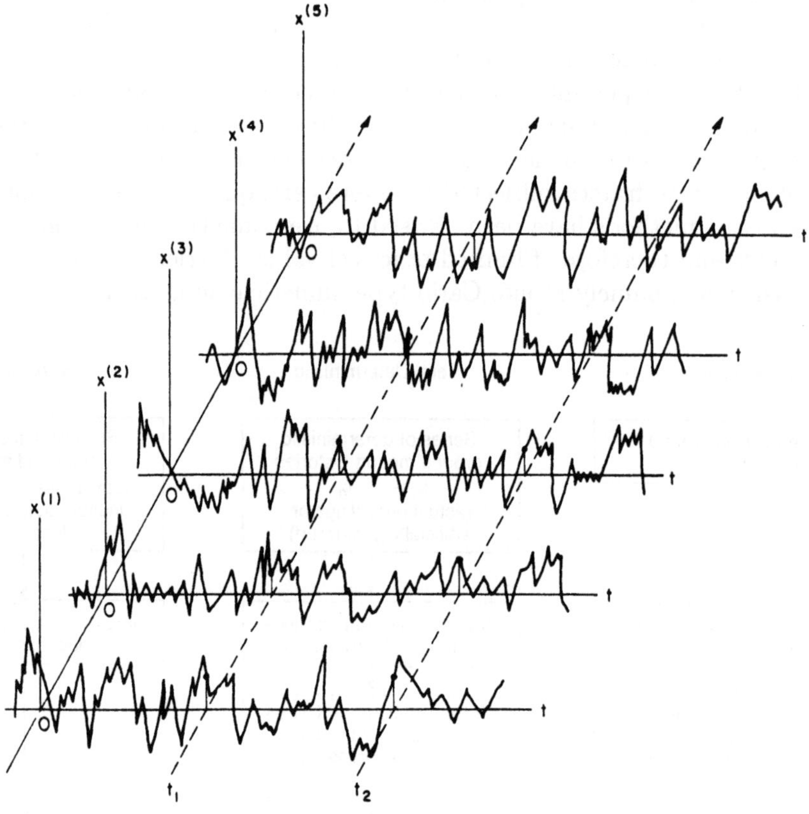

Fig. A.15 Ensemble of random sample functions

Although each individual sample function of a random process appears very unpredictable, certain regularities can be established, if the number of different functions is large or the duration of the functions sufficiently long. These regularities appear when we take averages, mean values, or expectations of the individual samples or their spectrum contents. This requires the introduction of a number of definitions:

The *ensemble average* is established by averaging the values of each sample at a given time t_1, i.e., $x^{(j)}(t_1)$, or simply x, i.e., $\frac{1}{n}\sum_{j=1}^{n} x^{(j)}(t_1)$. Suppose that with each sample x we associate a value $g(x)$, where g is a known function, then the average of $g(x)$ is $\frac{1}{n}\sum_{j=1}^{n} g(x^{(j)})$. If there are infinitely many samples, whose distribution is described by the probability density function $p(x)$, then

$$E[g(x)] = \int_{-\infty}^{\infty} g(x)p(x)dx \qquad (A.47)$$

where by definition, $\int_{-\infty}^{\infty} p(x)dx = 1$. When $g(x)$ is simply x itself, then the ensemble average becomes the *mean* or *expected value* of x,

$$E[x] = \int_{-\infty}^{\infty} x\, p(x)dx \qquad (A.48)$$

If we consider two different points in time, t_1 and t_2, and denote by x_1 and x_2 the ensembles of samples $x(t_1)$ and $x(t_2)$, then the *autocorrelation function* is defined as,

$$E[x(t_1)x(t_2)] = \int_{-\infty}^{\infty}\int_{-\infty}^{\infty} x_1\, x_2\, p(x_1, x_2)dx_1 dx_2 \qquad (A.49)$$

where $p(x_1, x_2)$ is the joint probability density function, i.e., $p(x_1, x_2)dx_1dx_2$ is the probability that the value of a random sample falls between x_1 and $x_1 + dx_1$ and between x_2 and $x_2 + dx_2$.

In order to be *stationary*, a process has to have a probability density function which is invariant under a shift of the time scale, i.e., $p(x)$, $E[x]$, etc are all independent of time t. If $p(x_1, x_2)$ is time invariant, it is a function of only the time lag $t_2 - t_1 = \tau$, i.e., $p(x_1, x_2) = p(t, t + \tau)$, and the autocorrelation function becomes,

$$E[x_1 x_2] = E[x(t)x(t + \tau)] = R(\tau) \qquad (A.50)$$

For a process to be strictly stationary, it can have no beginning and no end, i.e., each sample must extend from $-\infty$ to $+\infty$.

Next we shall introduce *temporal averages*, where we look not across the ensemble but at a single sample function $x^{(j)}$. For example, the temporal mean is defined as

$$< x^j > = \lim_{T\to\infty} \frac{1}{T}\int_{-T/2}^{T/2} x^{(j)}(t)dt \qquad (A.51)$$

Within the subclass of stationary random processes there exists a further subcategory known as *ergodic* processes, for which the ensemble averages are equal to the corresponding temporal averages taken along any representative sample function, i.e.,

$$E[x] = \int_{-\infty}^{\infty} x\, p(x)dx = < x^{(j)} > = \lim_{T\to\infty} \frac{1}{T}\int_{-T/2}^{T/2} x^j(t)dt \qquad (A.52)$$

An ergodic process is necessarily stationary, since $< x^{(j)} >$ is a constant, while $E[x]$ is generally a function of the time $t = t_1$ at which the ensemble average is taken, except in the case of a stationary process. And each sample function must be completely representative of the random process. For example, returning to the truck experiment: if the one driver with the one truck carefully drives "always the same" along the same route under similar traffic and weather conditions, each sample run would be statistically similar, and the ergodicity assumption would apply. If, however, k different drivers, each with their own trucks make the run along their own routes, then the individual samples will likely be so different that the process cannot be considered ergodic.

If a process can be assumed to be ergodic, the analysis is simplified significantly, because all statistical information can be obtained from a single sufficiently long sample.

Turning now to *spectral density*, we recall from engineering mathematics that a periodic function can be represented by a Fourier series,

$$f(t) = \sum_{n=1}^{\infty} a_n sinn\omega_o t, \quad \text{where} \quad a_n = \frac{2}{T}\int_0^T f(t) sinn\omega_o t dt \qquad (A.53)$$

The coefficients a_n indicate how much the n'th *harmonic* contributes to the total sum that represents $f(t)$. If $f(t)$ is a non-periodic function, it can similarly be represented, not by a discrete series, but by a continuous integral,

$$f(t) = \frac{1}{2\pi}\int_{-\infty}^{\infty} G(\omega)e^{i\omega t}d\omega \quad \text{where} \quad Q(\omega) = \int_{-\infty}^{\infty} f(t)e^{-i\omega t}dt \qquad (A.54)$$

where the period $T \to \infty$, and the frequencies are now represented by a continuous spectrum rather than the discrete values of Eq. A.53, and $Q(\omega)$ indicates the frequency content of the given function $f(t)$ over the entire frequency spectrum. Eqs. A.54 form a Fourier Transform pair, introduced already earlier (see Eqs. A.32 and A.33).

The *power spectral density function* is defined as

$$S_f(\omega) = \lim_{T\to\infty} \frac{1}{2\pi T}\int_{-T/2}^{T/2} |f(t)e^{-i\omega t}dt|^2 = \lim_{T\to\infty} \frac{1}{2\pi T}|Q(\omega)|^2 \qquad (A.55)$$

It defines the mean square of the function $f(t)$ as

$$< f(t)^2 > = \int_{-\infty}^{\infty} S_f(\omega)d\omega \qquad (A.56)$$

and, like $Q(\omega)$, is a measure of the spectral content in function $f(t)$. For example, white noise is defined as

$$S_f(\omega) = \text{const} \qquad (A.57)$$

which means each frequency component has an equal probability of being represented in the random process f.

A.12 Description of Dynamic Loadings

Even though most of the loads to which our structures are exposed are random, some are more random than others. Depending on the loads' severity, their randomness may play a varying role.

Earthquake ground motions are clearly among the most random loads. The ground acceleration histories actually recorded during earthquakes exhibit an extremely erratic nature, which derives from the irregularities of the various rock strata and soil layers through which the seismic waves travel, undergoing all kinds of reflections and refractions, amplification and attenuation along the way from the causative fault to the site under consideration. Since it is impossible to predict the actual expected ground motion at a given site due to a prescribed seismic event, we have to be content with estimating the spectral content of such motion, which is a little easier, if the properties of the soil layers and rock strata in the vicinity of the site of interest are known. These can be measured at least for small amplitudes, using acoustic-seismic experiments. An example of a widely used power spectrum is due to Kanai-Tajimi [A.9,A.10], Fig. A.16,

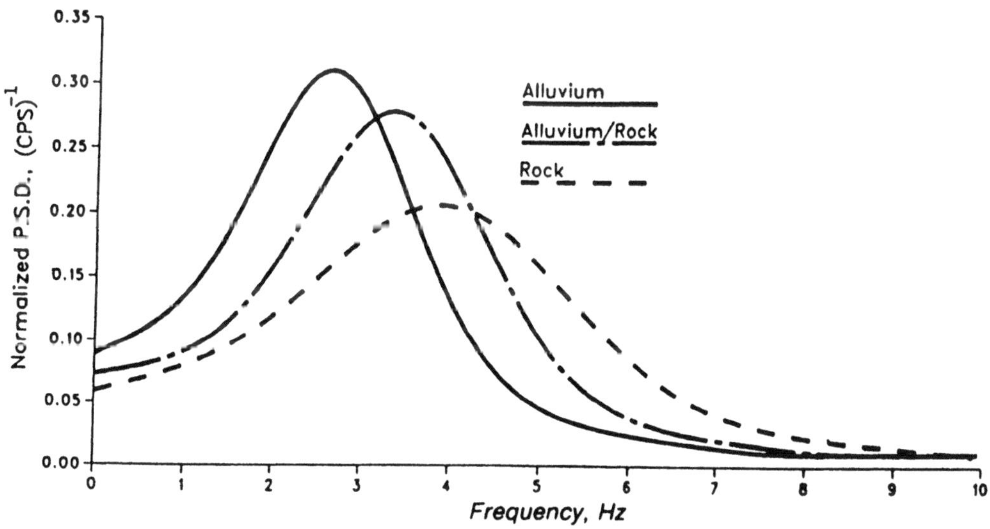

Fig. A.16 Kanai-Tajimi spectrum

$$S(\omega) = S_o \frac{1 + 4\zeta_g^2(\frac{\omega}{\omega_g})^2}{[1-(\frac{\omega}{\omega_g})^2]^2 + 4\zeta_g^2(\frac{\omega}{\omega_g})^2} \qquad (A.58)$$

where ω_g = characteristic ground frequency, ζ_g = predominant damping coefficient, S_o = intensity of Gaussian white noise over the range $-\infty < \omega < \infty$. As for the expected peak acceleration, it can be estimated from the epicentral distance, using any one of the attenuation laws described in the literature [A.11]. The question of seismic hazard involves also the probability of occurrence. For example, Gutenberg and Richter have postulated an empirical relationship between earthquake magnitude and probability of occurrence.

Strong earthquake ground motions are characterized by three phases, an initial build-up phase, followed by a strong-motion phase with relatively constant peak accelerations, and a final attenuation phase, Fig. A.17. Such motions clearly do not qualify as a stationary random process. Yet, the assumption is often made anyway, because of the dominant influence of the strong-motion phase on the structural response and damage potential. Likewise, earthquakes are typically assumed to be an ergodic process, purely because of computational convenience.

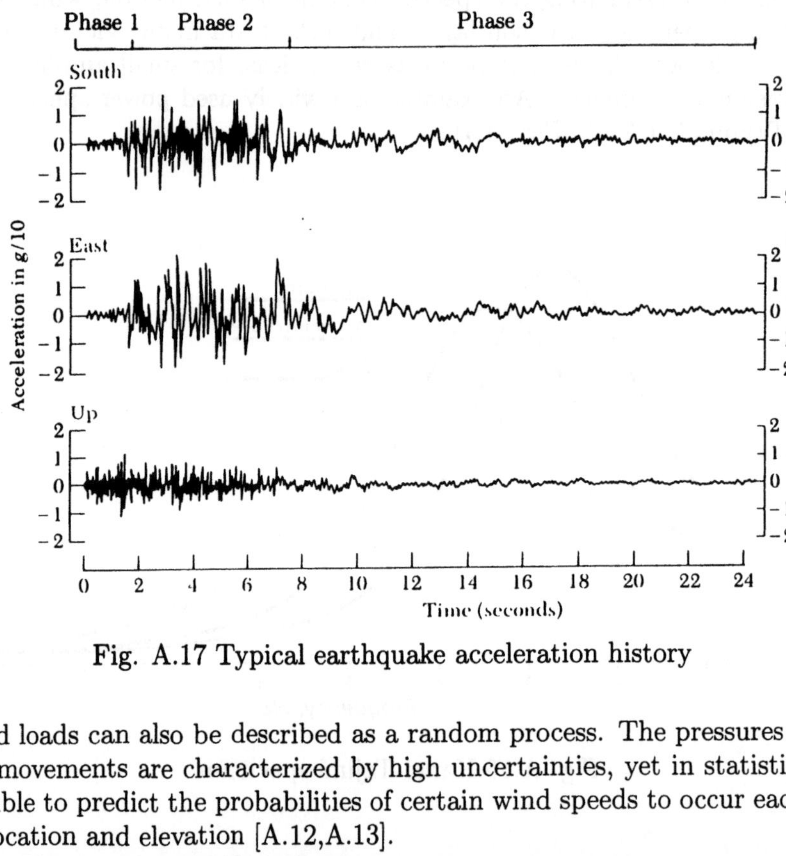

Fig. A.17 Typical earthquake acceleration history

Wind loads can also be described as a random process. The pressures resulting from air movements are characterized by high uncertainties, yet in statistical terms it is possible to predict the probabilities of certain wind speeds to occur each year at a given location and elevation [A.12,A.13].

Wind loads can be decomposed into a static or steady-state component due to the mean wind velocity and a dynamic component due to buffeting or wind gusts. For dynamic analysis this second component is of prime interest, especially if it is capable of exciting important structural modes of deformation. Various functions have been proposed to describe the frequency spectrum of such longitudinal turbulence, that is the fluctuations in the direction of the wind. Of similar importance for structural design purposes are across-wind fluctuations, which result from vortex shedding if the free air flow is obstructed by an object. It is characterized by the Strouhal number,

$$S = \frac{n_e D}{V} \tag{A.59}$$

where n_e is the frequency of the vortex shedding, D is a characteristic dimension of the body, e.g. the diameter of a cylinder, and V is the velocity of the incoming flow.

Other types of loading can also be represented as random processes, although for design purposes, codes specify safety or overload factors which are supposed to account for the stochastic nature of the loads. For example, the live load on a highway bridge could be measured by counting and weighing passing vehicles. A dynamic analysis could then be performed for an exceptionally heavy vehicle. But it is the rare exception that a random vibration analysis would be performed for civil engineering structures other than for wind or earthquake loading. In mechanical engineering and aerospace structures this is more common, e.g. for pressure loadings on an aircraft wing, or shock loading for a motor vehicle chassis.

A.13 Random Vibration Analysis

Most random vibration analysis methods are mathematically demanding. Here we shall restrict ourselves therefore to only a short sketch of one such technique.

Let us consider earthquake ground accelerations as the example random process and make the following assumptions,

a) the structure is represented by a single degree-of-freedom oscillator with linear elastic response;
b) the ground motion is a stationary process;
c) the ground motion is an ergodic process;
d) the earthquake power spectrum density function, $S_p(\omega_p)$, is given or assumed (e.g. the Kanai-Tajimi spectrum, Eq. A.58).

It can be shown [A.14] that the power spectral density function of the output process, i.e. of the structural response $x(t)$ is given by,

$$S_x(\omega_p) = |H(i\omega_p)|^2 \, S_p(\omega_p) \tag{A.60}$$

where $H(i\omega_p)$ is the complex frequency response function, Eq. A.27. The expected or mean value of the system response follows simply as,

$$< x > = \frac{1}{2\pi} \int_o^\infty |H(i\omega_p)|^2 \, S_p(\omega_p) dp \tag{A.61}$$

and the probability that some critical response value x^* is exceeded, is given by

$$Pr[x > x^*] = \int_{x^*}^\infty f(x) dx \tag{A.62}$$

where $f(x)$ is the probability density function of the response x.

As these results indicate, it is possible to obtain important structural response information without the need for a specific earthquake ground motion. Only its power spectral density function needs to be given, together with the structural system parameters (frequency and damping ratio). For further details, reference is made to the specialized literature, e.g. [A.1,A.8,A.14].

A.14 Monte Carlo Simulation

The limitation of most random vibration techniques to linear structural response is severe from a practical viewpoint. Also the representation of earthquake ground motions as a stationary random process is inappropriate, because for structures with degrading stiffness under cyclic load (such as reinforced concrete frames), the duration of the strong motion phase may become a critical factor. One solution to both of these problems is numerical simulation using Monte Carlo techniques [A.15].

Ground acceleration histories can be generated artificially to serve as input. By multiplying a stationary Gaussian process $g(t)$ with an envelope function $s(t)$, rather realistic acceleration histories are obtained, Fig. A.18. For the Gaussian process, the Kanai-Tajimi spectrum, Eq. A.58, can be used as power spectral density function. A sample function of the Gaussian process, $g(t)$ is then generated by a Monte Carlo technique,

$$g(t) \quad = \quad \sqrt{2}\sum_{k=1}^{N}\sqrt{G(\omega_k)\Delta\omega}\,\cos(\omega_k t - \phi_k) \qquad (A.63)$$

where ϕ_k is a randomly generated phase angle, uniformly distributed between 0 and 2π, $\omega_k = k\Delta\omega$, $\omega_u = N\Delta\omega$ is the upper cut-off frequency, and $G(\omega_k) = 2S(\omega_k)$ is the one-sided power spectrum. To generate an artificial earthquake, the following relationship between intensity S_o and peak ground acceleration PGA has been suggested,

$$PGA \quad = \quad p_o\sqrt{S_o\pi\omega_g(\frac{1}{2\zeta_g} + 2\zeta_g)} \qquad (A.64)$$

where ζ_g was the predominant ground damping coefficient, and p_o is an empirical factor, taken to be 3.0.

A complete analysis requires a nonlinear time history analysis of the mathematical model of the structure for several artificially generated ground motion histories. Our studies have indicated that for ductile moment resistant concrete frames ten sample histories should suffice to obtain meaningful statistical averages of the response quantities in question, such as damage indices.

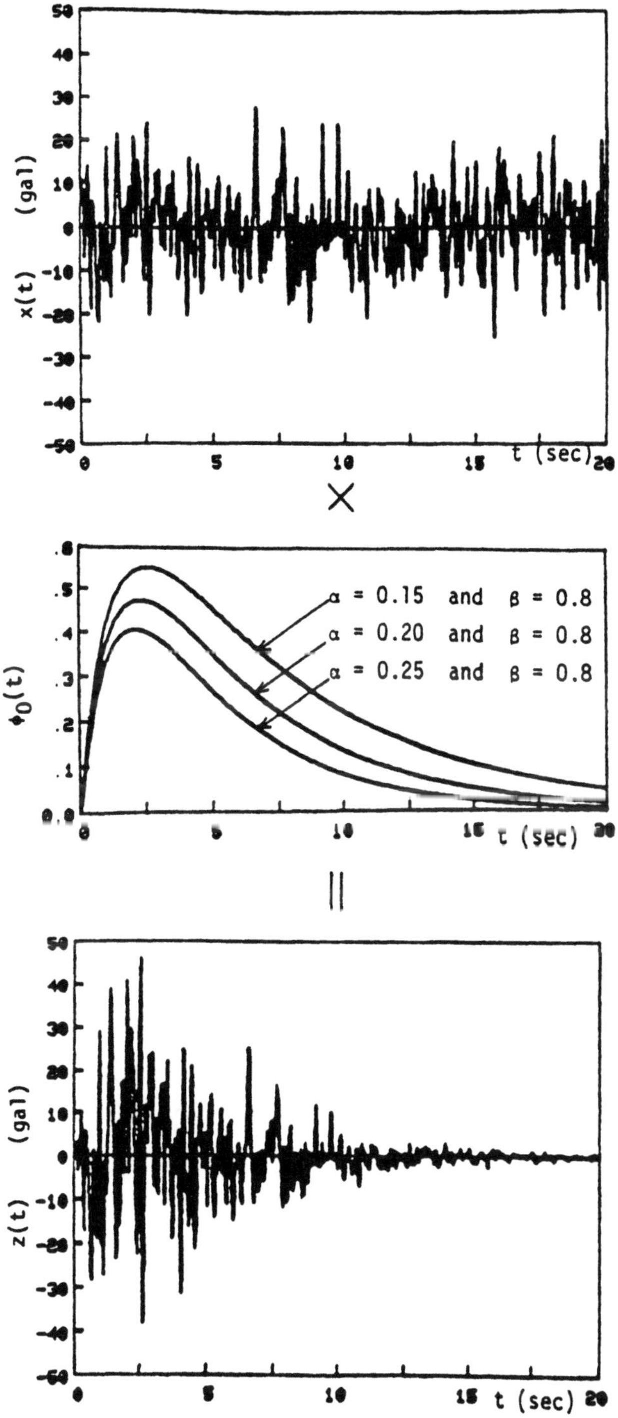

Fig. A.18 Monte Carlo simulation of non-stationary random process

A.15 References

A.1 Clough, R.W. and J. Penzien: Dynamics of Structures, 2nd Ed., McGraw Hill, Inc., New York, 1993.

A.2 Paz, M.: Structural Dynamics Theory and Computation, 2nd Ed., Van Nostrand Reinhold, New York, 1985.

A.3 Chopra, A.K.: Dynamics of Structures, Prentice Hall, Englewood Cliffs, NJ, 1995.

A.4 Nigam, N.C. and P.C. Jennings: Digital Calculation of Response Spectra from Strong-Motion Earthquake Records, Earthquake Engineering Research Laboratory, California Institute of Technology, Pasadena, Calif., June 1968.

A.5 Gupta, A.K.: Response Spectrum Method, CRC Press, Inc., Baton Rouge, 1993.

A.6 Wilson, E.L.: The static condensation algorithm, Int. J. for Num. Meth. in Engin. 8 (1974).

A.7 Bathe, K.-J.: Finite Element Procedures, Prentice Hall, Englewood Cliffs, NJ, 1996.

A.8 Crandall, S.H. and W.D. Mark: Random Vibration in Mechanical Systems, Academic Press, New York, 1973.

A.9 Kanai, K.: Semi-empirical formula for the seismic characteristics of the ground, Bull. Earthquake Res. Inst. 35, University of Tokyo, 1957.

A.10 Tajimi, H.: A statistical method of determining the maximum response of a building during an earthquake, Proc., 2nd World Earthquake Engineering Conference, Tokyo, 1960, Vol II, 781-797.

A.11 Mohraz, B. and F.E. Elghadamsi: Earthquake Ground Motion and Response Spectra, in: The Seismic Design Handbook (F. Naeim, ed.), Van Nostrand Reinhold, New York, 1989.

A.12 Simiu, E. and R.H. Scanlan: Wind Effects on Structures, 2nd Ed., John Wiley and Sons, New York, 1986.

A.13 Davenport, A.G., S. Mackey, and W.H. Melbourne: Wind Loading and Wind Effects, Tall Building Criteria and Loading, Volume CL, Monograph on Planning and Design of Tall Buildings, ASCE, 1980.

A.14 Newland, D.E.: Random Vibrations and Spectral Analysis, Longman Inc., New York, 1984.

A.15 Chung, Y.S., C. Meyer, and M. Shinozuka: Automatic seismic design of reinforced concrete frames, ACI Structural Journal, May-June 1990.

MIX
Papier aus verantwortungsvollen Quellen
Paper from responsible sources
FSC® C105338

If you have any concerns about our products,
you can contact us on
ProductSafety@springernature.com

In case Publisher is established outside the EU,
the EU authorized representative is:
Springer Nature Customer Service Center GmbH
Europaplatz 3, 69115 Heidelberg, Germany

Printed by Libri Plureos GmbH
in Hamburg, Germany